CISM COURSES AND LECTURES

Series Editors:

The Rectors of CISM
Sandor Kaliszky - Budapest
Mahir Sayir - Zurich
Wilhelm Schneider - Wien

The Secretary General of CISM
Giovanni Bianchi - Milan

Executive Editor
Carlo Tasso - Udine

The series presents lecture notes, monographs, edited works and
proceedings in the field of Mechanics, Engineering, Computer Science
and Applied Mathematics.
Purpose of the series is to make known in the international scientific
and technical community results obtained in some of the activities
organized by CISM, the International Centre for Mechanical Sciences.

INTERNATIONAL CENTRE FOR MECHANICAL SCIENCES

COURSES AND LECTURES - No. 365

DATA ACQUISITION AND ANALYSIS FOR MULTIMEDIA GIS

EDIT BY

L. MUSSIO
POLYTECHNIC OF MILAN

G. FORLANI
POLYTECHNIC OF MILAN

F. CROSILLA
UNIVERSITY OF UDINE

 Springer-Verlag Wien GmbH

Le spese di stampa di questo volume sono in parte coperte da
contributi del Consiglio Nazionale delle Ricerche.

This volume contains 207 illustrations

In order to make this volume available as economically and as
rapidly as possible the authors' typescripts have been
reproduced in their original forms. This method unfortunately
has its typographical limitations but it is hoped that they in no
way distract the reader.

ISBN 978-3-211-82806-9   ISBN 978-3-7091-2684-4 (eBook)
DOI 10.1007/978-3-7091-2684-4

PREFACE

The International Centre for Mechanical Sciences hosted from 12 to 16 June 1995 the Workshop on "Multimedia GIS Data" organized by the Technical Commission One of the International Society for Photogrammetry and Remote Sensing. On that occasion a tutorial on "Spatial Data Analysis: Theory and Algorithms" was also organized by the Working Group III/IV of the same Society.

The book contains the proceedings of both events and is made up of 27 papers, presented in 10 technical sessions of the workshop, on different fields of application, and 4 more papers given by invited speakers at the one-day tutorial.

The papers reported in the book and presented at the workshop can be ideally collected in five main chapters. In the first one, that can be defined "guidance and navigation", some experiences on data acquisition with low cost DGPS for road survey and an overview on Vehicle Navigation Systems (VNS) are shown. An original method for a mobile robot to explore an unknown environment is also reported.

The second chapter on "GIS data acquisition and evaluation" collects a sort of papers treating robust statistical techniques applied to preprocessing, analysis and testing for different kinds of GIS data.

Within the topic on "image acquisition and preprocessing" in particular some experiences on test and calibration of different scanners for GIS data acquisition are reported as well as some original approaches to the automatic DTM generation for cartographic and close range applications.

Finally some applications to the environmental monitoring and to the use of different kinds of geodetic data in multipurpose regional GIS, together with some examples of the applicability of multimedia technology to architecture and civil engineering are shown.
The most important aspects emerging from the workshop are perhaps the following:
- *data acquisition and update need to be automated to the highest level, not only for object geometry reconstruction (which is anyway not trivial in many*

applications to architecture) but also for object classification;

- *data quality and data currency evaluation is also crucial if the GIS is to provide meaningful information: testing strategies have been presented for geometric data, but something equivalent is needed also for thematic data;*
- *multimedia integration is at present rather limited, being applied mainly to help users getting into the system or in applications like describing historical buildings, where a combination of sound and images becomes very effective.*

Futhermore the book contains the 4 presentations given at the one-day tutorial. The goal was to highlight the current status on the conceptual aspects in designing GIS. In particular the problems of modeling and organizing data in structures, the processing techniques of GIS data for queries to the system and the so called Dynamic GIS have been reported in detail. In addition a lecture on Computer Graphics principles is enclosed, intended to grasp an impression of what's behind the graphics tools which are of great relevance, at least from the user interface side, also to GIS.

F. Crosilla
G. Forlani

WELCOME ADDRESS

It is always my pleasure to visit Italy and particularly to welcome attendees to an ISPRS sponsored event, I am very glad to have this opportunity to participate on behalf of the ISPRS Council in this ISPRS Commission I, Working group III/4 Tutorial on "Multimedia GIS Data."

This is a very timely Tutorial. Around us all, we find the technology and science of our disciplines trending toward all-digital: Digital Systems, Digital Processing and Digital Applications. A vanguard of this new environment is the emergence of many new ventures in commercial digital imaging from space and the value-added industries it will be spawning in its wake. We no longer are just serving the mapping industry, but rather we are placing our skills and ingenuity toward applications which will be greatly enhanced by our inherent knowledge for extracting and presenting semantic information. Multimedia is needed for working with the data and for spatial information presentation to the users.

Photogrammetry benefits from the semantic information embedded in digital multispectral images. Remote sensing, benefits from the metric quality inherent in digital photogrammetry. It is clear that the all-digital future will rely on both photogrammetric and remote sensing techniques. The success of the burgeoning geo-spatial information industry is totally dependent on our knowledge and skills. The specialists in this room and tutorial conveyors of advancements and innovations in our disciplines are positioning us all to meet the challenges of the future. I am sure that we all will be somewhat better equipped to address the needs of Society as a result of this Tutorial.

In closing, allow me to tell you that at our ISPRS Council Meetings in Vienna last week we discussed intently the future of our disciplines. It is very rewarding for ISPRS Council to note that Italy continues to lead in our disciplines and has signified its support by upgrading its category of leadership at the next ISPRS Congress in Vienna, July 1996. I am happy that I have the present opportunity to thank the Italian Society represented here by its President, Prof. Selvini and my dear friend of many years, Prof. Riccardo Galetto. It is an honour to be here with them and you all today.

L. W. Fritz
ISPRS Secretary General

FOREWORD

First of all let me thank the Academic Authorities of the Udine University and CISM (International Center for Mechanical Sciences) who have hosted this Workshop and supported it. Furthermore I want to say a special thanks to my colleague and friend, Prof. F. Crosilla, who has spent a lot of time to organize this Workshop very very well.

Regarding the Workshop, it will receive, in form of a Tutorial on Spatial Data Analysis: Theory and Algorithms, the contribution of WG III/4 and, for this reason, the Workshop has became an official intercommission activity. The Workshop on Multimedia GIS Data covers all topics of the Commission and its WG's and wants to be the last appointment for people interested in Commission I activities.

Finally, I must thank very much the convenor Prof. R. Galetto, the Secretary General of the ISPRS Dr. L. Fritz and the President of the SIFET (Italian Society for Surveying and Photogrammetry) Prof. A. Selvini, who have authoritatively opened the Workshop, and the secretary of TC I Prof. G. Forlani, who has been also the scientific supervisor of this Workshop.

Furthermore, I want to thank the lecturers and the chairmen of the session, who have actively contributed to fill in the Workshop of contents, suitably combining both scientific and technical aspects and proving that theory and applications should grow together. Last but not least, I thank all participants to this Workshop, for their active and qualified participation.

An additional comment can be done near to the end of the 1992-96 period, remarking that the participation in TC I has been increased offering many opportunities to exchange important experiences by means of the cooperating WG's. Therefore TC I is not a small, isolated and, maybe, useless Commission, but it plays an important role among several Technical Commissions increasing their cooperation and emphasizing the topics of primary data acquisition and evaluation. The philosophy of this opinion is that "cooperation" is better than "competition" and assures bigger and more stable advantages.

L. Mussio

CONTENTS

Page

MULTIMEDIA TECHNOLOGY
A NEW OPPORTUNITY FOR GISs IMPROVEMENT

R. Galetto and A. Spalla
University of Pavia, Pavia, Italy

ABSTRACT

With reference to the usual classification the GISs are assumed to be divided into two categories: those supporting decision makers (DSS: decision support systems) and those of the managerial type (utility company GISs). Only the latter, at least in Italy, have already achieved a notable level of success.

The DSS-GISs, on the other hand, haven't yet taken off, these being the ones which are of greatest interest to our scientific community. In fact, DSS-GISs see the territory not merely as a topographical reference to give a spatial link to technological and service networks, but as a real subject composed of soil, natural features, climatic elements, infrastructures, populations , activities, public services and rules and regulations.

The contributions of geodesy, topography, photogrammetry, remote sensing, image processing and so on are the determining factors in the DSS-GISs (which from now on we shall refer to only as GISs); those are therefore of most interest to us, and those to which we shall turn our attention in this paper.

1.   INTRODUCTION

GISs are like little plants which has difficulty growing even with the tender loving care, and green fingers of many people.

The reasons for this underdevelopment, which threatens to become chronic, are discussed in all meetings on the subject.

The most recent attempts to bring to life to this struggling shoot regard the planning aspect of the GISs. From a deterministic approach and through a period of positivism we have now reached an hermeutic one; this should partially counterbalance a pure salesman's policy, which flies in the face of understanding the needs of the user and his difficulty in expressing them coherently to a third party.

This attitude, which demands a long and costly period of investigation on the part of the planner, can bear positive results, but only with difficulty will it be able in a short space of time to bring new vitality to this sector which continues to lag behind.

Indeed, in order to improve things, it is too idealistic to adopt a philosophical approach and entrust ourselves to people's good will; better by far to look around and try to find ways and means that have had success in other fields.

In other words, it is useless to go on hoping against hope for a miracle; that those in power should recognize the need *to know the territory in order to take decisions about it,* which would stimulate demand for GISs. We need to apply a strategy which convinces them through the product's sexiness rather than through its usefulness.

So let's forget our natural inclination for original technical or scientific solutions, and concentrate instead on those characteristics that GISs have in common with multimedia products that are intended for the general public and for cultural and entertainment purposes.

A multimedia product (an encyclopaedia, an atlas, etc.) can be seen as a real information system, composed of data and related software to manage them, from which the user extracts information.

In some cases, for example atlases, where the data also have a geographical component, we can speak of a real GIS.

2.   PRINCIPAL ADVANTAGES FROM MULTIMEDIA TECHNOLOGY IN THE
      IMPROVEMENT OF GISs

The fact that the multimedia products are widely used, has already been of indirect advantage to the GISs.

One of these derives from the fact that multimedia products requires large mass memory storage and relevant computer power to manage the pictures and animation.

This has brought about an increase in the power of PCs and in the widespread use of CDROMs.

This increase in power allows the PCs to be used to run many types of GISs, favouring their direct use by those who know the technical problems.

But this acceleration in the development of the hardware will probably continue in order to satisfy new requirements of multi media technology techniques with the result that it will be easier to introduce the use of images in the GISs and also the use of digital photogrammetry as an everyday tool.

Another advantage consists of the fact that the multi media world is highly standardized while GISs is still in the slow process of standardization.
Taking advantage of multi media technology could therefore have a beneficial effect on this.

But the principal advantage to be had from multi media technology in the improvement of GISs is the assimilation of basic concepts and the techniques by which information is presented to the user.

The characteristic element in a multi media product as opposed to a traditional product of similar content is not that of increasing the quantity of information, but that of improving the user's access to it.

This increase in access is brought about by an exact technique, of which the following are the most relevant aspects.

2.1 Non-linear data access.

The software managing the data is based on the principal of permitting the user to move about in the data in a non-linear way, that is, not in the linear way by which information is acquired reading a book page after page.
On the contrary, it is as if all the pages of a book were open simultaneously before the user who can pass from one to another extremely easily.

2.2 Interactivity.

Interactivity is one of the characteristic principles of multi media technology and consists of giving an immediate answer to the user's action.
Interactivity is obtained through hypertext, balloons, buttons and similar devices.
By hypertext we mean a word which is connected with other information which can be accessed by the user, selecting that particular word with the mouse.
A typical use of hypertext is, for example, in *on line help* where in order not to make the text too heavy there is a direct link between some words in the text and other sectors of the menu which explain them or show related arguments.
Hypertexts are sometimes called *key words*.
The hypertext technique allows a very simple and immediate navigation in the data and this concept could be usefully taken from the reality of the GISs, creating hyperdrawings and hyperimages.
For example, selecting a building shown on a map we could have the photographic image of its front; then, by indicating one particular storey, we could have that storey's plans on the screen, at which point a set of buttons could appear capable of furnishing us with data about that floor (land register information, etc.), and so on.

Balloons are explanations which appear automatically when the mouse is pointed to a particular word, icon, or any other element on the screen.

In multi media technology the tendency is to eliminate the pop-up menu and to replace it with buttons, that is to say, button commands in icon form that clearly show their purpose.

This icon-driven technique as opposed to the menu-driven one, also called top-down, improves the user's knowledge of the navigation instruments' capabilities and encourages their use.

This aspect is obviously not of secondary importance to the GISs. In fact, in many of them it this user interface which creates difficulties and discourages further investigation into the use of the full range of capabilities.

The icon-driven technique together with other means that we shall examine later, helps focus the user's attention on the most interesting information or on the most significant functions to manage the data.

2.3   Easing navigation in the system.

By *navigation* we mean the movement of the user through multimedia application by the use of the interactive instruments at his disposal.

A greater availability of interactive instruments corresponds to a greater ease of navigation in the system.

The concept of navigation implies that the movement in the system is not rigidly structured but as random as possible. This prompts the user to explore the data and gives a sense of an *infinite* product.

In the GISs both the concept of easy navigation and interactivity should be more widely employed. Indeed, moving through a GIS application is often a very constricted operation based on a sequential series of operations with long periods of waiting which discourage the user.

2.4   Substituting speech for text.

It is well known that long text discourages the user from reading or, at least, they are tiring. Indeed, and perhaps for this reason, the written elements in GISs are quite limited.

It would be opportune in many cases to integrate maps and images with speech which gives detailed descriptions of phenomena which could not be easily detectable by the user.

The user could concentrate on maps and images and at the same time be directed by the spoken word to a greater understanding.

2.5   Use of animation

Animation consists of giving movement to a scene or a figure on a fixed background or to one or more lines of a graph.

The usefulness of this technique is in that it permits easier consultation of thematic maps or aerial pictures, where lines or other graphic features are shown using animation in order to highlight certain phenomena. Besides, if the use of speech is added to animation, complex phenomena become more readily understandable even to a non expert user.

This is particularly true of aerial pictures or images, the use of which is rapidly increasing in GISs, even if few users are sufficiently skilled to know how to interpret them.

A picture, whether it be an aerial or satellite image or an orthophoto, is extremely rich in information, but for a user who is not particularly expert it can be difficult to read.

The use of speech together with animation could therefore be a winning strategy in making images more fruitful for a greater number of users.

An other area in which animation could be of help is in the simulation of events that are connected to phenomena in which movement plays a role.

For example in a GIS concerning atmospheric pollution it could be very useful to be able to observe on a map or on a picture the simulation of the movement over a territory of clouds of smoke or other gases, connected to atmospheric changes.

2.6   Use of morphing.

Morphing is the technique by which an object is transformed into another through a process of animation.

Morphing could for example be used to represent in a dynamic way the change in shape of an area of restriction according to the variation of some parameters.

Instead of visualizing a certain number of situations that correspond to a series of values, we could vary the values through a continuous range and watch the area of restriction evolve in a dynamic manner.

This could be useful not only in giving more easily understandable information, but also in that it avoids the possibility of the omission of values which can produce unpredictable and unexpected effects.

The use of morphing can be used in conjunction with animation to make the study of a particular phenomenon more immediate.

For example, the extent of the area covered by the floodwater of a river can be represented on a graph by animation as a function of the value of the rainfall in its basin; at the same time the variation of that area could be visually represented on a map by morphing; and in addition the quantities of solid material brought down by the river can again be represented dynamically on a graph by animation.

And so on.

3.   CONCLUSIONS

Obviously we must not expect that multimedia techniques could easily or amateurishly applied to GISs; their use must be finalized in the project phase according to specific targets.

For instance, a wide use of the navigation concept is desirable, but it must fit the real user's needs concerning the queries on the available data.

The use of the various media and of the different techniques listed above, must, moreover, be carefully chosen and supervised by an art director who must collaborate with technicians in the project and in the realization of the GIS applications.

The contemporary use of text, images, maps, graphs, tables, speech, animation and morphing, give new and great possibilities of improving GIS capabilities. But all these tools must really facilitate the user getting information from data by means of a more friendly and powerful interface, avoiding confusing him with an excess of possible actions.

The multimedia techniques could also be profitably adopted not only for data processing, analysis and representation but also in the input phase or, more generally, to give to the user more tools than keyboard and mouse for communicating with the computer.

The most advanced techniques are in the direction of the use of speech, gaze and hand gesture jointly, in order to improve the interface in power and friendliness.

To conclude, we would like to bring the attention to a new concept which is connected to the world of multimedia.

That is to say, in the near future the current GISs will transform and evolve towards those which, in the world of artificial intelligence, are called Intelligent Multimedia Presentation Systems (IMMPS).

The philosophy of these systems is not that of adopting a rigid query structure to investigate data, but of profiting from the means within the system at the user's disposal in such a way as to enable him to pose a wide range of useful, if unexpected, questions.

Indeed, one of difficulties in the widespread use of GISs, comes from the fact that the commercial software in use offers a series of investigative tools which don't entirely fit the user's needs.

This isn't really as a result of incompetence on the part of the GIS designers, but is also in part due to the lack of certainty of the user as to his future needs. This is as a result of the fact that the needs of the decision makers who actually manage a territory with a GISs are difficult to foresee.

This means that the system is designed on the basis of the most predictable needs of the user and doesn't take into account any more original or unusual requirements.

Multimedia techniques and some principles of artificial intelligence could therefore be extremely useful in assisting at last the rapid development of GISs, promoting their use through increasing efficiencies, a more persuasive range of capabilities and a more alluring come hither look.

REFERENCES

Gould, M.D. (1994). GIS Design: A Hermeneutic View. PHOTOGRAMMETRIC ENGINEERING & REMOTE SENSING, VOL. LX, No. 9, pp 1105-1116

Ramage, P.(1994). Information in, rubbish out. MAPPING AWARENESS AND GIS IN EUROPE, VOL. 8, No 4, pp 34-36

Campbell, H., Masser, I.(1994). Implementing GIS: the organisational dimension. MAPPING AWARENESS AND GIS IN EUROPE, VOL. 8, No 2, pp 20-21

Tomlison, S.(1994). Virtual Reality or Virtually Useless. . MAPPING AWARENESS AND GIS IN EUROPE, VOL. 8, No 1, pp 28-30

Hadden, D.(1994). From Videodisk Geographic Information System to Multimedia Empowerment. INTERNATIONAL ARCHIVES of ISPRS, VOL. 30, Part 2, pp 391-396.

Dobson, J.E.(1993). A Conceptual Framework for Integrating Remote Sensing, GIS and Geography. PHOTOGRAMMETRIC ENGINEERING & REMOTE SENSING, VOL. LIX, No. 10, pp 1491-1496

Greve, C.W.,Kelmelis, J.A., Fegeas, R.,Guptill, S.C.,Mouat, N. (1993). Investigating U.S. Geological Survey Needs for the Management of Temporal GIS Data. PHOTOGRAMMETRIC ENGINEERING & REMOTE SENSING, VOL. LIX, No.10, pp 1503-1508

Raper, J. (1993). Multimedia and GIS: an interactive future?. MAPPING AWARENESS AND GIS IN EUROPE, VOL. 7, No 1, pp 8-10

Carter, J.R. (1992). Perspective on Sharing Data in Geographic Information Systems. PHOTOGRAMMETRIC ENGINEERING & REMOTE SENSING, VOL. LVIII, No. 11, pp 1557-1560

Welch, R.,Remillard, M.,Alberts, J. (1992). Integration of GPS, Remote Sensing and GIS Techniques for Coastal Resource Management. PHOTOGRAMMETRIC ENGINEERING & REMOTE SENSING, VOL. LVIII, No. 11, pp 1571-1578

"QUALITY PROJECT" IN SURVEY APPLICATIONS

G. Bezoari and F. Guzzetti
Polytechnic of Milan, Milan, Italy

ABSTRACT

This paper regards the world quality project in the area of surveying. Standards ISO 9000 and EN 45000 define a new common language and recognize elements that qualify each operation and organization. In the ambience of "Quality System at Politecnico of Milan" the "Laboratorio Gino Cassinis" determines operating precision of surveying equipment.
Standard deviations was chosen as a meaningful value for accuracy for levels, theodolites, optical diastimeters and EDM. Measures have to be repeated using particullary procedures, following schemes specific to each instruments.
Lastly are reported some critical remarks about the contents of DIN 18723 standards, used to define the instrumental standard deviation.

1. THE WORLD QUALITY PROJECT

The world quality project was born a few years ago in Japan.
Its basic philosophy is as follows: each activity, whether productive or not, has to be optimized; to do that, each step in the "productive process" must follow a specific design, planned to allow easy noise analysis, fast process modifications and direct links with the human element. The operator isn't a cog in a machine; he is the brain behind his specific tasks; he is an active collaborator with the designer to optimize the system.
International ISO 9000 were created for this purpose. All the countries of the E.U. use the even more restrictive EN 45000. These standards determine "Quality Systems" that each firm and research center has to design in order to follow the world quality project. They will have enormous repercussions, and not too far off, on world economic equilibrium. Consider for example the importance that long-term guarantees have gained in the automobile sector since they were introduced by Japanese. This is the result of a careful intensive design, revision and maintenance of quality that characterizes the Japanese car industry.
In the last two years national firms are also moving toward adjusting to the new situation. The most important companies have created new departments or quality offices, organized by capable managers, to draw up the internal quality plan. Little firms and craftsmen have autonomously adjusted their operations with help from outside consultants (in high demand).
Each supply, product or manufacture, each sale has to follow the internal quality plan, guided strongly by the regulations EN 45000. The staff has to obtain this sole objective.
For many operators this plan will be seen as yet another bureaucratic exercise of a very inelastic system. This opinion is mistaken, also if we recognize that the initial expense of time and energy to become familiar with the logic of quality is great and above all very bureaucratic. There is little doubt that in the end also little firms stiching to the quality plan will take advantages: a typical example, coming from Japanese car industry, is the management of products "out-of-tolerance", that in a process "in quality" has a failure rate of practically zero.
Above all, those who will not move toward normalization will find themself quickly outside of the market, from the legislative point of view (in a few months) or from the productive point of view (for the increased competitiveness that quality production could raise).
The problem is not affecting only the industrial and productive sectors. Also research centers (not directly productive) and the services sectors

need to assimilate the logic and rules of quality. All laboratories must in fact guarantee to follow the rules of ISO 9000 or EN 45000 standardization: machinery, software and hardware, outside suppliers, waste disposals, the secrecy of results obtained.

Having followed for almost two years now the quality project we are able to say that, while the industrial world has been for a long time at work getting good results, the research centers and the services sectors are slower to start working in accordance with the new rules.

2. SURVEYING WORLD AND QUALITY PROJECT

Standards ISO 9000 and EN 45000 involve the whole area of surveying. ISO 9000 aim to define a common language between sector employees and to be able to recognize elements that qualify an organization; they have been determined by the world organization for standardization (ISO is derived by the Greek "isos", which means "equal to"). These rules have been acknowledged and developed by CEN (Comitato Europeo per la Normalizzazione) in the two series EN 29000 and EN 45000 for organizations doing experimental laboratory work for research purposes or for determining the characteristics of a product or of a service.

To work in quality in surveying or photogrammetry instruments should be provided with a certificate of rectification, adjustment and guarantee approving their operative condition. To be able to participate in a call for tenders with respect to EN 45000 standards, you have to be able to use surveying instruments for precision tracing and controlling during construction and technical characteristic related with the work in plan.

For some time now the big foreign companies producing surveying instruments (Leica, Geotronics, Topcon, Zeiss, etc.) are subject to the new rules. The respective national agencies instead aren't yet recognized as authorized testing laboratories for EN 45000. To tell the truth, some of these agencies are setting up to obtain as soon as possible the certification of authorized laboratory from recognized organizations.

It is obvious that a large amount of work is required in our sector for the new rules to be effective; on the other hand, following a standard procedure is exactly what a good surveyor should do: instead of remembering to check an instrument once in a while, the new rules provide a schedule to follow for this operation, and the same operation has to be testified by a certificate by a laboratory, that in its turn has to be recognized as working according to the EN 45000 rules.

Parallel to the requests coming from the industrial world operating in the surveying sector, from Politecnico itself came the need to promote the realization of the "Quality System" inside the Politecnico di Milan, with the aim to normalize, according to the new procedure, the whole activity of research and outside contract done by various laboratories. The proposal seems very interesting and presents some notable possibilities for applied research and therefore the decision has been made to open the "Laboratorio Gino Cassinis" at Politecnico of Milan.

3. "LABORATORIO GINO CASSINIS" AND DIN 18723 STANDARDS

It's difficult to get a laboratory started without specific economic and staff endowments (that is, having to employ personnel from Politecnico already involved in other activities). It's long and laborious to integrate an already operating laboratory into the ambience of the "Quality System at Politecnico". To do both things at once is practically impossible if you don't have a very pragmatic attitude.
With this aim some procedures have been filed with international acknowledgment ready to be used to determine operating precision of surveying equipment. Therefore the DIN 18723 standards were agreed to and structured as follows:
- part 1 defines the statistic repeatability logic of the next parts; the data processing for each topographic instrument and the χ index are defined; the χ index is a function of the independent repetition number of each measurement and of the expected reliability confidence interval;
- part 2 prescribes the measurement and data processing procedures of the observations necessary for the theodolites standard deviation definition;
- part 3 prescribes the measurement and data processing procedures of the observations necessary for the levels standard deviation definition;
- part 4 prescribes the measurement and data processing procedures of the observations necessary for the optical diastimeters standard deviation definition;
- part 5 prescribes the measurement and data processing procedures of the observations necessary for the gravimeters standard deviation definition;
- part 6 prescribes the measurement and data processing procedures of the observations necessary for the EDM standard deviation definition;
The first operation consisted of translating with attention to technical details, all parts except part 5, regarding rarely used instrumentation.

Italian text is introduced to be part of the official references of "Quality System".

Along with the specification of technical procedures neceddary to determine standard deviation of surveying instrumentation subjected to tests, rules have been defined, regarding receipt of material, staff training, secrecy of results, in the ambience of the "Quality System at Politecnico of Milano". Much attention has been paid to professional requalification of the personnel employed in technical testing, with an effective revaluation of the personnel itself.

We ought at this point to examine the technical specification of various part of Din 18723 to highlight evaluate the scientific and research implications of this operation.

First of all we have to carefully evaluate what follows in part 1 of the above-mentioned DIN standards. Tests have to be carried out in non-extreme operating conditions, reproducing the normal ambience of angles, distances and differences of height. Measures have to be repeated, following schemes specific to each instrument, outdoors, by different operators, possibly on different days, trying to simulate all disturbances normally encountered in regular use.

Because the reliability of a measurement depends strictly on the instruments rectification, these rules can be used only if the instrument is rectified.

The measurement results are subject to random effects which are difficult to analyze: they are function of all the principal and additional instrumentation components, depend strongly on operator ability and are deeply influenced by environmental, meteorological and illumination conditions.

Over short time intervals, the observations made with the same instrument, on the same object and by the same operator with the same procedure, provide a repeatability sample. We obtain a comparability element using the same procedure, on the same object, by different operators and different instruments, in different environmental conditions.

Standard deviations was chosen as a meaningful value for accuracy.

For a single measurement series:

$$s_j = \sqrt{\frac{vv_j}{f_j}} \qquad (3.1)$$

where vv_j is the sum of the residual squares and f_j is the redundancy number in the j series.

For the whole procedure:

$$s = \sqrt{\frac{\sum_{j=1}^{\mu} f_j s_j^2}{f}} = \sqrt{\frac{vv}{f}} \qquad (3.2)$$

where μ is the number of the series each made of n measuraments, and f is the global redundancy number:

$$f = \mu \cdot f_j = \mu(n-1) \qquad (3.3)$$

The S value is different from the σ theoretical value due to the limited number of series that are really executed. Statistical analysis permits us to define an interval within which σ is contained with a probability equal to 95%.

According DIN 18709 part 4, this interval is named "confidence interval". The probability is:

$$P(\sigma \le C_{\sigma e}) = 1 - \alpha \qquad (3.4)$$

and

$$C_{\sigma,e} = s \cdot \chi_e = s \cdot \sqrt{\frac{f}{\chi_{f,\alpha}^2}} \qquad (3.5)$$

The χ_e values are listed as a function of f in table 1.

f	P = 90%	P = 95%	P = 99%	f	P = 90%	P = 95%	P = 99%
	χ_e	χ_e	χ_e		χ_e	χ_e	χ_e
2	3.08	4.41	9.97	15	1.32	1.44	1.69
3	2.27	2.92	5.11	16	1.31	1.42	1.66
4	1.94	2.37	3.67	17	1.30	1.40	1.63
5	1.76	2.09	3.00	18	1.29	1.38	1.60
6	1.65	1.92	2.62	19	1.28	1.37	1.58
7	1.57	1.80	2.38	20	1.27	1.36	1.56
8	1.51	1.71	2.21	25	1.23	1.31	1.47
9	1.47	1.65	2.08	30	1.21	1.27	1.42
10	1.43	1.59	1.98	32	1.20	1.26	1.40
11	1.40	1.55	1.90	36	1.18	1.24	1.37
12	1.38	1.52	1.83	40	1.17	1.23	1.34
13	1.36	1.49	1.78	50	1.15	1.20	1.30
14	1.34	1.46	1.73				

Table 1 · χ_e values for P = 90%, P = 95% and P = 99%, as a function of f

Part 2, which covers theodolites, describes the kinds of measurements to be executed, the surveying scheme and the kind of process to use (cf. fig. 1). Briefly, we have to execute four series of three layers to targets placed according to the given scheme at about 100-150 m, with sighting not far from the horizon. Measurement has to be executed for azimuth as well as for zenith in both telescope positions. Mean values are set at 0, in such a way as to determine with statistical exactness the differences for each direction or measured inclination. The sum of squared errors, integrated with the related index χ which considers independent observations, allows you to determine the standard deviation through an Excel work-sheet reproducing the process table provided by DIN 18723.

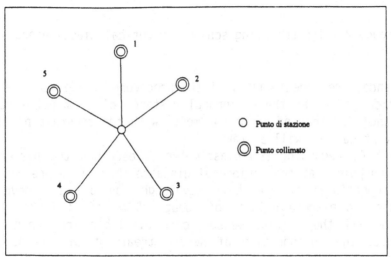

Figure 1 - The surveying scheme for theodolites

Part 3, which covers levels, provides two procedures for the definition of the standard deviation, either by the single difference of height or by the related kilometric value (usually more significant). In this case too we have to repeat a double leveling 5 times joining 5 bench marks through 4 tracts each 250 m long (normally with 3 sections for tract). Also in this case the standard deviation is valuated through the analysis of normalized errors in the measurements taken always considering the true independences obtained by the operating scheme adopted.
Part 4 covers optical diastimeters, that measures the distance through stadia hairs or through the use of a reduction tacheometer. The schedule provides for the determination of distance through the reading of the rod placed respectively at 25, 40, 75 and 100 meters (cf. fig. 2). For each

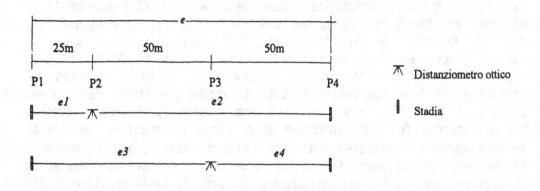

Figure 2 - The surveying scheme for optical diastimeters

basic distance, the repeatability of the procedure is set up on a series
of 11 station points at the reciprocal distance of 7 cm put across the
basic distance. In the end 44 measurements with the constant parallactic
angle and vertical rod will be taken.

Finally part 6, regarding EDM diastimeter, prescribes the placement of
seven station points at the reciprocal distance that, in a pre-determined
way, is proportional to the half wavelenght on a total development
(between the two extreme points) of about 400 m (cf. fig. 3). In this
occasion too all the partial measurements available from each station
point permits the introduction of severe treatment of the normalized
observations in such a way as to determine the related standard
deviation.

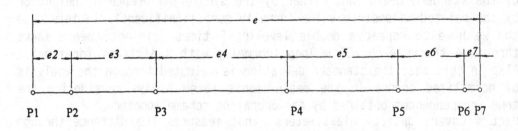

Figure 3 - The surveying scheme for EDM

It is easy to observe that the above mentioned procedures provide for a heavy amount of outdoor measurement and a statistical analysis on the differences generated by the set of normalized measurements. In fact, there are no estimates of variance propagation in the schedule nor rigorous least squares processing of taking advantage of the redundancy of the surveying schemes.

Adopting a pragmatic attitude in the effort to activate both the laboratory and seeking the general approval of "Quality System" we have until now followed almost to the letter the standards, trying only to optimize the number of independent repetitions (which determine the coefficient χ). This was to finf a balance between personnel employed in the laboratory and the needs of purchasers who find to minimize themselves temporarily having to work without their instrumentation because of testing and therefore needing speedy results.

During 1994 about a hundred technical tests were conducted, some concerning more than one instrumental characteristic. In 1995 the forecast is for a doubling of the number of samples submitted to test and the requests keep growing.

4. CRITICAL ANALYSIS OF DIN 18723 STANDARDS

After more than one year of laboratory activity it is possible to make some critical remarks about the contents of DIN 18723.

First of all it's useful to remember that DIN standards have been integrally used in an almost passive way just to start work "in quality" inside the "Laboratorio Cassinis"; it is just in the revision process of procedures provided by the rules themselves that we are working at the level of applied research.

The problem is to identify the outliers. The data analysis required by DIN 18723 part 1 is based, as already said, on determining the instrumental empirical standard deviation starting from the sum of the residual squares with respect to the mean value of each series of measurements.

It's common knowledge that least squares method doesn't distinguish between data and is therefore structurally prone to outliers. The anomalous values cause a strong distortion of the arithmetical mean value. The degree of dispersion. ie. the evalueted standard deviation, results also falsified by the outliers.

A statistically correct procedure should include a preevaluation stage with a robust estimator, to eliminate the outliers and then, finally, the least squares errors analysis. The quality procedures cannot however allow modifications in values measured in the field. In fact, the data sheet can't show any corrections or cancellations. All surveyed data must be used in the instrumental standard deviation evaluation.

Using a series of operative compensators and a rough but immediate measurement analysis, it is possible to deal with eventual blunders. The possible outliers must be investigated with a statistical approach.

The least squares principle is not adapted to seeking outliers. A robust estimator recommanded by many is obtained minimizing instead the sum of absolute residuals:

$$\sum_i |v_i| = \min \tag{4.1}$$

To be able to correctly perform this preprocessing it would be useful to build a data analysis for each procedure.

This analysis, while in itself feasible, has yet to be introduced because it entails a radical redefinition of field data evaluation and the consequent and indispensable education of personnel employed in technical testing.

The first experimental evaluation will be applied to the operating procedures used for levels. We are studing the possibility of obtaining instrumental standard deviation by measuring the difference of height between some datum points in a small levelling polygon. Each difference of height is measured with a single levelling observation, never longer than 50 m. The measurement can become very redundant and so can be elaborated with the above described double procedure.

If after testing a significant number of levels we obtain a strong coincidence between instrumental standard deviation obtained with DIN 18723 procedure and a posteriori instrumental least squares estimation, a modification of the Sistem Quality procedure is called for.

Already today, comparing some instrumental nominal standard deviation [1] with rigorous evaluation results (using the described polygon levelling), we note essential agreement in the results. The tests however must be conducted with good methodology, testing a sufficiently large number of instruments. As soon as the results are sufficiently proven, they will be detailed in a scientific journal.

Returning to the actually used procedures, it must be remembered that the high number of repetitions steadies the results automatically, smoothing laying and reading errors. Tests have been done aiming to optimize the number of repetitions to coefficient χ costs; reducing excessively the amount of repetitions tends to under-estimate the instrumental standard deviation and to be more affected by the randomness of disturbancies. Therefore statistical tests are required to verify the reliability of each measurement taken. This then is against what is stated by rules which don't allow correction of measurements. On the other hand to maintain the same number of repetitions provided by DIN 18723 creates concrete problems for the redelivery of the instruments to the owner firms. The number of independent repetitions has been checked analyzing testing reports done till now. It seemed appropriate to differentiate between the two procedures, for theodolites as well as for levels in relation to the category to which they belong (largely due to whether they have the plane parallel plate or not).

Other important analysis are in the executive phase, always starting from the statistical study "a posteriori" of a wide number of certified instruments. A simple example is the one related to the diastimeter constant of optical diastimeter. It is actually easy to evaluate the measurements executed on the processing electronic sheet if, inserting a different value from the one indicated from the producer, the related standard deviation is lower. For a certain category of instrument (all from the same producer) is suitable to use for the constant K the value 99.9 instead of the traditional 100.

5. OUTLOOK

Obviously from a scientific point of view, what is reported in the previous paragraph is worthwhile defining. In fact, if we could succeed in normalizing the double stage approach to the evaluation of instrumental standard deviation, paying attention to all the practical aspects connected (sometime conflicting with a correct statistical estimation) we could activate procedures that would be simpler and faster but still have the same performance.

In the same way, a deep statistical analysis "a posteriori" of the resulting standard deviations could be used for example as an indicator for firms. They would be able to qualify their instrument amongst those available, in such a way as to use that with the required precision to have a better quality result. Another consideration worth mentioning is the large amount of obsolete and low-precision instruments that are still used in jobs where high accuracy is required.

The choice of the appropriate frequency for certification has notable practical repercussions: it repeating the instrumental test every 2 or 3 years seems to be fair. In the meantime it would be right to define a series of practical procedures defining a quality workplace to verify at each on site the precision quality of the instrument employed, at least where a high level of accuracy is required.

Another problem to solve is that of those firms seeking similar
certificates for other categories of surveying instruments, specifically
GPS receivers: everything has still to be studied and defined to be able
to design a procedure for this purpose.
Certainly much work is still to be done, from the strictly theoretical
point of view, also considering the practical applications; the duty of
universities in the sector is in our opinion to keep up qualitatively
with the production world where in a little while quality will be the
basic element for confrontation.

REFERENCES:

1.   F. Deumlich: Surveying Instruments, Walter de Gruyter, Berlin 1982
2.   G. Bezoari, C. Monti, A. Selvini: Fondamenti di rilevamento
     generale, Hoepli, 1984, Milano.
3.   G. Forlani: Metodi robusti di stima in geodesia e fotogrammetria,
     in: Ricerche di Geodesia, Topografia e Fotogrammetria n.8, CLUP,
     Milano 1990.
4.   A. Capra, P. Russo, A. Selvini: Test di controllo e taratura per i
     teodoliti integrati: risultati di prove sperimentali condotte sul
     teodolite elettronico integrato Zeiss Modello Elta 3, Rivista del
     Catasto e dei SS.TT.EE., n. 2/91, Roma.
5.   F. Guzzetti, C. Monti: Prove di operatività e di precisione compiuti
     sul teodolite Topcon ET2, Costruzioni, anno XXXVIII, n. 40 aprile
     1989, Milano.

ROAD SURVEY FOR GIS BY MEANS OF LOW COST DGPS/DR

R. Cefalo and G. Manzoni
University of Trieste, Trieste, Italy

G. Skerl
University of Pisa, Pisa, Italy

ABSTRACT

The different GPS modes to be used for kinematic road surveying depend on the accuracy level required for the GIS. Several applications are possible: real time location of commercial vehicles, of accidents and landslide interruptions, real time navigation and location of rescue vehicles, cadaster of road and their belongings, cadaster of geometrical parameters of roads, location of statistical and real time enviromental parameter values.

Postprocessed and real time DGPS, integrated with a gyro header and an odometer have been experimented, as well as the interferential continuous kinematic GPS, both postprocessed and in real time.

An analysis of the various errors affecting the DGPS surveys is presented together with the prediction of the RDOP geometric statistic parameter .

1. Road GIS requirements

There are several users requiring Road GIS with different accuracy levels:
- real time location of commercial vehicles with an accuracy of 100 mt;
- real time location of accidents and landslide interruptions with an accuracy of 50 mt;
- real time navigation and location of rescue vehicles, taxis and parcel services with an accuracy of 10 mt;
- real time general aid to the downtown navigation;
- cadaster of roads and their belongings with an accuracy of 1 mt or better for property determination and fiscal charge;
- cadaster of the geometrical parameters of roads for safety, comfort, and maintenance, with an horizontal accuracy of 10 mt and a vertical accuracy of few centimeters;
- location of the statistical and real time enviromental parameter values like fog, visibility, rain, snow, ice, atmospheric pollution witha an accuracy increasing from intertown up to downtown roads, till 5 mt.

All the above-mentioned requirements can be satisfied by the GPS in its main modes:
- absolute, for real time navigation and location; in case of downtown navigation a map matching is used to correct the SA errors;
- differential, in real time for an accurate location in parcel services, taxi, etc, as well as for the cadaster of roads;
- interferential (kinematic) is only used when a very accurate road survey is needed, mainly in road outlining to detect irregularities;
- integrated with a dead reckoning (DR) navigation system, in urban canyons and tunnel, as well as in the mountainous region; the real time DGPS is probable even if not strictly necessary since the DR navigation system is usually a differential one, which is to say that the errors in GPS positioning, heading and velocity directly propagate into the DR; on the other hand, the real time DGPS requires a pseudorange correction broadcasting and receiving system, which could be not-available or very expensive. When the corrections are not available in real time, the DR is always useful since it smooths the GPS stand alone noise, while it is useless to correct the SA.

2. Cartography

A properly accurate cartography is used as background.
At present, various existing maps are used in Italy:

- IGM 1: 250000
- Touring Club 1: 200000
- IGM 1: 25000
- Regional Cartography 1: 10000 or 1: 5000, also in numerical form, when available.

Various Road GIS based on these maps are available from private Companies.

Difficulties have been found in overlaying the DGPS trajectories of vehicles running on the mapped roads; they can be due to:

- DGPS errors, mainly multipath and erroneous coordinates of the master receivers
- DR noise and drifts
- Geodetic networks errors
- Map detail errors
- Mechanical deformations of the maps
- too small scale of the original maps, then magnified by the computer.

In general, roads have to be surveyed by GPS methods in order to reference and update the existing maps, provided GPS masters are operated on national geodetic network benchmarks.

The road survey is in progress at European level by means of DGPS/DR equipped vehicles.

Various DR equipment is used, starting from the low cost Odometer/Piezolelectric Gyro till the very expensive Three Axis Laser Fiber Gyros and Accelerometers derived from air navigation systems.

Low cost DR is worthwhile for navigation. Its accuracy in road survey depends on the lenght of the satellite eclipsing intervals: an average accuracy (10 meters) can be obtained with lock losses of few seconds; longer losses can be associated to small scal maps only. Low cost DGPS/DR sensors can easily be mounted on any car. Care should be given to the integration between the GPS or DGPS and the Odometer and the Gyro in order to match the specific GPS noise of the surveyed site: the best compromise should be achieved between:

- the low noise and the high drift of the DR (see Figure 1 and 2, where the trajectory of a car in Trieste town centre has been done with DR stand alone); and

- the multipath noise and no drift of the DGPS (see Figure 3 in the same environment and with DGPS).

3. RDOP prediction

In various kinematic surveys, both pseudorange than interferential, divergences in the solution have been detected, due to particular constellation geometries.

This can be due to the complanarity of the visible satellites [2] or to some satellite configurations which generate singularities in the double difference observation equation system, in particular when no redundance exists (four satellites) and according to suddenly increases in the RDOP (Relative Diluition of Precision):

$$RDOP = \sqrt{\frac{tr\left(A^{T}\Sigma_{dd}^{-1}A\right)^{-1}}{\sigma_0^2}}$$

where: A is the coefficient matrix of the linearized double difference equations

Σ_{dd}^{-1} is the covariance matrix of the pseudorange or phase measurements

σ_0^2 is the unit weight variance

Let us consider the phase double difference equation:

$$\lambda \cdot \nabla\Delta\varphi = \nabla\Delta\rho - \nabla\Delta d_{ion} + \nabla\Delta d_{trop} + \nabla\Delta N \cdot \lambda + \varepsilon$$

$$\nabla\Delta\rho = \lambda \cdot \nabla\Delta\rho - \nabla\Delta N \cdot \lambda + \nabla\Delta d_{ion} - \nabla\Delta d_{trop} - \varepsilon$$

for two stations i, j and two satellites k, m :

$$\nabla\Delta\rho = \rho_i^k - \rho_j^k - \rho_i^m + \rho_j^m$$

$$= \left((X^k - X_i)^2 + (Y^k - Y_i)^2 + (Z^k - Z_i)^2\right)^{1/2} - \left((X^k - X_j)^2 + (Y^k - Y_j)^2 + (Z^k - Z_j)^2\right)^{1/2}$$

$$- \left((X^m - X_i)^2 + (Y^m - Y_i)^2 + (Z^m - Z_i)^2\right)^{1/2} + \left((X^m - X_j)^2 + (Y^m - Y_j)^2 + (Z^m - Z_j)^2\right)^{1/2}$$

Linearizing the equations in $\underline{x} = \underline{X}_j - \underline{X}_{j_0}$ and leaving the terms containing the unknows on the right – hand side, the left – side becomes

$$\left(\frac{X^m - X_{j_0}}{r_{j_0}^m} - \frac{X^k - X_{j_0}}{r_{j_0}^k}\right) \cdot x + \left(\frac{Y^m - Y_{j_0}}{r_{j_0}^m} - \frac{Y^k - Y_{j_0}}{r_{j_0}^k}\right) \cdot y + \left(\frac{Z^m - Z_{j_0}}{r_{j_0}^m} - \frac{Z^k - Z_{j_0}}{r_{j_0}^k}\right) \cdot z$$

with a four satellites k, l, m, n constellation, the design matrix A becomes:

$$A = \begin{bmatrix} \dfrac{X^m - X_{j_0}}{\rho_{j_0}^m} - \dfrac{X^k - X_{j_0}}{\rho_{j_0}^k} & \dfrac{Y^m - Y_{j_0}}{\rho_{j_0}^m} - \dfrac{Y^k - Y_{j_0}}{\rho_{j_0}^k} & \dfrac{Z^m - Z_{j_0}}{\rho_{j_0}^m} - \dfrac{Z^k - Z_{j_0}}{\rho_{j_0}^k} \\[3mm] \dfrac{X^n - X_{j_0}}{\rho_{j_0}^m} - \dfrac{X^l - X_{j_0}}{\rho_{j_0}^k} & \dfrac{Y^n - Y_{j_0}}{\rho_{j_0}^m} - \dfrac{Y^l - Y_{j_0}}{\rho_{j_0}^k} & \dfrac{Z^n - Z_{j_0}}{\rho_{j_0}^m} - \dfrac{Z^l - Z_{j_0}}{\rho_{j_0}^k} \\[3mm] \dfrac{X^n - X_{j_0}}{\rho_{j_0}^m} - \dfrac{X^k - X_{j_0}}{\rho_{j_0}^k} & \dfrac{Y^n - Y_{j_0}}{\rho_{j_0}^m} - \dfrac{Y^k - Y_{j_0}}{\rho_{j_0}^k} & \dfrac{Z^n - Z_{j_0}}{\rho_{j_0}^m} - \dfrac{Z^k - Z_{j_0}}{\rho_{j_0}^k} \end{bmatrix}$$

A prediction of the RDOP values during the desired survey period, could be useful to plan the observation so as to avoid divergences problems. A software to do this has been developed at our Department, starting from the satellite Almanac.

In Table 1. and 2. there is a comparison between the RDOP values obtained with an "epoch by epoch" kinematic pseudorange processing of a static baseline using a 4 satellite constellation (sats. N. 5, 7, 9 and 12), and the values obtained using this software for the two values pointed out in the first Table (574679 and 574874 GPS time of week).

As an example of what the consequence in kinematic GPS could be, in Figure 4 a divergence of the elipsoidal heights in meters is shown together with the corresponding RDOP values. The trajectory is part of an Antarctic glaciological survey [3]and, in the figure, the heights are plotted versus the run distance. The plot has been obtained in the postprocessing with a four satellite constellation. Neither cycle slip nor satellite set occurred in the considered time interval. Figure shows the same trajectory processed with the addition of one further satellite.

These divergences are also dangerous in the RTK applications: they can cause the transition from the "fix" to the "float" solution, or even to the pseudorange one.

4. Enviromental parameter location

GPS is a unique tool for locating the values of the environmental parameters measured by means of sensors mounted on the vehicle. We have proposed this method for monitoring the atmospheric pollution due to the traffic[4]. Experiments have been done in tunnels and in the urban canyons of Trieste (see Figure 5) by means of a carbon oxide sensor. In order to map the CO contour lines, it is necessary to repeatedly drive the vehicle along the streets. An accurate positioning guaranteeing the repeatability of the trajectories is therefore nedeed. Hence, the DGPS/DR is required.

REFERENCES:
1. Manzoni G., Cefalo R., Fonzari S., 1994 - Urban and suburban road survey by means of real time direct DGPS/DR and inverted GPS - DSNS 94, Londra 18-22 Aprile 1994, Proceedings.
2. Betti B., Crespi M., Sansò F., 1989 - GPS, geoide e sistemi di riferimento - in Ricerche di Geodesia Topografia e Fotogrammetria, Miscellanea per il 70° di Giuseppe Birardi, Milano, Clup, 1989
3. Cefalo R., Manzoni G., Tabacco E., 1995 - Kinematic processing of GPS trajectories around DomeC and to and fro DomeC and Dumond Durville - in printing on Terra Antarctica
4. Camus R., Cefalo R., Fonzari S., Manzoni G., Ravalli R., 1995 - Esperimenti di monitoraggio mobile dell'inquinamento atmosferico a Trieste - Rassegna Tecnica del Friuli Venezia Giulia N. 1, Gennaio-Febbraio 1995, Anno XLVI

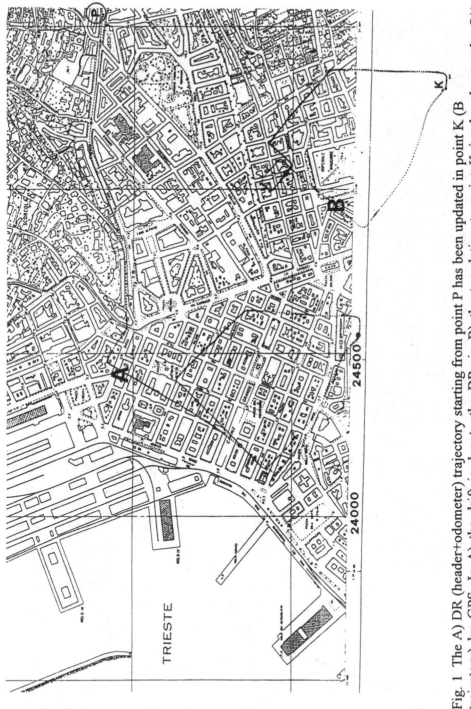

Fig. 1 The A) DR (header+odometer) trajectory starting from point P has been updated in point K (B trajectory) by GPS. In A) the drift is due to the DR, in B) the translation to point K is also due to the S/A

Fig 2 The B) DR trajectory of Figure 1 is superimposed to the 1: 5000 cartography

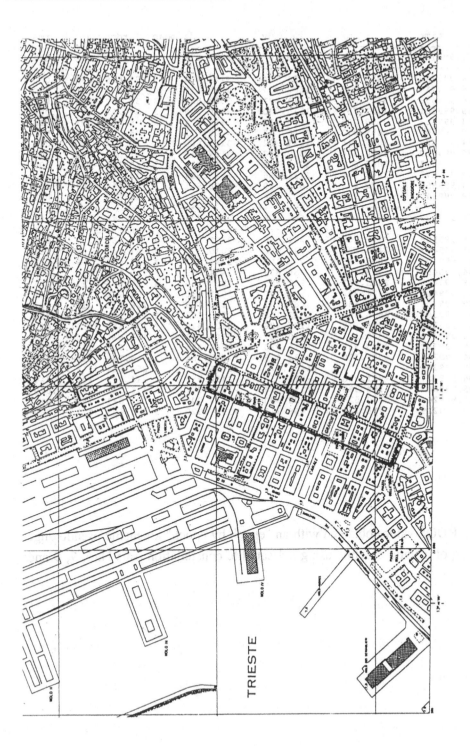

Fig. 3 A DGPS/DR trajectory in Trieste town center

time	dx	dy	dz	meas	rdop	epoch	rms
574679.999753	3396.724	1556.406	-3006.228	3	28.293	93	.002
574694.999735	2826.239	1233.704	-3124.480	3	29.618	94	.000
574709.999717	3454.337	1583.707	-2985.117	3	31.043	95	.000
574724.999699	3139.530	1411.591	-3063.530	3	32.583	96	.000
574739.999681	3321.751	1515.820	-3020.313	3	34.250	97	.000
574754.999663	2649.266	1135.958	-3165.274	3	36.060	98	.000
574769.999646	2859.232	1256.538	-3126.491	3	38.028	99	.000
574784.999628	2647.215	1131.382	-3161.505	3	40.177	100	.000
574799.999610	2450.304	1012.806	-3197.643	3	42.530	101	.000
574814.999593	2868.632	1255.144	-3115.133	3	45.113	102	.000
574829.999575	3647.368	1711.566	-2971.614	3	47.960	103	.001
574844.999557	4272.755	2082.189	-2858.911	3	51.114	104	.000
574859.999540	3120.335	1402.226	-3069.891	3	54.627	105	.001
574874.999522	3564.024	1666.658	-2992.704	3	58.549	106	.000
574889.999505	2868.508	1254.870	-3119.072	3	62.961	107	.001
574904.999487	3496.501	1632.090	-3010.395	3	67.943	108	.001
574919.999470	2035.615	752.108	-3249.497	3	73.626	109	.002
574934.999453	718.081	-58.873	-3454.552	3	80.145	110	.003
574949.999435	2323.337	922.905	-3199.350	3	87.668	111	.003
574964.999418	3471.222	1621.979	-3019.630	3	96.462	112	.002
574979.999401	2916.688	1288.391	-3110.166	3	106.875	113	.001
574994.999384	3638.643	1732.108	-3003.163	3	119.339	114	.001
575009.999366	2779.527	1195.326	-3121.310	3	134.540	115	.002
575024.999349	1373.953	310.747	-3304.703	3	153.431	116	.005
575039.999332	3283.930	1514.930	-3060.645	3	177.400	117	.006
575054.999315	6684.770	3686.405	-2657.354	3	208.788	118	.007
575069.999298	3189.837	1458.437	-3074.882	3	251.904	119	.013
575084.999281	-501.016	-918.653	-3480.519	3	314.212	120	.040
575099.999263	1184.385	156.318	-3278.971	3	411.281	121	.022
575114.999246	2490.529	1003.880	-3145.001	3	583.841	122	.017
575129.999229	255.325	-466.312	-3339.520	3	975.623	123	.065

Table 1. RDOP values obtained with an "epoch by epoch" kinematic pseudorange processing of a static baseline using a 4 satellite constellation (sats N. 5, 7, 9 and 12)

1995　　1　　14　　15:38:00　　1

		X (m)	Y (m)	Z (m)	local azimut (°)	local zenit (°)	pseudorange (m)
Yes	5	+22697031.56	-9064348.66	+10508662.43	+243.46	+43.98	21778600.57
Yes	7	+6349327.08	+16988341.16	+19616283.51	+66.75	+41.87	22046893.10
Yes	9	+15368297.29	-1888329.17	+21616306.11	+311.14	+69.73	20528997.10
Yes	12	+11720901.98	+4222856.16	+23527209.03	-+10.70	+68.10	20614360.63
No	23	+15729537.99	-19383161.21	+8826839.76	+265.88	+18.51	23756524.08
No	26	+25531533.14	+6652464.80	+4311605.64	+177.72	+43.56	21931930.76

Constellation　　5　7　9　12

PDOP = 284.59　　　RDOP norm. **28.21**

1995　　1　　14　　15:41:15　　14

		X (m)	Y (m)	Z (m)	local azimut (°)	local zenit (°)	pseudorange (m)
Yes	5	+22525816.69	-8855810.59	+11043295.88	+244.59	+45.32	21692057.15
Yes	7	+5890352.41	+16839222.53	+19882499.11	+65.01	+41.28	22085773.82
Yes	9	+15450708.71	-1360527.67	+21600136.56	+312.94	+71.09	20491998.27
Yes	12	+11599676.17	+4786514.39	+23492245.53	+14.92	+67.84	20634415.10
No	23	+15928207.39	-19464390.55	+8254019.58	+264.58	+17.73	23826193.45
No	26	+25600728.72	+6734017.99	+3707881.36	+177.57	+41.99	22034617.34

Constellation　　5　7　9　12

PDOP = 584.81　　　RDOP norm. **58.17**

Table 2. RDOP values obtained using the previsional RDOP software produced at our Department: the comparison has been made for the two values pointed out in the Table 1 (for 574679 and 574874 GPS time of week).

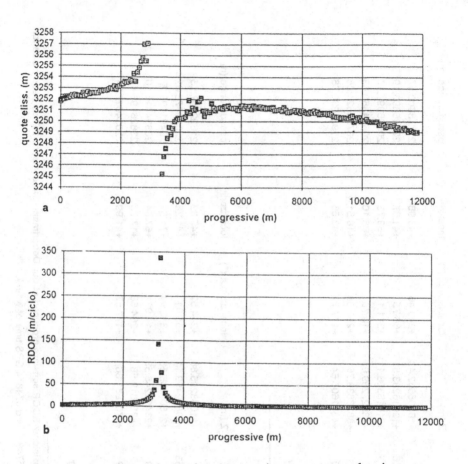

Fig 4   a) divergence of the solution in a kinematic postprocessed trajectory
          b) the corresponding RDOP values

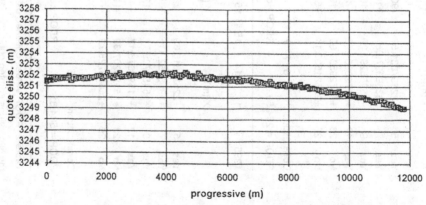

The same trajectory processed adding a fifth satellite

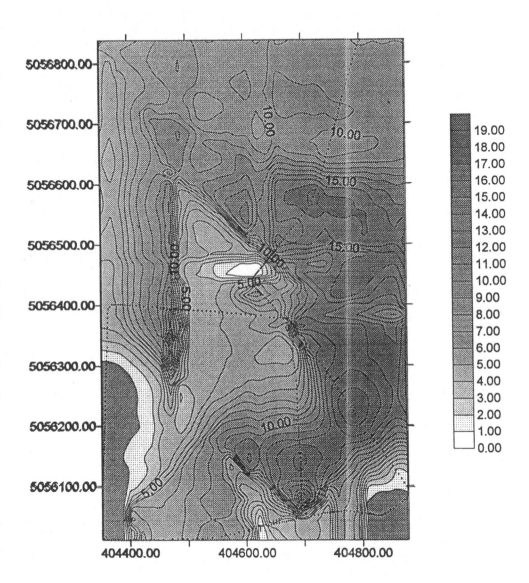

Fig. 5 A part of a DGPS/DR trajectory with the contour levels of the atmospheric CO concentration in ppm measured in real time by a sensor mounted on the vehicle

Fig. 5. A part of a DCU/SDU trajectory with the conforic levels of the same three
CO concentrations in ppm measured in real time by a sensor mounted on the vehicle.

DIGITAL ROAD MAPS AND PATH OPTIMIZATION APPLIED TO VEHICLE NAVIGATION SYSTEMS

M. Wieser

Technical University Graz, Graz, Austria

Abstract: This paper first gives an overview on Vehicle Navigation Systems (VNS), which are - at the time - intensively discussed within technical science in general and geodetic science in particular. Starting with a short description of VNS-components we briefly discuss the various types of systems available nowadays.

After dealing with numerous methods for the vehicle's location we shall focus on the digital road map as an essential instrument to handle all the geometrical, topological and thematic information of the road infrastructure by using the well known mathematical model of a graph. Novel topological elements and norm definitions for edges and edge sequences are introduced in order to meet the high degree of specializations within the thematic information. The digital road map also serves as an excellent experimental field for path optimization, route planning and other applications which all are related to combinatorial optimization. In this respect one has to investigate a lot of problems which occur together with the huge amount of digital road data and together with time consuming optimization procedures.

1. VNS ... components and types of systems

At the time we are confronted with a lot of projects, prototypes and commercially applied systems as a proof for the fact that there is a worldwide acceptance and usage of vehicle navigation.

From the geodetic point of view most of the systems contain the following main components which cover a wide range of scientific disciplines such as positioning, digital cartography, geoinformatics and optimization theory, the most of which more or less related to geodesy:

- **Vehicle location and position finding:**

  This major component should provide us with the vehicle's coordinates within the reference frame defined by the database. Starting at an initial point an autonomous system uses the so-called dead reckoning method, which is often used together with map-matching systems.

  In addition to these autonomous components we consider systems supported by infrastructural devices (e.g. transmitting and receiving facilities on beacons), radio communication or satellite positioning, such as the Global Positioning System (GPS).

- **Digital road map and geographic database:**

  The description of the geographic space concerning the roads and the traffic environment is performed by mapping the road system to a vector type structured digital road map which is done by using the mathematical model of a weighted (=valuated) graph. Within this structure of nodes and directed edges geometry, topology and thematics as the components of our digital model are handled in a way to meet the diverse types of information as road classification, capacity of roads, influence of topography, traffic restrictions, one way roads, prohibited turns and much more.

- **Path optimization and route planning:**

  Path optimization from a starting location to a given destination is an essential component of a VNS acting as a routing system. A shortest path serves as an optimal route between two given points, optimal either in the sense of geometric distance or time. Thematic information on the traffic environment has its influence on the results of the optimization procedure, which is handled with well known tools of graph theory. In the case of special types of a VNS questions may arise which are related to combinatorial optimization and are again based on the optimal path problem. Besides "spanning trees" and site planning, the traveling salesman problem, vehicle routing and the pickup and delivery problem are the most famous examples.

As shown in figure 1 the digital road map (a geographic database included) plays the role of a link between vehicle location and path optimization. Pure positioning is completed by route guidance and therefore leads to a fully integrated navigation system.

With respect to the kind of application of the above mentioned components one distinguishes between the following types of systems:

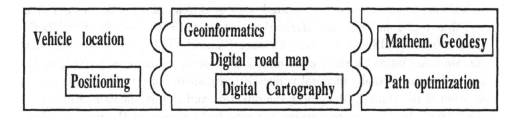

Figure 1: VNS-components related to geodetic topics

- **Autonomous systems:**

  An autonomous in-vehicle system is primarily used to answer the question "where am I ?". The aim is to position the vehicle basically using onboard devices as distance and heading sensors and to plot the calculated position on a display using a digital road map in order to visualize the vehicle's location relative to a desired destination. The main feature of that kind of a stand-alone system is that there are no external signals (maybe with the exception of GPS to support positioning), no control center, no communications link, no traffic information and that path optimization is only optional.

- **Dispatching systems:**

  A navigation system with the vehicle's position transmitted to a central office deals with the question "where are you ?" and is often called a dispatching system. In this case the navigation system turns to become a kind of GIS in the sense of a decision support system, which can be seen as an application to commercial vehicle operations for fleet management of commercial and emergency vehicles to accelerate deliveries and to increase operational efficiency. Besides GPS autonomous positioning of the fleet's vehicles is supported by Terrestrial Radio Frequency (TRF) techniques.

- **Routing systems:**

  In case one is interested in the question "how do I get there ?" so-called routing systems are taken into consideration, that means we are talking about vehicle navigation as the guidance of a vehicle from a starting point to a destination. A traffic control center collects actual traffic data with the help of beacons mounted on the sides of the streets, whereas the individual vehicle is anonymous to the control center, what is not the case within dispatching systems. Besides onboard path optimization the driver is provided with instructions how to choose an optimal route to a destination point with respect to a given set of constraints and criteria in the sense of an advanced traveler information system.

In addition to automatically generated route selection, taking into consideration real-time traffic and weather information transmitted with the help of the Radio Data System (RDS) on a Traffic Message Channel (TMC), such systems may contain information on service stations, gas stations, restaurants, parking facilities and so on. Visually or aurally, questions are answered concerning the actual location of the vehicle, concerning the direction and the line-of-sight distance to the driver's destination, concerning the road to be chosen at the next crossing, etc. In the future advanced vehicle control systems should apply additional technologies to vehicles, like radar instruments, night-vision sensors, and automatic braking in order to identify obstacles and adjacent vehicles to increase safety avoiding road accidents.

- **Inventory systems:**

  If one intends to collect data relevant for geoinformation systems a navigation system can be used as an inventory system.
  Used for data acquisition to create a thematic database of a specific area the kinematic or semi-kinematic system combines the position coordinates with information gained by an onboard surveying equipment supported by CCD- or video cameras. Very often the coordinate and time related data refer to objects of the road infrastructure as buildings, traffic signs, etc., see [1].

Not mentionend here are a lot of desktop systems, which cover the range of path optimization on the basis of digital road maps but do not have any component for vehicle location as it is typical for navigation systems. Because of the huge amount of already realized products worldwide we do not present a list for examples of VNS in this paper but rather give you an impression, how the various types of systems are used in selected regions of the world. As shown in the following table out of [2] dispatching systems are intensively distributed all over North America and Europe, whereas autonomous and routing systems are predominant in Japan, whereas inventory systems are rather rarely used all over the world, because of the relative high costs compared to the other types of VNS.

		Autonomous	Dispatching	Routing	Inventory
North America	%	18	62	4	16
Japan	%	55	3	39	3
Europe	%	22	50	26	2
worldwide	%	27	46	18	9

Table 1: Summary statistics on the types of 150 VNS (1975-1993)

2. VNS ... from the geodetic point of view

2.1. Vehicle location

As already mentioned in the previous chapter the determination of the vehicle's location or position plays an important role within a navigation system and it goes hand in hand with the computation of the vehicle's trajectory in an defined reference frame. The most frequently used techniques are listed and explained below:

- **Dead reckoning:**

  The dead reckoning method provides the actual position adding vectors gained by measured increments of distances and directions and requires wheel sensors which are generating pulses. These wheel sensors may be applied as a differential odometer as well as a heading sensor combined with magnetic compasses.

- **Map matching:**

  As a complement to the "hardware facilities" map matching compares the trajectory of the vehicle to the digital map database on a storage medium like a CD-ROM in order to correct the position and to give a visual information of the vehicle's actual location.

- **Global Positioning System (GPS):**

  GPS is the well known satellite positioning system developed by the US Department of Defense which is available to civil users under certain accuracy and reliability constraints. On the basis of code measurements the vehicle's absolute position using differential GPS (DGPS) can be gained in real time with an accuracy of $\pm 10m$, although one has to be aware of the visibility problem in a highly masked urban environment.

- **Terrestrial Radio Frequency (TRF):**

  Radio frequency signals are received from a number of transmitters distributed within the area of interest. The exact position of the vehicle is determined by the intersection of the incoming signals.

- **Beacon Systems:**

  Infrared or microwave devices which are mounted on sign posts or beacons on the sides of streets are able to receive and transmit data from and to vehicles passing by and being equipped with suitable transceivers.

A representation relative to the total number of navigation systems within the already mentioned continents is given in table 2, which again is derived from statistics compiled by Krakiwsky on the basis of 150 VNS between 1975 and 1993. The column headed by "others" counts inertial systems on the one hand and recently developed satellite systems combining positioning and communication tools on the other hand. Generally spoken one can say that the high usage of GPS in North America goes hand in hand with the fact that dispatching systems are strongly represented there. In Japan and Europe, where you have a lot of autonomous and routing systems dead reckoning, map matching and beacon systems are frequently used.

	Dead Reck.	Map Matching	GPS	TRF	Beacons	Others
North America	30	11	70	21	5	6
Japan	68	55	48	3	32	0
Europe	55	24	38	19	19	5

Table 2: Statistics on types of positioning systems

One important thing remains to be mentionend in this context: to avoid disadvantages of the individual positioning techniques one is forced to combine several methods, although mathematical tools of signal processing have to be intensively applied in addition when smoothing, filtering and combining different signals to calculate the vehicle's position. In this respect dead reckoning, map matching and GPS are often combined recently (see [3]): because of measurement and digitalization errors the vehicle's trajectory cannot be directly identified with the equivalent path in the digital road map and is therefore map matched to it; in the surroundings of road intersections it may happen that the trajectory does not uniquely result in one digital path and one needs additional tools for a decision-making; besides the information out of a path optimization, that one should be on the suggested road, absolute positioning in general and GPS in particular can be very helpful.

2.2. Digital Road Map

At first it has to be mentionend that general aspects of any digital data handling are of course valid with respect to a digital road map. Essential importance is attached to data acquisition including information updating, data quality and data exchange. What data acquisition is concerned Europe is on the way to a digital road map, designed to be to a high degree accurate and complete with respect to data quality. Besides several national activities on road data acquisition the European Digital Roadmap Association (EDRA) as a cooperation of Bosch (Germany), Tele Atlas (The Netherlands) and Etak (United States) has started to produce a large scale digital street map database for all of Europe in response to a great demand for a uniform digital road map of our continent. These plans go hand in hand with the effort to generate a standardized data exchange format for road data, what has been done by the task committee TC 278 on "Road Transport and Traffic Telematics" within CEN (Comité Européen de Normalisation) in a first step by generating a standard format which is called GDF (Geographic Data Format); for more details see [4].

Let us now discuss model aspects and the graph model representing the digital road map. We use a valuated graph, which is of course finite and in addition directed and strongly connected, because for every pair of nodes (A,B) there should be a path from A to B and a path from B to A. In general the graph is cyclic (i.e. with at least one cycle) but it is not complete (not every pair of distinct nodes is joined by an edge) and not regular (nodes have different degrees (degree=number of edges incident with a node)).

As shown in figure 2 geometry is usually defined by two- or threedimensional coordinates of nodes (node=vertex) representing all intersections of roads. In addition one has polygonal points to describe the connection between two nodes in case it is not a straight line and special sites to handle objects slightly off the link between two nodes. Basic topology is defined by edges and by novel topological elements, which will be called traverses in the sequel. While an edge describes a direct link between two nodes, the traverse is responsible for a pair of links or a group of three nodes respectively. The crucial point is that thematic information can either affect nodes (e.g. traffic lights), edges (e.g. road classification, traffic conditions, etc.) or traverses (e.g. traffic rules at intersections). Special informations (so-called thematically based topology) such as one-way restrictions or forbidden turns are handled by oriented edges (arc from A to B) and oriented traverses (arc from A over B to C) establishing a directed graph.

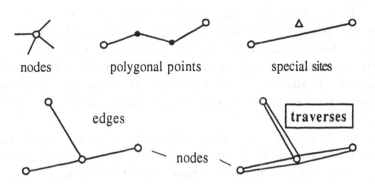

Figure 2: Geometrical and topological elements

2.3. Path Optimization

In connection with path optimization and route planning the question arises how to perform the above mentionend valuation of the graph's topological elements, or in other words, which norm is to be used in order to meet reality.

As far as the edge is concerned one may introduce the pure geometrical length l_{ij} (on the basis of the L_2-norm) if an optimal path with respect to distance is looked for. The definition of a valuation number b_{ij}, describing traffic conditions in the sense of a resistance value (the higher the value the lower the probability to choose the edge for an optimal path with respect to time), may lead to a hybrid norm $t_{ij} = b_{ij} * l_{ij}$. As in random walk theory this "thematic" length t_{ij} ($0 < t < \infty$) may be replaced by a probability number p_{ij} ($0 < p < 1$, $t_{ij} = -\log p_{ij}$). Even for the traverse a valuation number b_{ijk} ($1 < b < \infty$) may be introduced.

As explicitly stated in [5] it is a matter of experience to choose a thematic length for a path as a sequence of arcs in order to meet reality. A suggestion for a hybrid norm, which has been tested in practice and has reproduced suitable results may be expressed as follows: take the sum of the thematic lengths of all directed egdes and multiply it with the average of all valuation numbers belonging to the traverses along the path.

With the help of the above norm definitions one easily gets an optimization criterion how to proceed from A to B on an optimal path. In principle we minimize the sum of thematic lengths or maximize the product of all probability numbers along the path:

$$\phi b_{ijk} \sum_s t_{ij} \stackrel{!}{=} Min \quad or \quad \left(\prod_s p_{ij} \right)^{\phi b_{ijk}} \stackrel{!}{=} Max$$

What is needed now is a suitable path algorithm, which we have to apply with respect to our norm definitions. On the one hand combinatorial optimization offers a lot of algorithms (FORD-MOORE, DANTZIG-DIJKSTRA, etc.), on the other hand one can use descriptive methods based on the definition of a sufficient set of optimal routes within the network. In addition one must not forget the concept of "spatial reasoning" which uses hierarchical structures to simplify wayfinding on the basis of a cognitive map; see [6].

Besides the optimal path problem there are a lot of other applications of combinatorial optimization which are useful in navigation systems applied as a kind of decision support systems. One of the most famous example in this respect is the so-called traveling salesman problem, which provides us with a number of positions and the objective to visit each of the sites and then to return to the starting point while minimizing the total distance (or time) traveled. In a more general context these problems are handled as "vehicle routing" or as "pickup and delivery" problems. In addition to that one may come in contact with site planning, that means the search for suitable positions for schools, bus stops etc. Looking for shortest routes to positions within a given distance from a common site leads to a so-called spanning tree which can be used to display zones of certain distance areas. The example shown in figure 3 is generated by the automobile information system (AIS) which has been developed at our Department of Mathematical Geodesy and Geoinformatics at the Technical University in Graz.

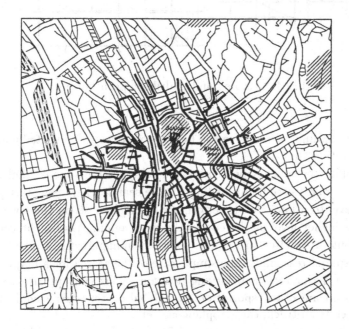

Figure 3: Example for a spanning tree

When optimizing within digital road maps one must be aware of the fact that time consuming algorithms are combined with a huge amount of geometric and thematic data. With the help of intelligent path algorithms and under certain conditions it is possible to reduce computation time, which originally grows with the order of $n * n$, computing a path from A to B (n being the number of nodes within a complete graph); computation time is significantly reduced in the case of digital road maps which are usually close to so-called planar graphs (a planar graph can be imbedded in a plane such that it edges intersect only at its nodes, see [7]).

Another possibility is to introduce modern strategies, two of which are briefly introduced in the sequel and are shown in figure 4:
Developing the search tree to reach a destination point one can introduce higher weights for edges correlated with the line of sight between the starting and the destination point. What you get is an elliptic search tree instead of a circle, and what you do in the sense

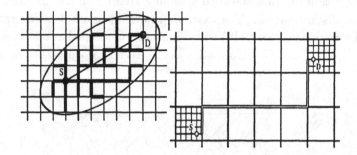

Figure 4: Sectored and multilevel path finding

of "sectored path finding" is reducing the number of nodes to be visited and therefore
reducing computation time.

A second possibility to reduce node density and computation time is "multilevel path
finding", in the simplest case performed as a triple search where you start proceeding
from a higher level to a lower level, performing the main task on the lower level and
returning to the higher level at the end of the path again. You might get a suboptimal solution but in practice this equals - corresponding to the already mentioned
GIS-concept of spatial reasoning - the simple case of the search for a major road, proceeding along the major road and leaving it again. This concept also works if you
consider different means of transportation, in a global context this is equivalent to the
combination truck-train-truck, car-aeroplane-car, etc.

A last thing to be mentionend in context with way finding is the address conversion
in case your route planning refers to house addresses rather than to node numbers. In
this connection one has to be aware of the fact that the introduction of a node per
address increases your data set and your computation time tremendously. Because of
that reason either introduce precomputed additional nodes for addresses in case they
are used or just find the accessory node by interpolation between road intersections
where addresses have to be known.

References

1. Schwarz, K.P., H.E. Martell, N. El-Sheimy, R. Li, M.A. Chapman, and D.
    Cosandier: VISAT - A Mobile Highway Survey System of High Accuracy. VNIS
    Conference, Ottawa, October 1993.

2. Krakiwsky, E.J.: Comparison of IVHS Navigation Systems, in: Proceedings of the 3rd International Conference on Land Vehicle Navigation, Dresden, 1994, 5-12.

3. Kreft, P.: Multipath Map Matching (M^3), in: Proceedings of the 3rd International Conference on Land Vehicle Navigation, Dresden, 1994, 199-206.

4. Ad Bastiaansen: Status and Directions of Uniform Digital Road Maps in Europe, in: GIS in Transport und Verkehr (Ed. N. Bartelme), Mitteilungen der geodätischen Institute der Technischen Universität Graz, Folge 80, 1995,71-75.

5. Wieser, M.: Efficient use of algorithms for path optimization in GIS: problems and requirements, paper presented at the 8th Conference of the European Consortium for Mathematics in Industry, to be published by Teubner 1995.

6. Car, A. and A.U. Frank: Modelling a Hierarchy of Space Applied to Large Road Networks, in: IGIS'94: Geographic Information Systems (Ed. J. Nievergelt et al.), Springer, Berlin Heidelberg 1994, 15-14.

7. Chachra, V., P.M. Ghare, and J.M. Moore: Applications of Graph Theory Algorithms, Elsevier North Holland, New York 1979.

MAINTAINING CONSISTENT GEOGRAPHIC DESCRIPTION OF THE ENVIRONMENT OF AN AUTONOMOUS MOBILE ROBOT

G. Borghi

Polytechnic of Milan, Milan, Italy

and

CNR-LADSEB, Padua, Italy

D. Brugali

Polytechnic of Milan, Milan, Italy

and

Polytechnic of Turin, Turin, Italy

ABSTRACT

In this paper we present a method for a mobile robot to explore autonomously an unknown environment. many issues are involved by such a task. In particular we present a model to represent geographic knowledge, based on an extension of the "Diktiometric representations"; this kind of representations broaden the topological model to include the geometric relations between places. The model we propose is a network of 2D geometric descriptions connected by arcs with geometric relations between nodes. We paid special attention to the maintenance of this model, providing mechanism to allow the consistent fusion of sensory observations. To keep low uncertainty inside each node, we introduce in the fusion process "internal relations", whose measurements are affected only with the uncertainty of the sensor system and not on the robot location.

1. INTRODUCTION

Many tasks for mobile robots require a detailed description of the environment.
A goal for many researchers is programming the robot to acquire autonomously robust and consistent descriptions of its surroundings using sensory information; the robot must be able to operate in unstructured environments with little a priori information. This kind of task is commonly called exploration of an unknown environment (see for example [3]).
The exploration task involves many problems, including the robot self-location, the sensor data integration, the exploration strategies, the path planning, the obstacle avoidance.
The issue of representing sensory information in a complete and consistent manner is one of the most challenging problems faced by the research community. In recent years, several approaches to the robot exploration have been proposed.
The geometric approach to world representation attempts to build a detailed metrical description of the environment from sensor data (see for example [2], [6]). This kind of representations has a reasonably well-defined relation to the real world, but is highly vulnerable to metrical inacurancy in sensory devices and movement actuators.
The "Occupancy Grid" [5] framework represents a fundamental departure from traditional geometric approaches. It uses a stochastic tessellated representation of spatial information maintaining probabilistic estimates of the occupancy state of each cell in a spatial lattice.
Other works [3], taking a more qualitative approach, show great promise of overcoming the fragility of purely metrical methods: they consist in a topological description of the environment. The model is a network of nodes, where nodes represent distinctively recognizable places in the environment, and arcs control strategies which take the robot from one place to another.
DIktiometric representations [1] broaden the topological representation to include the paths' shapes, i.e. the geometric relations between places. In [1] only "point-like" places are considered.
The aim of this paper is to present an extention of the "diktiometric representation" model of Engelson and McDermott [1] to represent geographic knowledge; our extension broadens the "diktiometric representation" to include not only the geometric relations between places, but also the geometric description of every place.
Furthermore we explain how this extension is advantageous to cope with the main issues involved by the exploration task, such as reducing the robot position uncertainty and maintaining the consistecy of the representation model.
The whole representation is a network of 2D geometrical representations with the characteristic that the uncertainty about features is locally bounded.
When new features have to be merged in one of the geometric descriptions of the graph, their parameters must be expressed in the correspondent reference frame. This needs relations (called "external relations") that are affected even with uncertainty on the robot location.
To keep low uncertainty we introduce in the merging process "internal relations". Internal relations specify distances between corners or angular widths between edges belonging to

the same visual scene; therefore their measurements are affected only with the uncertainty of the sensor system and not on the robot location.

Measurements about internal and external relations are used to obtain a new estimate of the parameters.

Some hypotheses about the world limit the range of applicability of the approach: the robot is able to explore an indoor environment where the obstacles are polyhedral objects. Furthermore the world is static: the position, shape and size of the objects do not change.

The theory outlined in this paper is implemented in a project called EXPLORER; a first prototype of EXPLORER has been implemented and is currently under assessment.

The paper is organized as follows. Section 2 introduces the theoretical approach and the basic features of the representation model. Section 3 focuses on our version of the "diktiometric representation" and presents a new approach to the problem of maintaining the consistency of the model. Section 4 draws some conclusions.

2. THEORETICAL APPROACH AND INSTRUMENTS

In order to cope with intrinsic limitations of the sensory system and with the uncertainty of the provided observations, representation models must be able to express uncertain, partial and often spurious information.

We chose to represent geometric uncertainty in a probabilistic way.

The robot acquires sensory observations using a telemeter device composed of a laser range finder and a rotary table that allows the laser finder to rotate around a vertical axis.

Each telemetric reading provides the distance measurement of the nearest obstacle in front of the laser device; this measurement, together with the sensor orientation, provides the location of an obstacle point in polar coordinates inside the reference frame tied to the rotary table.

A telemetric scanning provides therefore the location of a set of points belonging to the edge of visible objects around the robot.

This set of points is then clustered in subset, in order to obtain line segments by fitting a staight line to each subset. The estimation of the line parametres is a typical linear regression problem: the least square method is used. This metod supplies the parameters of each line segment and the associated covariance matrix.

Once this process is concluded a set of corners, related to the robot reference frame, are obtained by geometric intersection of corresponding pairs of line segments.

Corners are the basic elements of a robot visual scene.

Since vertices can be end-points of edges representing part of an object boundary, they do not necessarly corrispond to true corners in the real world. It is essential to distinguish true corners from "dummy" ones. Only true corners are recognizable from different points of view.

3. THE DIKTIOMETRIC REPRESENTATION

The world representation is built merging a sequence of consecutive 2D visual scenes, where a visual scene is a geometric description of the environment built using sensory data acquired from a single robot position. A visual scene is related to the robot reference frame and henceforth it will be called "local view".

The main problem in such world-modelling is to compute the relative position of features (i.e., corners) observed from different points of view in order to merge them in a consistent manner.

Two types of information may be used to compute the relative positions: the matching of geometric features from visual scene to visual scene, and the best estimate of the current position of the robot as given by the odometric system.

EXPLORER doesn't make use of visual landmarks and it doesn't rely on the odometric system as in [5]. EXPLORER uses only geometric characteristics of detected objects to find correspondeces among their features and then their relative position.

The proposed self-location method (Section 3.1) shows the weakness to depend on the amount of available information characterizing the environment, that is, the robot is obviously not able to estimate how far it traveled along a corridor, where it sees just a couple of parallel walls.

Since this weakness is common to every approach attempting to build a geometric description of the world referred to a global reference frame, we face this limitation proposing an extension of the "diktiometric model" of Engelson and McDermott [1].

Engelson and McDermot deal with a network of places being small regions which can be treated as single points. In our "diktiometric representation" places may be either single "local views" or "extended views", that is sets of merged local viewes referred to a single reference frame attached to the place.

Fig. 1 shows an example of our representation: each reference frame corresponds to a node of the diktiometric graph.

The displacements between frames are the measure associated to the arcs of the graph.

The robot begins by wandering in a new territory with a null or partial description of the environment. Whenever the robot reaches a new position, it builds a local view, related to the robot reference frame, using the information provided by the laser device.

Afterwards it tries to self-locate by matching the geometric features of the new local view with the geometric features belonging to one place of the graph so far acquired; i.e. the robot tries to estimate the displacement between the current reference frame and one attached to a place of the graph.

When the displacement is provided, the local view is merged into the correspondent place obtaining a new extended view (frames E1, E2, E3, E4 in Fig. 1).

The robot currrent position is related to the reference frame attached to that place. It is clear that in a diktiometric representation the robot absolute position is meaningless.

As said above, when the local view is lacking in information the robot cannot self-locate. In such a case the local view is not allowed to be merged into any place of the graph.

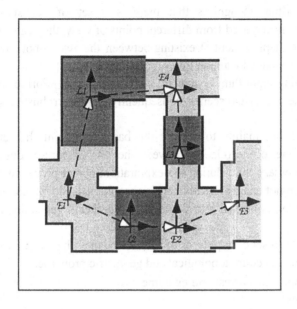

Fig. 1, The diktiometric representation

A new place in the diktiometric graph is inserted (frames L1, L2, L3 in Fig. 1), and an arc will be created between the new place and the place to which the robot previous position was related, representing the rigid displacement between their reference frames. The displacement is computed taking into account the odometric estimation.

This way the geometric uncertainty inside a place is kept low (the robot can self-locate) and it is not propagated among different places.

2.1 Matching and robot self-location

The exploration of an unknown environment requires the robot to change continuously its position in the environment.

It is essential for the model consistecy to know as well as it is possible the robot position in the real world.

The odometric system supplies an estimation of the robot position/orientation, but it is not possible to rely on it.

Therefore we need instruments and methodologies in order to cope with the uncertainty on the robot position and reduce it.

The followed approach is based on comparisons between two visual scenes: the last local view and an extended view attached to a place of the diktiometric graph.

This comparison supplies corrispondeces among the corners of the two visual scenes.

In our representation the matchng process has to compare different measurements of corners taken from different robot positions.

If the matching algorithm recognizes that there is a set of corrispondences among measurements of corners acquired from different points of view, they can be used to reduce the uncertainty of the displacement D existing between the two robot positions. This is done by applying the Extended Kalman Filter (Ayache-Faugeras [4]).

The proposed matching algorithm is able to work without a priori estimations of the relative position of the two visual scenes, consequently it shows robustness in presence of serious motion errors.

The matching process has mainly to satisfy the basis constraint that an hypotesis of correspondence is a one-to-one relation between the sets of corners describing the two visual scenes. It is necessary that, during the exploration, object properties do not change, that is object shapes do not change, relative positions of objects do not change, etc.

The process of matching is broken into two parts: hypothesis generation and hypothesis verification.

The algorithm generates a set of hypoteses of correspondances between corners of the two visual scenes, taking into account topological and geometric properties of their objects.

- Topological properties: corresponding corners have to join corresponding edges in the two visual scenes.
- Geometric properties: corresponding corners must have similar angle width in the two scenes, and similar distances from adiacent corresponding corners.

Fig. 2 shows an example of hypotheses generation, where several corresponding geometric properties have been individuated.

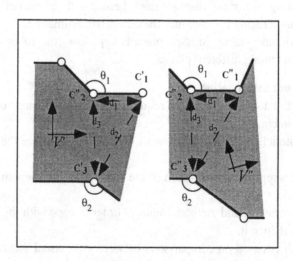

Fig. 2, Matching of two visual scenes

Each hypotesis of correspondences is verified: a subset of corrispondences between corners belonging to an hypotesis is used as measurements in the EKF to estimate the visual scene relative position.

This estimate is therefore used to verify the consistence of the other correspondences of the same hypotesis. If the hypothesis is validated all its correspondences are used to estimate the visual scenes relative position.

In case of particular symmetries several equivalent hypoteses can be validated; it is necessary to use the odometric estimate to disambiguate them.

In the next section we show how corrispondences between two visual scenes and estimates of the displacement between their reference frames are used to allow the fusion of those visual scenes.

2.2. Integrating sensory observations

A real environment can be very complex; therefore, because of the limited range of action of the sensors, the robot should be able to integrate sensor observations acquired from different points of view.

When the matcher successfully individuates a consistent set of corrispondences, among the corners of a new local view and the corners of an extended view of the diktiometric graph, the merger has to integrate the local view into the extended one.

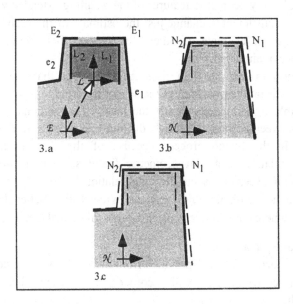

Fig. 3, Views integration. The local view L is merged with the extended view E obtaining the extended view N

For example, Fig. 3.a shows an extended view, related to the reference frame E, and a local view, related to the reference frame L. The matcher has identified the corrispondences {(E1, L1), (E2, L2)}.

The classical approach to the fusion problem would compute, using the EKF, a new estimate of the parameters of each common corner, taking into account the two available estimates, with associated uncertainty, and the estimate of the displacement D between frames E and L. Although we use a "good" estimate of the displacement D, serious fusion problems may occur.

Let's suppose that in the real world there are only mutually orthogonal edges, and that our sensorial system is able to measure the angular width of the corner with high precision. Let's suppose that in the so far acquired extended view, due to uncertainty propagation, there are edges e1 and e2 only roughly parallel.

Following the classical approach, the fusion of E1 with L1 and of E2 with L2 would produce an extended view as showed in Fig. 3.b in which only the parameters of these corners have been updated.

The result is unacceptabie, because the positions of corners N1, N2 was computed without taking into account the good confidence of the measurement of the angular width of the corresponding corners L1 and L2 in the local view and E1 and E2 in the extended view.

We argue that the key to obviate such a shortcoming is to introduce relations, among corners, that are indipendently from the displacements D and to distribute the uncertainty onto the parameters of every geometric features of the resulting extended view.

In other words, it is not sufficient to change just the estimate of the reobserved features; it is necessary to change the estimate of others features so that the consistency of the geometric description is maintained.

The proposed solution to this problem rises from the fact that in a local view the uncertainty only depends on the precision of the sensor device.

Therefore it is possible to individuate inside the local view relations between geometric features (distances between corners, angular width between edges, etc.), which have to be taken into account in the fusion process because of the low uncertainty on their parameters. We call relations between geometric features, such as distances between corners, angular width between edges, etc. "internal relations".

Following this approach we obtain a new extended view with a better defined relation to the real environment, where consecutive walls are quite orthogonal (see Fig. 3.c).

2.3 Estimating feature parameters

Starting from the corrispondences provided from the matcher every corner of the new extended view is set in corrispondence with a corner of the old extended view, a corner of the local view or both.

Subsequently the parameters of every features belonging to the resulting extended view are estimated, taking into account all new and old relations among them.

Following our approach to the sensory observation fusion, the problem of estimating the corners parameters is simply treated as a classical problem of linear regression.

Let be c_1, c_2, ... , c_i, c_{i+1}, ... , c_{2k} a set of 2*k parameters to be estimated where k is the number of corners belonging to the new extended view and (c_i, c_{i+1}) are the coordinates (x, y) of the i-th corner; given furthermore a set of N linear equations of the form

$$y(t) = c_1 * u_1(t) + ... + c_{2k} * u_{2k}(t)$$

representing relations (internal and external) among the corners of the new extended view, where t = 1,2,...,N and $y(t)$, $u_1(t)$, .. , $u_{2k}(t)$ are the measures of 2k+1 real variable obtained in the old extended view, in the local view or both.

The estimate of the 2k parameters c_i is computed using the Markov's weighted least squares method. Let be

$$\theta = \begin{bmatrix} c_1 \\ ... \\ c_{2k} \end{bmatrix}$$

and $\hat{\theta}$ the estimate of θ,

$$\Phi = \begin{bmatrix} u_1(1) & ... & u_{2k}(1) \\ ... & ... & ... \\ u_1(N) & ... & u_{2k}(N) \end{bmatrix}, \quad y = \begin{bmatrix} y(1) \\ ... \\ y(N) \end{bmatrix}, \quad v = \begin{bmatrix} v(1) \\ ... \\ v(N), \end{bmatrix}$$

and V the covariance matrix of v, the following relation holds:

$$y = \Phi\theta + v$$

the Markov's estimator and its covariance matrix are:

$$\hat{\theta} = \left[\Phi^T V^{-1} \Phi\right]^{-1} \Phi^T V^{-1} y, \quad \text{var}\left[\hat{\theta}\right] = \left[\Phi^T V^{-1} \Phi\right]^{-1}$$

External relations are used to specify corner positions in the new extended view, knowing the position of corrispondent corners in the old extended view, in the local view or both.
External relations involve the displacements between the local view reference frames and the new extended view reference frame; therefore their measurements are affected even by uncertainty of the robot location .
Internal relations specify distances between corners or angle widths between edges belonging to the same visual scene; therefore their measurements are not affected with the uncertainty of the robot location .
Let's consider again the examples of Fig. 3.
The local view L and the extended view E are to be merged into the new extended view N.
The old extended view and the new one are referred to the same reference frame; furthermore it is known the displacement D between this frame and the local view.

The merger, sets in corrispondence the corner N2 with the corner E2 trough an external relation, whose coefficients are represented in the first two rows of matrix Φ; furthermore rows 3 and 4 of Φ represent the coefficients of the external relations setting in correspondences N2 with L2:

$$\Phi = \begin{bmatrix} 0 & 0 & 1 & 0 & ... & 0 \\ 0 & 0 & 0 & 1 & ... & 0 \\ 0 & 0 & \cos(\theta) & \sin(\theta) & ... & 0 \\ 0 & 0 & -\sin(\theta) & \cos(\theta) & ... & 0 \\ & ... & & & ... & 0 \end{bmatrix} \quad y = \begin{bmatrix} E2_x \\ E2_y \\ L2_x + a*\cos(\theta) + b*\sin(\theta) \\ L2_y - a*\sin(\theta) + b*\cos(\theta) \\ .. \end{bmatrix} \quad \theta = \begin{bmatrix} N1_x \\ N1_y \\ N2_x \\ N2_y \\ .. \\ Nk_y \end{bmatrix}$$

a, b, θ are the parameters of the displacement D.

The merger can furthermore identify some internal relations, such as the distance between corners (N1, N2), whose measurement is obtained in the local view computing the distance between corners (L1, L2).

In a concise way this relation can be written as follows:

$$DIST_{L1-L2} = |N2 - N1|$$

Obviously these relations are to be linearized; the corrispondent coefficients are inserted in new rows of the matrix Φ and the correspondent measurements in the vector y. Relations such as the distance between not consecutive corners provide constraints on angular width between edges.

5. CONCLUSION AND RESULTS

We have presented a new approach to world-modeling based on an extension of diktiometric representations.

The use of internal relation allow us to introduce in the merging process information such as angular width and distances from corners, information affected only by sensor uncertainty. Diktiometric representation and internal relations allow us to obtain a network of geometrical representation in which the uncertainty is locally bounded.

Fig. 4 shows snapshots of the user interface during the EXPLORER activity. Fig 4.a shows the simulated environment. Fig. 4.b shows the diktiometric representation: the dashed lines individuate the different extended and local views. A high degree of consistency between the diktiometric representation and the simulated environment can be observed.

The presented approach was tested in simulation. The simulation results show that the robot may succesfully represent an indoor environment. Currently we are developing real experiments with one of the mobile robot of the PM-AI&R Laboratory.

The present implementation of the EXPLORER project runs on a workstation SUN.

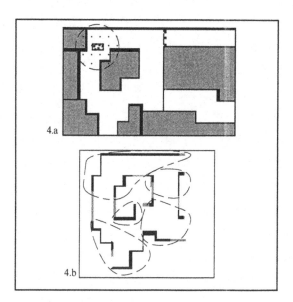

Fig. 4 - The simulated environment 4.a and
the diktiometric representation 4.b

6. REFERENCES

1. Engelson S.P and McDermott D.V. : "Error Correction in Mobile Robot Map Learning", Proc. IEEE International Conference on Robotics and Automation, Nice (1992), pp. 2555-2560
2. Durrant-Whyte H.F. : "Integration, Coordination and Control of Multi-Sensor Robot Systems", Kluwer Academic Publishers 1988
3. Kuypers B.J, Byun Y., : "A robot exploration and mapping strategy based on a semantic hierarchy of spatial representations", Robotics and Autonomous System 8 (1991), pp. 47-63
4. Ayache N. and Faugeras O.D. : "Maintaining Representations of the Environment of a Mobile Robot", IEEE Trans. on Robotics and Automation, 5(6) December 1989, pp. 337-350
5. Elfes A : "Sonar-Based Real-World Mapping and Navigation", IEEE Journal of Robotics and Automation, Vol RA-3, NO 3, June 1987
6. Preciado A., Meizel D. et al. : "Fusion of Multi-Sensor Data: a Geometric Approach", Proc. IEEE International Conference on Robotics and Automation Sacramento, California - April 1991, pp. 2806-2811

ROBUST TECHNIQUES FOR DATA PREPROCESSING

M.A. Brovelli, F. Migliaccio, L. Mussio and O. Sharif
Polytechnic of Milan, Milan, Italy

ABSTRACT

A summary on numerical cartography and digital photogrammetry, taking into account also some aspect of GIS's and image processing, is presented, in order to list the principal problems of spatially referenced data analysis.

As it is very well known, reliability and robustness are important properties, that may be assured in the data analysis by using suitable advanced techniques, but their introduction in the collocation method is at present problematic.

An attempt to introduce robust estimators in the collocation method, taking into account the different steps in which it is split, has been done and the results are here presented and discussed.

1. A VIEW ON GIS WORLD

A GIS requires information from image processing, obtained by photogrammetric or remote sensors, as well as by means of hardcopy scanning of previously existing maps. Furthermore a GIS exchanges information with more general data bases.

A GIS is often object oriented and these objects represent entities or entity classes. They get a spatial reference, while the most commonly used data banks do not get it.

Attributes of various types are associated to the entities: they supply information about the entities both from the geometrical and semantic point of view. The entity-attribute relations are characterized, if binary, by four cardinal values that represent the smallest and the largest numbers of entities in each direction of the relation.

The geometrical entities are regions, lines and points, while their attributes may represent different characteristics and they may constitute particular subentities.

The georeference of each object allows for spatial (topological and geometrical) analysis. In this way data integration and data fusion are possible and suitable mathematical models can be applied to study particular phenomena of different physical nature and semantic contents.

The dimension of the data bases of a GIS requires a suitable organisation, with special regard to the structure of the data bases. A suitable methodology to implement the data bases and the set of entities of a GIS, that is used also in non-spatially referenced informative systems, is composed by four phases:
- spatial;
- conceptual;
- logical;
- physical.

The first phase consists of a spatial object description; successively this description becomes abstract, so that the real world is described in a complete, non-ambiguous and of minimum redundancy way. This set of formal rules represents the conceptual scheme of the data, that is followed by the complex process of data modelling.

Data exchange becomes quite immediate from the logical point of view, which gives some advantages, like the capability to supply information, relatively independent from hardware platforms and operative systems (software). Thus the physical implementation is only the last phase in the implementation of a GIS.

Many other problems are to be solved during the phases of GIS implementation and, among these, the most relevant are: data quality control, data security and facility in management, so that the system becomes user friendly.

The structure of the spatially referenced data bases often shows vector characteristics and is classified into the following types of structures:
- hierarchical;

- network;
- relational.

The first type of structure goes over the rigidity imposed by electronic sheets; it is very easy to be defined, but a lot of non-hierarchical relations are ignored.

The second structure is more general and efficient than the first one, but it requires a lot of pointer arrays and therefore its implementation is too complex.

The last structure goes over the constraints of both the first and the second structures, although it can be time consuming in the search of information. It merges two topological graphs, while the hierarchical structure is represented by a tree and the network structure by a graph only.

An example confirms the validity of the relational structure. In fact given a geometrical object constituted by regions, lines and points, it is immediate to find for each line the two end points, but not the two adjacent regions.

On the other hand applying the topological relations of duality:
- regions-nodes;
- lines-arcs;
- points-zones;

a new abstract dual graph is obtained, where it is immediate to find for each arc the two end nodes, i.e. the two adjacent regions.

Suitable combinations of these structures, especially when object oriented, are also possible. Among these, structures capable to export and import data are very important, because in the external phase particular operations can be performed. Indeed they, because of their dimension and complexity, can not be usually defined inside the system and require special packages, highly qualified.

2. SUMMARY OF NUMERICAL CARTOGRAPHY

A GIS supplies information in form of maps and, if necessary, of auxiliary images. Indeed its aim is to establish relations among objects and users, by means of dedicated hardware and software, in order to get data analysis and information distribution/circulation.

Geographic data combine semantic attributes, related to different properties of the investigated objects, and geometrical components, that supply information about the form of the objects and their positions, related to the positions of other neighbouring objects.

The main structure of the input/output primary data is a raster.

Data compression is possible by using the technique of quadtrees, i.e. the same technique of image pyramids. This alternative structure is represented by a tree, where each node has four branches.

Data compression is also possible, avoiding the rigidity of quadtrees, by means of techniques to store sparse matrices (i.e. design matrices or normal matrices of geodetic networks, as well as block matrices) or, more recently, by means of form descriptors.

The geometrical data, that supply information about forms, dimensions and positions of the objects, have different characteristics for altimetry and planimetry.

In the first case a DTM involves methodologies for surface reconstruction or object reconstruction, while in the second case the main problems of numerical cartography regard form descriptors and map matching.

A DTM allows for the evaluation of different parameters:
- heights (DEM);
- slopes;
- bendings;
- volumes;
- height changes;
- movements of well identifiable points;
- (surface or object) deformation.

The altimetric output can be given by:
- contour lines and break lines;
- sections or profiles;
- grid data.

The planimetric output belongs to the map production procedures and involves photogrammetric restitution (by means of automatical or semi-automatical techniques), as well as orthoimages, both with the superimposition of vector elements:
- skeleton lines and geodetic vertices;
- main texture lines;
- administrative bounds;
- toponyms and conventional symbols.

The rules in a raster output are:
- frequency;
- importance;
- center.

and they must be applied in the given order.

It can be recalled that a planimetric representation needs geocoding and map projections, when objects occupy a space larger then 10-20 km's; furthermore elevation models require earth curvature correction, but in case of a very close range (less than 1 km), and this involves geoid undulations, when the space occupied by the objects is again larger than 10-20 km's.

For these reasons, although numerical cartography and digital photogrammetry, taking into account important contributes given by GIS's and image processing, are going from the field of earth sciences to the field of information technology, geodesy still plays an important role in the surveying and mapping disciplines.

3. TOWARDS DIGITAL PHOTOGRAMMETRY

In recent years, the methodologies of photogrammetric surveying are going in the direction of image processing, which concerns a set of new technologies tending to evaluate images, as primary data, for qualitative and quantitative documentation and representation of objects.

In this frame the primary data are:
- digital images;
- digitized analogical images.

The hardware, necessary in all the phases from data acquisition until restitution, is composed by:
- digital cameras, with linear arrays (e.g. the SPOT and the three-line cameras) or with areal arrays (e.g. the CCD cameras);
- scanners, of high resolution, with linear arrays and diffused light;
- PC's or Workstations, with ram of the order of megabytes and cd-rom of the order of gigabytes.

The computers must be provided with graphic equipments and act as servers for special instrumentation:
- analytical plotters;
- scanners;
- solid state memory data recorders;
- plotters;
- orthoimage printers.

The modern trend intends analogical images and digitizing processes as an intermediate phase in the transition towards direct acquisition of digital images.

In any case primary data are grey level matrices (or colour scale matrices) i.e. raster data, where the pixels of the grid supply geometrical and semantic information in terms of position and radiance.

The software is therefore the most important tool in data management and processing:
- data acquisition;
- geometrical and radiometric data correction;
- data transformation;
- visualisation on screen;
- data representation.

In image processing, the passage from the image space to the object space requires not only geometrical algorithms, but also radiometric ones, that are necessary in semantic interpretation of morphological characteristic and image matching.

The procedure can be performed and repeated by using image pyramids, as proposed by Ackermann. In this case a hierarchical sequence of image levels substitutes the image itself, after reducing the number of pixels by filtering and resampling. The procedure begins at the lowest level and continues sequentially up to the highest level. Each passage reduces the investigated area and at the same time refines the accuracy in positioning.

The main operations of digital photogrammetry are:

- preprocessing, in which filtering separates the signal from the noise in the images;
- feature extraction: it is the most critical part of image processing from the point of view of a fully automatic implementation, because it concerns semantic interpretation and during it morphological characteristics, like points, lines or figures, are derived;
- image matching, where homological correspondences between two images (or more, for the sake of reliability and robustness) are established: they involve points in case of area based matching, lines in case of features based matching, figures in case of relational matching;
- restitution.

The last operation is connected to numerical cartography procedures, as briefly shown in the previous paragraph.

4. MAIN TOPICS IN IMAGE PROCESSING

Image processing begins, as already mentioned, with preprocessing procedures in which filtering separates the signal from the noise in the images. This operation is very important to eliminate elements or zones of disturbance or uncertainty before coming to the successive and more sophisticated phases.

In the second phase, features extraction is performed by applying different operators:

- Foestner operator, applied to points, in order to identify the maximum grey level variations in at least two directions;
- Roberts, Sobel, Prewitt, etc. operators, applied to lines, that are enhanced by the modulus of the gradient of the grey levels;
- Marr and Hildreth, Kirsch, Chen, Deriche operators, again applied to lines, that are here enhanced by the Laplacian multiplied by a Gaussian filter of the images;
- clustering of pixels and parsing, applied to figures; with the second operation, starting from the clustered elements, graphs with nodes and arcs or, better, principal nodes, polylines and circuits are formed.

Clustering operation are to be performed, because of the anisotropy of the clusters, by suitable techniques:

- line following;
- region growing;
- split and merge.

The technique of split and merge allows for the iteration of the first two techniques until the expected features are achieved. The same technique should be used at low level image processing, in the operations of thinning, erosion and dilatation.

In the third phase image matching can be performed by:
- maximizing the correlation among homological elements;
- minimizing the least squares norm of the observation equations, after establishing a correspondence among homological elements;
- minimizing a robust norm under the same conditions.

As already mentioned, image matching is called:
- area based matching, when the correspondence is defined among points;
- feature based matching, as proposed by Ballard and Brown, when the correspondence is defined among lines, by using:
  - epipolar geometry (Gruen),
  - a priori knowledge of the object and form descriptors (Ebner, Wrobel);
- relational matching, as proposed by Shapiro and Haralich, when the correspondence is defined among figures.

The search of the optimal solution for the relational matching requires to explore different combinations, starting from a suitable tentative solution, after defining a risk or gain function to be minimized or maximized during the search. At present relational matching has not operative solutions.

In the other two cases the mathematical model of the above mentioned correspondence is defined as follows:
- collinearity equations, when the matching is made in the object space on the:
  - groundels , i.e. projections of the pixels on the object (Helava),
  - voxels, i.e. elementary parts of the object (Schenk);
- complanarity condition, when the matching is made in the image space on the pixels, obtaining the same result of the relative orientation;
- relations among the coordinates of the homological elements in the image space and among their grey levels; these relations may be:
  - local, as shifts, S-transformations, affine-transformations,
  - global, where a black box model defines the transformation on the whole image.

In case of image matching in the image space, the absolute orientation should be performed successively. Notice that it is much more complicated to do it in a fully automatic way, because of the difficulties in obtaining a mathematical description of the semantic contents of the control points.

5. SPATIALLY REFERENCED DATA ANALYSIS

Data related to objects are, as already mentioned, spatially referenced data. A general classification distinguishes them as:

- constant;
- time depending;
furthermore , considering their nature, they can be diveded into:
- point functions;
- point position difference functions.
  Point functions are related to approximation-interpolation problems:
- line interpolation, section/profile reconstruction, time serie analysis;
- surface reconstruction;
- form descriptors, i.e. contours of figures or surfaces of objects;
- figure matching (images, maps), 3D matching (models, objects).
  Point position difference functions are referred to network adjustments in a broad sense; there exist in fact:
- geodetic networks, at different scales and for various kinds of tasks;
- spaceborne, airborne and terrestrial image blocks;
- different nature networks, like circulation/distribution/communication networks.
  In these cases it is necessary to fix a reference system, because the point position difference function observables are invariant for variations of the reference system, unlike point function data.
  The same solution given by constraints is achieved, as it is known, by using pseudo-observations with suitably high weights. This method is however more general, because it is able to process pseudo-observations with every kind of weight, like:
- previous information;
- auxiliary data;
- regularization condition.
  The design of these problems may be performed in an optimal way or, at least, in a suboptimal one, by using advanced techniques:
- sampling of the point function data;
- optimisation of the point difference function observables.
  In the first case variance analysis allows for:
- progressive sampling, when the explanation is too poor;
- selective sampling, when the amount of information is too large and the data are processed sequentially.
  In the second case a joint strategy of:
- simulation, where one or more tentative configurations are prepared;
- optimisation of the configurations themselves and of the weights of the observables;
leads to an improvement of the expected solution. It can be recalled that, because of the non-linearity of the simulation/optimisation strategy, this procedure should be linearized and repeated until convergence is obtained and controlled, iteration by iteration, for the sake of reliability of the whole scheme.
  Observation equations, pseudo-observation conditions and constraints form the functional models. A general classification distinguishes them as:

- grey box models, when physical or geometrical laws exist, whose parameters are to be estimated;
- black box models, when appropriate laws are missing.
  In this case it is necessary to adopt ad hoc mathematical models given by:
- polynomial interpolation;
- finite element method or spline interpolation;
- Fourier analysis;
- wavelet analysis;
- stochastic process approach.
  Together with the functional model, a stochastic model must always be defined. The easiest stochastic model is the Gauss-Markov model, but if the choice of the weights seems to be problematic it could be refined by assuming the more general Gauss-Helmert model, after suitably clustering some groups of weights.

  The most general stochastic model involves correlations too, yet it produces unreliable estimates, except for a large amount of data, like in the stochastic process approach.

6. RELIABILITY AND ROBUSTNESS

The least squares principle is the main estimation method. In fact it produces, as it is known, unbiased estimates of minimum variance; on the other hand it requires homogeneity among the variances of the observables, as it happens for normally distributed data.

Any way the redundancy of the problems, measured in term of ratio observations-parameters, should be large enough, so that the corresponding systems would be generally and locally reliable. It can be recalled that a reliable system is able to identify and locate errors.

Furthermore it is obvious that the same systems should be generally and locally well- conditioned, avoiding to estimate unobservable parameters.

Unfortunately in the data there exist gross errors, i. e. blunders, leverages and outliers, that produce a departure from the hypothesis of normal distribution. Therefore non-normally distributed data give distorted estimates, because the least squares method is unable to reject gross errors. An alternative to least squares is supplied by robust estimators. Galileo introduced them in the Dialogo dei Massimi Sistemi; successively they were formalized in terms of qualitative robustness by Boscovich and Laplace. The modern concept of mathematical robustness refers to the definition of:

- influence function of gross errors;
- break down point, consequent to the introduction of rejection point in the influence function;

introduced by Huber and Hampel respectively. At present robust estimators are not only able to identify and reject gross errors, but also capable to set up a different explanation, i. e. to adopt a different function model when they, after rejection, show a homogeneous behaviour.

After the estimation, hypothesis testing formulates a judgement on the quality of the obtained estimates.

The whole procedure runs according to the rules of statistical inference:
- formulate a null hypothesis;
- choose a significance level;
- assume a sampled distribution of probability;
- split it, in general, in three regions, where the critical zones are represented by the tails on both sides.

Notice that, in case of rejection of the null hypothesis, before accepting an alternative hypothesis it is necessary to verify the power of the test, so that the two hypotheses would be significantly different.

There exist two classes of significance tests:
- classical tests, suitable only for the normally distributed data;
- distribution free tests, for all kinds of data: their power is less than the power of the first class.

Both classes of tests evaluate under different conditions:
- significance of parameters;
- quantity of residuals or noise;
- correlations or other forms of dependence, i.e. regression or connection, against independence;
- goodness of fit of probability distributions.

The choice of second order models extends the capacity of analysis and understanding both of the estimation and of the hypothesis testing. In fact:
- deterministic (or quasi-) phenomena following precise laws have correlation equal to one (or near to-);
- random (or quasi-) phenomena are sets of independent events and they have correlation equal to zero (or near to-);
- other phenomena have intermediate behaviour and if their correlation is high enough, a memory remains in their evolution/distribution.

Phenomena with a high enough correlation, under a stochastic process approach (i.e. considered as a realisation of a stochastic process, after assuming a suitable invariance group for the definition of the stochastic process itself) permit the definition of an internal law by using the data themselves.

This internal law is called covariance function and allows for the separation, by filtering, of the signal from the noise. The same law allows for the prediction of the signal and, by covariance propagation law, of its functionals.

On the other hand the stochastic process approach suffers, as any generalized least squares adjustment, from the lack of reliability and robustness according to the above mentioned limitations of the least squares principle. Unfortunately robust estimators in this field have not a complete theory and their application is very rare; therefore methodological proposals have to be formulated and pilot experiments have to be performed.

7. ON ROBUST COLLOCATION

An attempt towards the introduction of robust estimators in the collocation method could be done according to the following approach, that splits the collocation method in different steps and modifies them in order to get robustness. As a matter of fact, the main steps of the collocation method are:
- preprocessing and trend removal;
- covariance estimation and covariance function modelling;
- filtering and prediction;
- crossvalidation and multivariate testing

and each of them should be suitably modified improving the capacity of robustness.

Collocation filtering and prediction require gross error free data, therefore blunders, leverages and outliers should be already identified and eliminated in the preprocessing step. To reach this aim an easy procedure is to compute the mid point of a median absolute value of a set of data neighbouring a given point. By repeating this operation for each point in a data set, the data in the tails will be automatically rejected, because their discrepancies with the estimated mid points are larger than a small multiple of median absolute values.

Preprocessing is followed by trend removal, which consists in applying robust estimators to trend parameter estimation. In this way, solution of weighted least squares adjustments are repeated until convergence to robust estimates is obtained and controlled in accuracy and reliability by the computation of robust norms.

Covariance estimation is a very important step in the collocation method. Indeed bad covariance estimates do not allow for a good separation of the signal from the noise. For these reasons gross errors must be eliminated before covariance estimation or, alternatively, covariance estimation should be done by using Spearman's rank correlation coefficients instead of empirical covariance estimates.

It can be recalled that the Spearman's rank correlation coefficient is a function of the sum of rank difference squares, where the ranks are assigned separately to each component, from the smallest argument to the largest one, and then compared, element by element, in terms of their difference.

Concerning covariance estimation, the two components are the data and the mid points of median absolute values of sets of data neighbouring given points. In this case the global median absolute value substitutes the variance.

Covariance estimation is followed by covariance function modelling which is performed in the classical way, considering that the selection of a covariance model from a library and the interpolation are not best fitting, but suitably investigate characteristic features: the behaviour in the origin, the number and the position of the zero-points, the maxima and minima, the correlation length and the size of the correlation.

Filtering and prediction use the results of covariance estimation and covariance function modelling and can be performed applying robust estimators to obtain the separation of the signal from the noise and the signal prediction respectively. In both cases, since the quality of covariance estimates is very high, they are computed by repeating weighted least squares adjustments until convergence is obtained. The weight change is determined on the basis of the magnitude of the residual noise

To this purpose sequential algorithms could be introduced to compute the innovation of the standard solution and the accuracy. On the other hand, remembering that raster data are processed by means of special algorithms for regular structures and that also irregular but dense fields are processed by means of iterative algorithms, the sequential processing is till now problematic.

Cross validation and multivariate testing is the classical procedure to analyze the results obtained by means of the collocation method. To this aim, since robust estimators imply the departure from the hypothesis of normal distribution of observations and estimates, distribution-free tests must be used.

There exist many different distribution-free tests and, among them, the following ones are important:
- Kruskall-Wallis' (or Friedman's) test for independent (or correlated) means, or variances;
- modified Wilcoxon-Wilcox's test for Spearman's rank correlation coefficients. They allow to verify, respectively:
- the behaviour of the signal, or the magnitude of the residual noise: both problems involve distribution-free variance analysis;
- the independence against dependence, i.e. the quality and quantity of the correlations, that look as those of coloured or white stochastic processes.

On the other hand, although distribution-free tests do not require heavy hypotheses and their application is quite immediate, they waste information and tend to be too conservative.

8. EXAMPLES

The procedures, explained in the previous paragraph, forming an approach to robust collocation have been preliminarly tested by using simulated test examples. The reason of this choice is that a correct judgement on the procedures can be obtained only if the magnitude of the perturbations, resulting in a departure from normality, are a priori controlled.

Therefore after the acquisition of a grey level profile of a SPOT image from the OEEPE Working Group on SPOT-image Test, it has been analyzed and evaluated by means of some elementary robust estimators. They represent a sample of moving medians, a sample of moving m.a.v.'s and a sample of moving Spearman rank correlation coefficients, computed after splitting the whole profile in subsets of small size.

Figure 1 shows the grey level profile; figures 2, 3 and 4 show the sampled statistics, while figure 5 shows the autocovariance function of the whole profile. The last product is, as it is known, necessary in the collocation filtering and prediction: their goodness depends strongly from the capacity of the autocovariance function to model the behaviour/distribution of the phenomena.

Therefore considering the identification and rejection of some isolated blunders as standard and assuming that a downweighting procedure in the collocation (similar to a sequence of generalized weighted least squares adjustments) is suitable for small outliers only, big outliers and leverages have been simulated and introduced in the grey level profile. A study of the quality of the auto covariance function and of the capacity of the elementary robust estimators in gross errors detection has been the aim of the proves.

The first simulation modifies of 100 tones the grey levels of a small subset of the grey level profile. The sampled moving medians immediately enhance the simulated perturbation.

The second simulation modifies of 10 times the discrepancies with respect to a local median in the same small subset. Again, the sampled moving m.a.v.'s immediately enhance the simulated perturbation. Figures 6 to 10 and 11 to 15 show the same results for both simulations (see figures 1 to 5 for the original data).

On the contrary, the third and last simulation, which locally modifies the correlation among the elements of the same small subset (e.g. bringing it from a large correlation to zero), does not give the expected results. Indeed since Spearman rank correlation coefficients for small subsets range, in the original data, from one to minus one, it has been really impossible to prove that a zero value in the sampled moving Spearman rank correlation coefficients is a perturbation. On the other hand, its effect on the autocovariance function vanishes and figures 16 to 20 seem to be (exactly) the same as figures 1 to 5 for the original data.

A comment can be done after the results of the simulations: well identifiable perturbations, like changing the center and/or the dispersion of a subset of data, are easily detected by elementary robust estimators, which must be used before covariance estimation. On the contrary, a more complex perturbation, like changing the dependence (e.g. the correlations) in the same subset of data, is not detected by elementary robust estimators, because of a lack of consistency of the

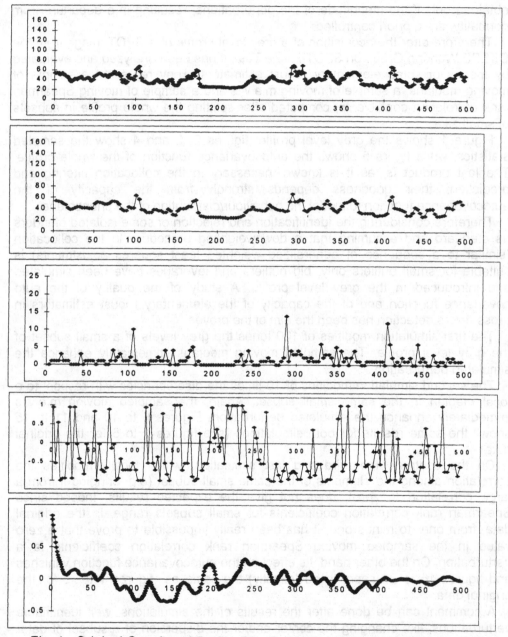

Fig. 1 - Original Grey Level Profile. Fig. 2 - Median Profile (Original Data).
Fig. 3 - Mav Profile (Original Data). Fig. 4 - Spearman Profile (Original Data).
Fig. 5 - Covariance Function (Original Data).

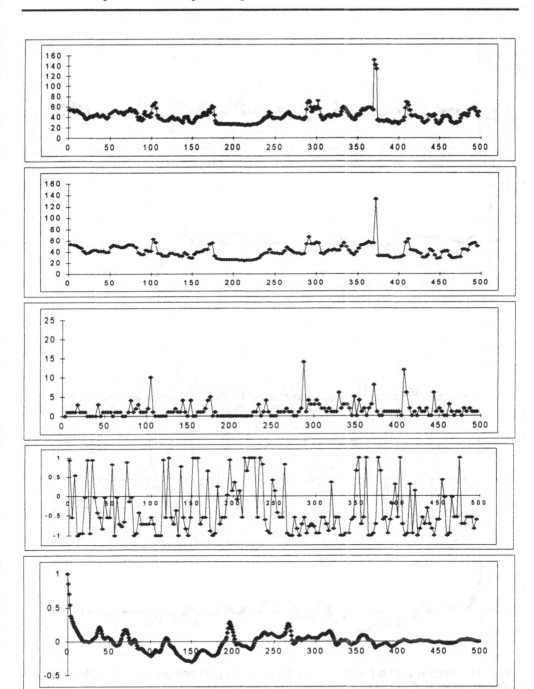

Fig. 6 - Profile with a Local Perturbation. Fig. 7 - Median Profile (Perturbed Data).
Fig. 8 - Mav Profile (Perturbed Data). Fig. 9 - Spearman Profile (Perturbed Data).
Fig. 10 - Covariance Function (Perturbed Data).

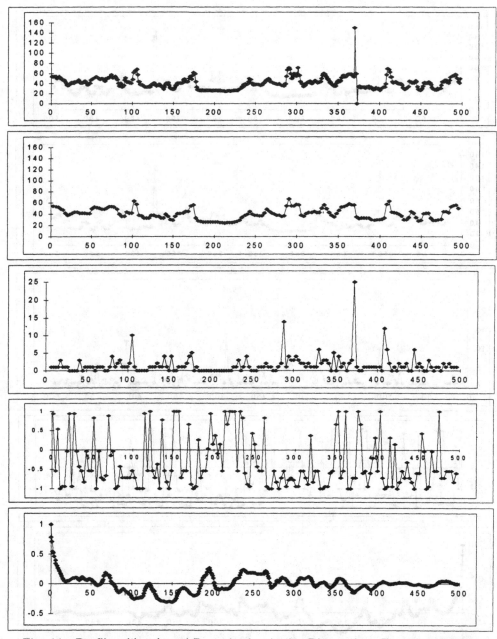

Fig. 11 - Profile with a Local Perturbation in the Dispersion. Fig. 12 - Median Profile (Perturbed Data in the Dispersion). Fig. 13 - Mav Profile (Perturbed Data in the Dispersion). Fig. 14 - Spearman Profile (Perturbed Data in the Dispersion). Fig. 15 - Covariance Function (Perturbed Data in the Dispersion).

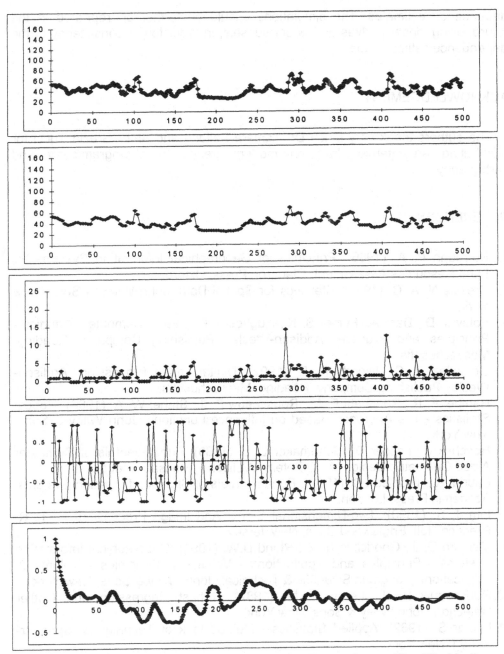

Fig. 16 - Profile with a Local Perturbation in the Dependence. Fig. 17 - Median Profile (Perturbed Data in the Dependence). Fig. 18 - Mav Profile (Perturbed Data in the Dependence). Fig. 19 - Spearman (Perturbed Data in the Dependence). Fig. 20 - Covariance Function (Perturbed Data in the Dependence).

dependence estimates. The simulations should be further on repeated in the future, using longer profiles and larger subsets, in order to get consistency of the dependence estimates too.

ACKNOWLEDGEMENT

The authors thank very much Dr. Tamara Bellone and Arch. Grazia Tucci for distributing an internal draft on modern trends in photogrammetry and cartography.

REFERENCES

1. Burrough P. A. (1986): Principles of Geographical Information Systems for Land Resources Assessment. Claderon Press, Oxford.
2. Cressie N. A. C. (1991): Statistics for Spatial Data. John Wiley & Sons, New York.
3. Foley J. D., Dam A., Feiner S. K., Hughes J. F. (1990): Computer Graphics - Principles and Practice. Addison-Wesley Publishing Company, Reading, Massachusetts.
4. Gonzalez R. C., Woods R. E. (1992): Digital Image Processing. Addison-Wesley Publishing Company, Reading, Massachusetts.
5. Hampel F. R., Ronchetti E. M., Rousseeuw P. J., Stahel W. A. (1986): Robust Statistics - The Approach Based on Influence Functions. John Wiley & Sons, New York.
6. Hoschek J., Lasser D., Schumaker L. L. (1989): Fundamentals of Computer Aided Geometric Design. A. K. Peters, Wellesley, Massachusetts.
7. Laurini R., Thomson D. (1992): Fundamentals of Spatial Information Systems. Academic Press, London.
8. Lim J. S. (1990): Two-Dimensional Signal And Image Processing. P. T. R. Prentice Hall, Englewood Cliffs, New Jersey.
9. Maguire D. J., Goodchild M. F., Rhind D.W. (1991): Geographical Information Systems - Principles and Applications - Volume 1: Principles - Volume 2: Applications. Longman Scientific & Technical, John Wiley & Sons, New York.
10. Rousseeuw P. J., Leroy A. M. (1987): Robust Regression and Outlier Detection. John Wiley & Sons, New York.
11. Lothar S. (1982): Applied Statistics - Method look of Techniques. Springer-Verlag, New York.
12. Schowengerd R. A. (1983): Techniques for Image Processing and Classification in Remote Sensing. Academic Press, New York.

A PACKAGE FOR THE MANAGEMENT OF DATA QUALITY IN GEOGRAPHIC INFORMATION SYSTEMS

G. Pillirone

University of Udine, Udine, Italy

D. Visintini

INSIEL S.p.A., Trieste, Italy

Abstract

This work describes a software package for the evaluation of the accuracy of a digital mapping. It implements a testing procedure based on the application of statistical tests, both parametric and non-parametric. Automatic tools for data quality assessment are not yet common; for this reason, the development of a computer program for accuracy testing is an interesting result.

1 Introduction

The literature about Geographic Information Systems has pointed out that the quality of spatial data and data bases is a major concern for developers and users of these systems. The kind of transformations and the tools of analysis available in these systems require that the user be aware of the error affecting the result of the analysis. To give only an example, map scale changing and map overlaying are standard operations in the use of geographic information systems which can introduce or propagate the uncertainty affecting data. With traditional techniques the experience of the user was enough to take into account the presence of error, but the use of computers and the way results are presented may give the illusion that inaccuracy and error are absent [1].

Coming to the field of multimedia GIS, the problem becomes even more important, since data collected from different sources have a different "quality". The user gets no warning from the system if he merges information coming from different maps; the ease in doing computations encourages the user to perform operations which could not be compatible with the nature of the data involved.

When we talk about quality we refer to the standard quality descriptors proposed in the literature [2]: accuracy of geometry and attributes; time currency; lineage or information about the origin of data; completeness; logical consistency.

This work is mainly interested in the accuracy of geometry. It has been proposed to add to every to every data set in a GIS some additional information, a kind of meta-information, giving a measure of the quality of the data set. The starting point in modeling uncertainty in a GIS is the evaluation the geometric accuracy of the maps contained in the system, because these systems are always built upon a digital mapping, which is a database of coordinates, the geometric reference of the information contained in the system.

A procedure for evaluating the accuracy of geometry was already proposed by the group who carried out the present project [3, 4] and will be recalled in the next sections. The theoretical results have supported the development of a reliable procedure to compute the values for a metric accuracy descriptor. The natural step to take was to implement the testing procedure in a software package. Besides a certain number of theoretical results and proposals, in fact, it can be noticed how the literature still presents very few real systems including a quality management module. The program described in this work tries to fill this gap and aims to become part of a real GIS.

2   The testing procedure

The testing procedure implemented in the package will now be described. It applies mainly to the testing of digital mappings, but it works also for testing the accuracy of any data base of coordinate values, such as a digital image coming from the rectification of a photograph of a building. Particular emphasis will be given to the most recent upgrades that have permitted to obtain a useful method for real practice.

The standard quality descriptors mentioned in Section 1 define the accuracy as the degree of similarity between the value assigned to an attribute in the data set to be tested, and the value (or its best estimation) of the corresponding attribute in the real world. For point-coordinate values, which are continuous-valued attributes, this can be measured by the differences between assigned and real-world values.

Since it is not possible to do this comparison for every point in the map, it is performed in a statistical way. A certain number of point coordinate values, say $p_1, \ldots, p_n$, are extracted from the data file of the digital map to be tested. The coordinates of the same points are then measured "on field" with high precision geodetic techniques, giving the values $q_1, \ldots, q_n$. Field measurements provide a reference data set to which the map is compared.

In statistical terms, the p_i and the q_i are two random samples coming from two distributions F_p and F_q; we want to know if there is enough evidence to say that the two distributions are the same. If this is the case, it means that the map shows a good similarity with the real world it wants to describe.

Two values are given for each point coordinate, reference value and tested value. This situation is called *paired comparison*. The equality can be checked computing the differences between the two values of each point and testing the statistical hypothesis that their distribution is centered about zero. The three components of the point coordinate, x, y and z, are tested separately, assuming statistical independence between the components.

One possibility to perform the paired comparison is using the well known t-test, based on the statistic $t = \frac{\bar{d}}{s/\sqrt{N}}$ (where N is the size of the sample, $\bar{d} = \frac{1}{N}\sum_{i=1}^{N} d_i$ is the sample mean and $s = \frac{1}{N-1}\sum_{i=1}^{N}(d_i - \bar{d})^2$ is the sample standard deviation). Acceptance or rejection of the null hypothesis depends on the value of the test statistic: if t falls outside the interval $(-c_{\alpha/2}, c_{\alpha/2})$, where $c_{\alpha/2}$ denotes the $\alpha/2$-th quantile of Student's t distribution, then H_0 is rejected, otherwise it is accepted, α being the probability of first kind error.

The application of the t-test and of other techniques related to it, requires the assumption that the differences between reference coordinate values and map coordinate values follow a normal distribution. Several analyses of the distribution of differences of coordinates for real examples coming both from cartography and from architectonic photogrammetry have been performed, and the normality of the distribution was not always accepted. The most frequent departures from normality were in asymmetry, but some cases also showed bimodality. For situations such as these, the t-test is no more reliable, and a non-parametric approach is proposed instead. Non-parametric methods are statistical procedures that do not require assumptions about the form of distribution of the data.

For the paired data setup such as the one we have described, and for the null hypothesis of equality of the populations giving the two samples, a good non-parametric test is the Wilcoxon signed rank test [6]. This test is based on *ranks* rather than on the observed values. A random sample of N paired observations, $(p_1, q_1), \ldots, (p_N, q_N)$ reference and tested, must be given, as before. The differences between observed and reference value $d_i = p_i - q_i$ are computed. The N differences are ordered according to their absolute value and a rank is assigned to each: rank 1 to the smallest absolute value, rank 2 to the second smallest, up to rank N to the biggest. The statistic W is then computed, as the sum of the ranks corresponding to positive differences: $W = \sum_{i=1}^{N} \text{sgn}(d_i) R_i$ where R_i denotes the rank of the i-th difference d_i and $\text{sgn}(x) = 1$ if $x > 0$, $\text{sgn}(x) = 0$ otherwise.

If the two distributions are the same, the ranks of the differences are expected to have as many positive as negative signs. Otherwise there will be many positive or negative ranks, and the statistic will take a very big or very small value.

The null hypothesis is accepted if W falls inside the critical region (c_1, c_2) given by the quantiles of the Wilcoxon-signed rank distribution, such that $\text{Prob}[c_1 < W < c_2] \approx \alpha$; α is the probability of first kind error, and can only be approximately satisfied because the distribution is discrete, and does not take any value between 0 and 1.

Up to this point we have proposed a simple way to measure the accuracy of the digital mapping, as a statistical test performed on the population of the errors in point coordinates, computed as differences between values in the data set and reference values. The most appropriate test to use, parametric or non-parametric, is chosen by a preliminary analysis of the form of the distribution of the differences. Several normality tests are known, such as the one based on the χ^2 statistic or the one based on the sample coefficients of asymmetry and kurtosis.

3   Sample size considerations

The most important upgrade to this testing procedure regards the power of the test. As is well known, the acceptance or rejection of the null hypothesis in a statistical test is done with a certain probability of error. Rejecting the null hypothesis when it is true is called first kind error, and is controlled by the critical region of the test. Conversely, accepting the null hypothesis when it is false gives the second kind error. Controlling the probability of second kind error is not easy, because it requires the knowledge of the distribution of the test statistic when the null hypothesis is false, which is not always available. On the other hand, the explicit consideration of the second kind error is important because it allows sample size calculations. First kind error, second kind error, and sample size are linked together by the power function of the test, which is the probability of rejecting the null hypothesis as a function of the alternative.

For Student's t test the distribution of the test statistic under the alternative hypothesis can be obtained with the so called non-central t distribution [5], and this allows to compute the power of the test.

Let's take into account the power of the Wilcoxon signed rank test. To reason about the power function we need to know the alternatives of interest when the null hypothesis is rejected. The alternative to the equality of two distributions is any difference between the two, and this is too general. We restrict our attention to a specific kind of alternatives, corresponding to a pure translation of one distribution with respect to the other. This is called shift model, and corresponds to investigate the presence of a systematic shift in the coordinate values of the tested map. In the shift model the alternatives are expressed as amount of the shift between the centers of the two compared distributions, called Δ. An interesting power expression for the Wilcoxon signed rank test in case of shift alternatives, is the following [6]:

$$\Pi(\Delta) \approx \Phi\left[\frac{N(N-1)\,g_1(0) + N\,g_0(0)}{\sqrt{N(N+1)(2N+1)/24}}\Delta - z_\alpha\right] \tag{1}$$

Here $\Pi(\Delta)$ denotes the power as a function of the location of the alternative hypothesis; we can see the explicit presence of the sample size N. The only problem with this formula are the quantities g_0 and g_1 which are respectively the density in zero of the distribution, say G, of the differences of coordinates, and the density in zero of the sum of two independent variables each distributed as G. The estimation of these quantities is not easy, since it requires the estimation of a density. It has been solved using the bootstrap [7], which is a statistical procedure based on the resampling of an available data set.

Only a brief overview of the bootstrap will be given. If $\mathbf{x}$ is a random sample, the bootstrap algorithm draws a high number, say B, of samples from $\mathbf{x}$ with replacement. B ranges from 200 to 2000 depending on the kind of parameter to be estimated. This gives raise to B new samples, denoted as $\mathbf{x}_i^*$, each giving a replication of the estimation of a certain parameter of interest. From the variability over the bootstrap replications, information about the variability of the parameter over the population can be derived. This works satisfactorily in many situations.

The bootstrap has been used to estimate the densities in zero, producing an empirical cumulative distribution from the bootstrap replications and computing its derivative in zero.

$$G(z) = \text{Prob}(Z \le z) = \frac{(\text{number of } x_i^* \le z)}{n} \quad ; \qquad g(0) = \frac{G(\varepsilon) - G(-\varepsilon)}{2\varepsilon} \qquad (2)$$

In this way we have all the information required to solve the power expression (1).

4 Implementation of the package

The theory described in Sections 2 and 3 can be schematically summarized in the following way:

1. Collection of data;

2. Normality test, in order to choose between parametric and non-parametric procedures;

3. Test of the null hypothesis of no systematic error;

4. Power computation, which can be solved with respect to N or with respect to Δ.

These are exactly the steps implemented in the "C/1" package that is going to be described. The source code is written in C language under Microsoft Windows, and in AutoLISP for some routines running under AutoCAD. The Windows environment was chosen in order to have quickly a first prototype release to submit to a few users to check their reaction. The availability under Windows of a package such as AutoCAD, moreover, makes the step of input data collection quite easy, as will be seen.

A schematic view of the architecture is shown in Figure 1. Two main parts can be seen: the statistical core of the program, and some input and output interfaces, which are filters between the user and the program.

Input data to the program are contained in disk files. The importance of the data collection module is so high, that it is explained in a separate section, together with the description of the output produced by the program. The remainder of this section is about the part that performs the statistical computations.

We will not give a complete description of the statistical routines, because this would be too long and not really interesting. Only a few remarks are worth making:

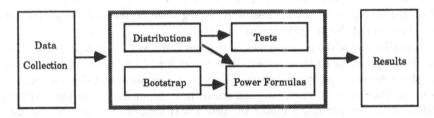

Figure 1: *Main components of the "C/1" package.*

- The routines are written trying to achieve portability, so that the inclusion inside a real GIS is made easier. The formats of the input text files are general and easy to change, and the C language has been used trying to avoid tricks or constructs that could be non-standard.

- Efficiency has been maximized. The program may be applicable in a "field" use, in the sense that with a portable computer, the testing can be conducted starting from few reference points, 15 or 20, then doing the power calculation and verifying if the achieved power and the location of the alternative are as required. If this is not satisfied, the program can be iteratively run, collecting a few more testing points, for instance 10 or 15 more, until the test shows the desired performance. For the purpose of efficiency, statistical distributions are calculated from series expansions or iterative methods. The only distribution for which the calculation is not quick enough is the Wilcoxon distribution, so that it has been stored in a file that is looked up as a statistical table.

- The implementation of the bootstrap algorithm was interesting. It does not present particular programming difficulties, while it gives good practical results. With the speed of modern computers, the generation of even a high number of bootstrap samples is not a problem.

- The user interface is a part that must be carefully designed in any application. In this case, particularly, it becomes a matter of great importance to give the user an understandable environment, since the subject of data quality in general, and the testing procedure we propose, especially for what concerns power computations, are not yet standard components of an analysis based on GIS. The Windows-style graphical interface that was developed for "C/1" is still a prototype; future work will be addressed to the realization of an interface in which it would be clear, in any moment, what is going on and what's the meaning of the results presented.

5 Geometric Data Collection

What makes this software package really useful, from a practical point of view, is the module for data collection, that is the part that prepares the samples of coordinate values for the testing. A very good statistical procedure for which the preparation of the data were cumbersome, would be soon abandoned by the users. The "real time" approach that was described, moreover, would not be possible if adding some points to the available samples of a preceding run of the program were a long process.

The input format of the program simply requires two text files, of a proper format, one storing the sample of points to be tested (file name with a .TST extension) and one storing the sample of reference points (extension .RIF). At the moment the files are read by the program, the differences are computed and stored in an input file (extension .INP). In the most simple case the user could supply the two ASCII files, or even the INP file directly, but this is not always possible or easy to do: the geometric data of the GIS to be tested are represented by a digital mapping stored in some vector format, such as the common DXF (Drawing eXchange File), whose structure is a text file of graphical primitives used

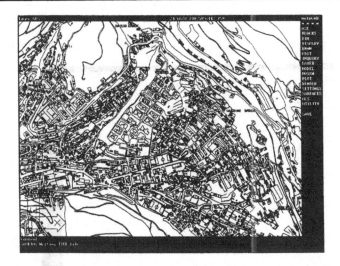

Figure 2: *A display of the tested area of a digital map; testing points are drawn in a different color, here not visible.*

to define all the objects to be drawn; the problem is in the size of these files, that for each element of the map can be as huge as 20 or 30 MB. The extraction of the test coordinates from these files would not be simple and not reliable.

To solve this problem, the program includes a semi-automatic data collection procedure that exploits the capabilities of the AutoCAD environment. The user can choose it from a submenu that starts the AutoCAD application from within "C/1". The procedure requires that the RIF file is created first. The structure of the RIF text file is simply a four-column array of coordinates; each row is a different point, the first column stores its identifier (its name) and the other three columns the East coordinate, the North coordinate and Height.

The RIF file will most often come from an existing file of reference coordinates, having a different layout, obtained from topographic survey or GPS measurement. Of course there are so many different formats that it is impossible to write a translation procedure to the RIF format for each of them. At present, a conversion filter reading the output file of the SKI software, for Leica's GPS system, has been developed; it automatically writes the RIF file of cartographic coordinates in the Italian Gauss-Boaga system. Other translators can be easily written and included in the program, but the RIF file is a minor problem, compared to the TST file.

The TST file has the same layout of the RIF file, a list of rows containing the name and the coordinates of the tested points. To identify these points in the DXF file, the digital map is imported and drawn in AutoCAD space; the points in the RIF file are then read by an AutoLISP routine and drawn over the map (see Figure 2); in this way the user can see what he is testing and what the control area of the test is.

Another AutoLISP routine produces a zooming, centering the display on each reference point, waiting that the user chooses the corresponding test point, one by one, for all the sampled points. Two circles are automatically drawn centered in every reference point, so that the operator can immediately see the accuracy of each test point or detect blunders

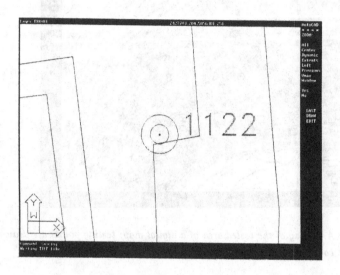

Figure 3: *Automatic zoom on one reference point, named 1122.*

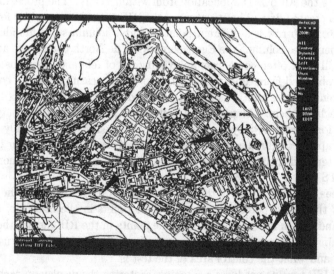

Figure 4: *Planimetric error vectors. The vectors are magnified by a user-selected amount.*

(see Figure 3). The radius of the circles was fixed taking as example the digital technical cartography of the Friuli-Venezia Giulia region (Italy), where the mean and maximum admitted planimetric errors are fixed at .45m and .90m. This procedure, requiring an operator, was preferred to the automatic selection of the tested point as the nearest point to the reference one, that had too much ambiguity existing when more than one possible test point is near enough to a reference (see Figure 3).

Also some utility procedures for the AutoCAD environment have been written, such as one that permits to re-select single "bad collimated" test points and one that draws the vectors "planimetric error" for each point (see Figure 4).

Another feature of the data collection module must be mentioned: sometimes the area corresponding to the RIF file spreads over several DXF files from which to extract the TST file (e.g. when a single GPS survey is extended on more maps); other times more RIF files cover one DXF file (e.g. when more GPS surveys are included in an unique map); in both cases the package supports the operator to create correctly the INP file, doing the work of identifying which DXF file contains each point.

A few words about the output produced by "C/1". The results of the testing procedure are saved in a text file containing the statistical responses: summary statistics such as mean and variance, normality of the sample, an "accept/reject absence of systematic error" answer and a power computation. In particular, the power of the test can be saved as a power table for various locations of the alternatives and sample sizes. No experiment was made to display the results of the test graphically; as it was said the matter of representing the concept of error and uncertainty inside a GIS is still an open problem.

6 Practical applications

We briefly report the application of the procedure to a real example, to which Figures 2 to 4 refer. It is the testing of 5 contiguous maps of Carta Tecnica Regionale Numerica (digital technical regional cartography) at 1:5.000 scale of the Friuli-Venezia Giulia region: it is a tridimensional vectorial cartography obtained from aerial photogrammetry and it may be the geometric data-base for a GIS. Every map has a geographic squaring of 2'30" in longitude by 1'30" in latitude that correspond about 3.200 by 2.800 m, and an area of about 900 hectares.

coord.	mean	med.n	st.dev.	norm.	syst. err.
East	−0.056	−0.089	0.405	Y	N
North	−0.045	−0.136	0.362	N	Y
Height	0.281	0.187	0.374	N	Y

Table 1: *Results of the testing of the Friuli-Venezia Giulia digital mapping. Lengths in meters.*

The reference sample has been obtained by GPS measurements of 24 points randomly positioned on such maps. The results obtained from "C/1" are summarized in Table 1, where the first three columns report some statistical summaries about the sample of differences, column four the result of the normality test, and column five the result of the systematic error test. A tabular form of the power function for the Wilcoxon test applied to the North coordinate is shown in Table 2. This component had the null hypothesis rejected, meaning that it is affected by a systematic error. The table shows the power of the test for various alternatives Δ and sample sizes N. From exact computation, it results that fixing $\alpha = 0.05$ and with 27 points, a power of approximately 0.90 is achieved for an alternative centered on 0.239m. If the user wanted to detect a smaller systematic error, for example an alternative located on 0.15m, the same α and the same power would have required about 40 points.

	N = 20	40	60	80
Δ = .05m	0.2108	0.2989	0.4245	0.5287
.07m	0.2888	0.5177	0.6715	0.7682
.10m	0.5184	0.6714	0.8614	0.9893
.15m	0.7988	0.8941	0.9697	0.9999
.20m	0.8777	0.9899	1.	1.
.30m	0.9319	1.	1.	1.

Table 2: *Power of the Wilcoxon test applied to the Friuli-Venezia Giulia digital mapping, North coordinate.*

Acknowledgments
We would like to thank INSIEL S.p.A. that has funded the present work.

References

[1] Goodchild, M. and S. Gopal (eds.): Accuracy of Spatial Databases, Taylor & Francis, London 1989.

[2] European Committee for Standardization: Data Description: Quality, Technical Report TC287 WWG2 N15, 1994.

[3] Crosilla, F. and C Garlatti: Un primo contributo analitico per una nuova proposta di collaudo informatico di una cartografia numerica. Boll. SIFET, 3 (1991).

[4] Crosilla, F. and G. Pillirone: Parametric and nonparametric procedures for testing the metric quality of a digital map, Manuscripta Geodaetica, 20 (1995), 231-240.

[5] Bickel, P and K. Doksum: Mathematical Statistics. Basic Ideas and Selected Topics, Holden Day, San Francisco, 1977.

[6] Lehmann, E.L.: Nonparametrics: Statistical Methods Based on Ranks, Holden Day, San Francisco 1975.

[7] Efron, B. and R. Tibshirani: An Introduction to the Bootstrap, Chapman & Hall, New York 1993.

PATTERN RECOGNITION IN DIGITAL CARTOGRAPHY AND GIS

V. Casella

University of Pavia, Pavia, Italy

Abstract

The present work describes an attempt to realize *digital maps automatic classification*. In particular the aim is to select the *built zone*, that part of the territory essentially characterized by the presence of buildings.

Different selection methods are taken into account: local and non local, with one, two or three parameters. The pattern recognition approach is explored. Enhancing techniques, such as outliers elimination, are also considered.

What is presented here could be useful also in the GIS world: data analysis is often and effectively synthetized by a thematic map. Less often the methods used for such elaborations allow to quantify their reliability. For this goal, sometimes, techniques similar to those described here could be used.

1 The goal and the method of the research

The present work is a part of a wider research on data quality in digital cartography, that deals with the digital map of the town of Verona, with a nominal scaling ratio of 1/2000, realized in the period 1989-91.

1.1 The goal: to outline the built zone

We had the necessity of selecting the built zone, that part of the digital map characterized by the presence of buildings. In particular we ideally overlapped a grid to the cartography and wanted to classify the status of the cells so that each of them could be: *red* (that means: in the built zone) or *blue* (that means: out of the built zone).

The first problem was to have a sharp idea of the built zone and this in not as easy as it could seem: indeed, looking to a map, there are zones that would be probably judged in the same way by many different people, but there are also ambiguous zones, that need a careful and difficult tuning of the decision criteria.

Once the built zone concept was defined, it was necessary to translate it in a quantitative criterion. But this is very difficult to do in an *a priori* approach.

1.2 The method: starting from examples

So we decided to follow a more complex way: to draw the classification of a test area by hand and to look for procedures that, learning by examples, could carry out the classification work on the same area with a good degree of fidelity to the example. This is a *pattern recognition problem*.

1.3 The sample area

We realized plots of three regions of 2400x1500 square meters that cover all the typologies of the built zones; the plots are at the 1/2500 scaling ratio and buildings are pointed out by colors. They represents, more or less, the 10% of the whole map surface and we called them *sample area*.

1.4 The sample classification

We overlapped the plots with gridded transparent sheets, with 50 meters cells. In the present work, cells dimension is a constant and not a parameter still to be tuned. But during the long try and retry time we used 25, 50, 100 and 200 meters cells, concluding that 50 meters is the best compromise.

On the transparent sheets we drew by hand, with three different colors, the status of each cell: *red* -built zone cell-, *blu* -unbuilt zone cell- and *yellow*, that means unclassified cell. We were obliged to introduce the third color because it was impossible to classify a 5% of the cells. Our work results are what we called *sample classification*.

Before starting the techniques description, it is useful to underline again the importance and the role of the sample classification.

- Sample classification can be used to tune automatic classification tecniques in such a way they learn from the example.

- Sample classification can be used to determine the efficiency of different criteria. Applying a generic criterion to the sample area, we obtain a thematic map that we can compare to the sample map; we can estimate the *misclassified cells rate* (**mcr** starting from now), that is the percentage of cells that have a different status in the calculated classification and in the example one.

2 Methods for the automatic selection of the built zone

2.1 Local threshold criteria

Historically we first looked for *local threshold criteria* that is, for example, selecting, as belonging to the built zone, cells that have inside not less than one or two buildings. This criterion can also be formulated in terms of *building covered surface* that is the surface which, in a certain cell, is covered by buildings.

Thanks to the digital nature of the map considered, it is possible to calculate, for each cell, by means of specific software procedures, the number of the inside buildings and the extension of the surface covered by them.

We tested these criteria for different values of the threshold and the best values for **mcr** are 7.87 for building number criterion (with the threshold fixed to 1) and 7.24 for covered surface criterion (with the threshold fixed to 63.56).

It is clear that local criteria (local underlines that such criteria consider only what happens inside the cells) have at least two kinds of problems:

- the problem of the isolated house in the country, which make local criteria decide, wrongly, that the corresponding cell is in the built zone;

- the problem of the squares in the cities that, on the contrary but always wrongly, make local criteria decide that the corresponding cells are out of the built zone.

2.2   Non local criteria

To overcome these problems it is necessary to abandon local criteria that is, we have to consider not only what happens inside each cell, but also what happens in the neighbourhood. We introduced two parameters S, the building covered surface in a cell, and S_3, the building covered surface in the eight nearest cells, measured in square meters. So each cell is characterized by two numbers and the sample classification can be represented on a plane. Blue and red point clouds (represented here by plus signes and dots, respectively) are separated, even if not completely.

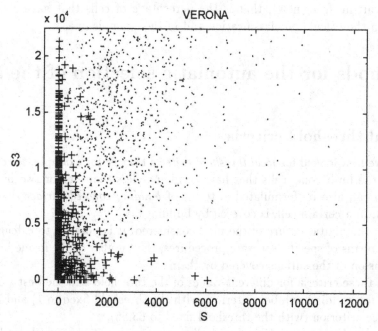

It seems possible to separate the two sets in some way, for an example with a line, so that blu points are under the line and red points are over.

2.3   Linear separators

Once we have determined the line that separates the sample points in the best way, we could use it to classify cells. Indeed, the generic unclassified cell is characterized by a couple of values S and S_3, and can be represented by a point on the plane: if the point is on the lower halfplane (respect to the line) we say that the cell belongs to the unbuilt zone; if, on the contrary, the point is on the upper halfplane, we say the cell belongs to the built zone.

So a line could be an useful classification tool: it remains the problem of finding the line that separates the two point sets in the best way.

2.4   Parametric methods

The red and blu points distributions can be considered as extractions from bidimensional random variables. If their distributions had a known shape, it would be possible to discuss the separation problems in a rigorous statistic way. We worked out a normality χ^2 test with significance level $\alpha = 0.01$ for blu and red points, always with the same result: the normality hyphotesis is unacceptable.

2.5   Non parametric methods

So we looked for non parametric methods and a *loss function* approach was chosen. In this case it is necessary to consider a $U(a,b)$, function of the line parameters. U must be written in such a form that *U reaches its minimum in two values $\hat{a}$ and $\hat{b}$ that identify an optimal separator line*. Optimal separator means that the line, used as a classifying tool on the sample area, shows a behaviour near the sample classification one. In other words: the highest possible number of blu sample points has to be under the line and, at the same time, the highest possible number of red sample points has to be above the line.

3   Loss functions for line separators

3.1   Linearity

The loss function can be expressed as the sum of $n+m$ terms where n is the number of the unbuilt zone sample points and m is the number of the built zone sample points.

We can write

$$U(a,b) = \frac{1}{n}\sum_{i=1}^{n} U_i^u(x_i, y_i, a, b) + \frac{1}{m}\sum_{i=1}^{m} U_i^b(x_i, y_i, a, b)$$

where U_i^u indicates the contribution of the i-th unbuilt zone point, while U_i^b quantifies the contribution of the i-th built zone point. Starting from now, x and y will be used, instead of S and S_3, for semplicity.

3.2   Some possible forms for the loss function

At this point it is necessary to give a form to U. Let's consider a generic line: if a certain red sample point is over the line, it is on the right halfplane and we translate this making the point give no contribution to U. On the contrary, if the point is under the line, on the wrong halfplane, we translate this wrong position making the point give a positive contribution tu U_i^b. Thus we can write

$$U_i^b(x_i, y_i, a, b) = \begin{cases} d^b(x_i, y_i, a, b) & y_i < ax_i + b \\ 0 & y_i \geq ax_i + b \end{cases}$$

For a blu sample point we can argue in a specular way and write

$$U_i^u(x_i, y_i; a, b) = \begin{cases} d^u(x_i, y_i; a, b) & y_i > ax_i + b \\ 0 & y_i \leq ax_i + b \end{cases}$$

where the d distance functions somehow measure the distance between the outlying point and the line.

Now the problem becomes to find useful distance measures. We could assume, as a measure, the vertical distance between the outlying point and the line, $ax_i + b - y_i$, that we called d_1

$$d_1 = ax + b - y$$

In this case U_i takes the form

$$U_i^b(x_i, y_i, a, b) = \begin{cases} ax_i + b - y_i & y_i < ax_i + b \\ 0 & y_i \geq ax_i + b \end{cases}$$

$$U_i^u(x_i, y_i; a, b) = \begin{cases} y_i - ax_i - b & y_i > ax_i + b \\ 0 & y_i \leq ax_i + b \end{cases}$$

We could consider also the square of d_1, d_2

$$d_2 \;=\; (ax + b - y)^2$$

It is possible to use the geometric distance between the outlying point and the line. Therefore we can write the distances d_3 and d_4

$$d_3 \;=\; \frac{ax + b - y}{\sqrt{1 + a^2}}$$

$$d_4 \;=\; \frac{(ax + b - y)^2}{1 + a^2}$$

3.3 Results

For each form of the distance function d, there is a different form of the loss function U, whose minimization gives different lines. Minimization can be worked out by iterative software procedures. A first indication of the quality of the obtained separator is given by the *residual* $U(\hat{a}, \hat{b})$ that quantifies the separability of the two point sets. Infact if blu and red point sets were completely separated, we would have $U(\hat{a}, \hat{b}) = 0$; on the contrary, if the two sets were highly overlapped, we would find a high value of the residual.

The only way to choose the best separator among the four we found, is to calculate their **mcr**.

Distance fn	$\hat{a}$	$\hat{b}$	U_0	mcr
d_1	-5.498	10710.88	220.98	5.70%
d_2	-4.015	11056.78	$1.96\ 10^6$	6.11%
d_3	-18.893	15434.39	20.85	5.00%
d_4	-17.418	17781.79	$2.86\ 10^4$	**4.95%**

The distance functions d_3 and d_4 give similar performances, better than d_1 and d_2 do. The best misclassified rate is equal to the 63% of the best rate obtained with the building number threshold method.

Next pictures show a part of sample area classified: (i) by hand (it is the sample classification); (ii) by means of the building number threshold method; (iii) by means of the best separator line. Black squares represent built zone cells, while white squares represent unbuilt zone cells; gray color represents unclassified cells. It easily seen that the third picture is much more similar to the sample than the second one.

3.4   Error estimation

Once we find the best separator line, we obviously want to apply it to unclassified areas and it is very important to estimate the percentage of the errors it will commit. The value of **mcr**, that is the error rate calculated on the sample area, can be considered as a general estimation of the error rate, in relation to any unclassified zone. This identification is as reasonable as the sample area is significant.

4   Improvments of the loss function technique

The results described in the previous section represent the basic starting point. This section deals with the possibility of improving line separators performances. We explored more than one direction.

4.1   Parabolic separators

We tried to reach lower values for **mcr** using more sophisticated separators. We considered parabolas, with the analitical form

$$y = ax^2 + bx + c$$

Results are not good: next table compares the best parabola **mcr** to the line one, both referred to the d_1 distance function.

Distance fn	$\hat{a}$	$\hat{b}$	$\hat{c}$	U_0
d_1 - line	-5.498	10710.88		**220.98**
d_1 - parabola	0.0016	-78.685	6372.15	2475.98

If the parabola residual is so much bigger than the line one, there is no hope of obtaining good results in this way. Moreover we met many difficulties due to the instability of the solutions: changing the starting point of the iterative method makes the result change greatly.

4.2   Plane separators

Another possible improvement is to consider larger neighbourhoods. For every cell we considered S, S_3 and S_5, that is the building covered surface in the eighteen second nearest cells (they are the sides of a 5x5 cells square). Therefore the status of each cell is characterized by three numbers and can be no more represented on a plane, but in a volume.

This is an important question to consider: being the sample points number constant, passing from a bi-dimensional environment to a three-dimensional one, causes a decreasing of the information density. As a consequence, a plane separator will be less well determined than a line separator.

Moreover it is necessary to verify if the S_5 parameter is significant. To control all these things, we determined the plane separator, with the equation

$$z = ax + by + c$$

with d_1 distance (d_1 is the simplest to implement, so we always started from it), obtaining the following results

Distance fn	$\hat{a}$	$\hat{b}$	$\hat{c}$	U_0	mcr
d_1 - line	-5.498	10710.88		**220.98**	**5.70%**
d_1 - plane	-18.72	-0.10	28191.47	1109.39	7.97

Both U_0 and **mcr** are worse than in the line case; therefore there was no reason to go on.

4.3 Coordinate transformations

It is well known, from cluster analysis, that coordinate transformations can increase separation. At the same time it is not clear what to do in a specific case, if it is possible to enhance data and in what way.

We explored many coordinate transformations and it seemed that, taking the square root or the cubic root of the representative point coordinates, gave good results. Next table shows **mcr** calculated for line separators on original coordinates -those used in section 3- (Case 1), on square root of coordinates (Case 2) and on cubic root (Case 3).

Distance fn	Case 1	Case 2	Case 3
d_1	5.70%	4.88%	4.88%
d_2	6.11%	5.12%	5.00%
d_3	5.00%	4.49%	4.52%
d_4	4.95%	**4.47%**	4.59%

The transformation from first coordinates to their square root gives always a gain; the next transformation between square and cubic root has a variable effect which is, in any case, small.

4.4   Outliers removal

Without any doubt, during the sample classification work, we committed errors, so we tried to remove some of them.

Unfortunatly, as already remarked, the distributions of the sample points are not known. Therefore we estimated mean and variance of each component of blu and red points distributions and used Tchebycheff's inequality to determine 0.95 probability intervals. Then we discarded all those points having at least one component out of the above-said intervals. In particular the test discarded 3 of the 1812 built zone points and 30 of 2327 unbuilt zone points . The interesting results are synthetized by the following table containing the values of **mcr** calculated on the cleaned data, for the line separator.

Distance fn	Case 1	Case 2	Case 3
d_1	4.33%	4.19%	4.43%
d_2	4.50%	4.55%	4.63%
d_3	4.31%	**3.82%**	3.90%
d_4	4.33%	4.21	3.99%

It is noticeable that, with cleaned original coordinates (Case 1), all the distance functions give almost the same results. The joint application of coordinate transformations and outliers removal gives a **mcr** of 3.82, that is one half of what we obtained with the building number threshold method.

4.5   Further possible improvements

Though reached results are satisfactory, we think it is possible to do something better in bi-dimensional problems using, as separators, piecewise defined functions, such as splines.

5   Looking to the future: applications to GIS and reliability enhancement

5.1   GIS and optimal separators

Grid analysis is very diffused in GIS softwares: they allow to represent many quantities in different grid planes; it is possible to make selections in each grid with interval conditions (a cell is selected if its value is in a certain range); it is possible to combine corrisponding cells of different planes with boolean operators. All this operations belong to local criteria.

The present work shows that certain thematic -grid based- work can reach a good reliability degree only by the use of non local techniques. In such situations it could be useful to adopt pattern recognition tools that, starting from sample classification, learn from it to riproduce the same classification, allowing to estimate their work accuracy.

5.2   Reliability enhancement

It should be a good thing to be able not only *to decide the status of a cell, but also to specify the reliability of this classification.* This would be easy, as usual, if it were possible to make some hypothesis on the distribution form of the points to classify. Differently, and this is the case, everything is more complicated. However we are working out something that will probably be the subject of a next paper.

6   Conclusions

The paper analyzes different techniques and different data-enhancing ways for selecting the built zone.

The best result is a classification tool that, applied to the sample, has an error rate of 3.8%. This is a very good value, thinking that the starting point is, in any case, a hand-made classification. Indeed this underlines the high accuracy of the sample classification work.

7   References

COMINCIOLI, *Metodi numerici e statistici per le scienze applicate.* Casa Editrice Ambrosiana, Milano, 1992.

MEISEL *Computer oriented approaches to pattern recognition.* Academic Press, New York and London, 1972.

MENDEL, FU *Adaptive, learning, and pattern recognition systems: theory and applications.* Academic Press, New York and London, 1970.

VAPNIK *Estimation of dependences based on empirical data.* Springer Verlag, New York Heidelberg Berlin, 1982.

DATA ARCHIVING BY CLUSTERS AND GRAPHS

T. Bellone
Polytechnic of Turin, Turin, Italy

L. Mussio
Polytechnic of Milan, Milan, Italy

C. Nardinocchi
University of Ancona, Ancona, Italy

ABSTRACT

A summary of cluster analysis techniques and graph theory algorithms is presented and their application to data archiving is discussed. A specific system of programs and a preliminary example show that the procedures prove to be effective.

1. FOREWARD

Data compression, archiving and understanding are very important tools in the GIS world. Indeed information acquired by images and/or maps, as well as different kinds of spatially referenced data, have to be organized, evaluated and stored in order to get specific data bases for geographic information systems.

There are many different methodologies on texture analysis and pattern recognition, useful for data acquisition, evaluation and archiving. They involve some mathematical, statistical, numerical and computer science procedures and algorithms, at different complexity levels, which are presented and explained in the scientific reference.

Data classification, clustering and explanation, in the preliminary sense, are derived from the application to the data of cluster analysis techinques. A more refined explanation, taking into account structural topological relations underlaying the geometry of objects, is obtained by using graph theory algorithms.

A system of programs able to organize elements of objects in clusters and recognize structural relations, to be formalized as graphs inside the clusters, has been implemented and tested by means of a small synthetic example. It plays an important role in a package of methodologies on morphological feature extraction and form descriptors, that respectively precede and follow the presented procedures.

2. CLUSTER ANALYSIS

Cluster analysis collects different statistical procedures and algorithms, able to classify data, to gather them in sets and to give a preliminary explanation about their behaviour.

The aim of cluster analysis is to minimize the internal dispersion of a group of data (cluster), respect their center (cluster point), and to maximize at the same time the distance among the above defined clusters.

The procedures of cluster analysis could be subdivided, taking into account the starting point, the collection strategy and the end point respectively in:

• agglomerative or divisive techniques

• hierarchical or not- techniques

• clustering or clumping.

Qualitative cluster analysis could be performed e.g. by MEDOIDS algorithm, while quantitative cluster analysis could be achieved by ISODATA algorithm.

The second algorithm is very important for data compression, archiving and understanding, when data are supplied by images and/or maps, as well as different kinds of spatially referenced data, providing suitable data bases for geographic information systems.

The main steps of ISODATA algorithm are the following:

1. selection of preliminary cluster points and thresholds for the iterations

2. clustering in a cluster the elements neighboring a cluster point

3. division of clusters, whose internal dispersion is greater than the chosen threshold

4. new clustering

5. fusion of clusters, whose distance is less than the chosen threshold

6. starting a new iteration from the 2nd step, or stopping the algorithm if a reproducing point is achieved.

The steps of clustering could be done in a different way. When the objects are well configurated, the Euclidean norm or the Manhattan distance help cluster the surrounded elements.

On the contrary, when the objects are ill configurated, these norms show evident difficulties to cluster the elements.

Hence specific techniques of parsing, like line following, region growing, to be iterated by means of split and merge, provide the expected results.

The parsing is a methodology of Linguistics, which has already been applied to Artificial Intelligence, Image Processing and Robotics.

A special subroutine has been implemented in order to get the clustering of ill configurated objects; it is called PATTERN and its main steps are the following:

1. selection in a cluster of the element which is the closest to its cluster point

2. selection in the same cluster of (if any) elements slightly more distant from the cluster point, thus taking into account possible bifurcations

3. attribution of the role of cluster point to the just selected elements

4. reiteration of the 1st step

5. reiteration of the 2nd step

6. reiteration of the 3rd, 4th and 5th steps as long as selection of elements is possible

7. selection of a possible not yet selected element as a new cluster point; otherwise stop the algorithm

8. reiteration, for the new cluster point only, of the 1st...5th steps as long as a selection of elements is possible.

When preliminary cluster points are not available, the algorithm begins flatly at the 7th step and it proceeds to form isolated clusters. Notice that the construction of clusters each after the other, instead of all together, is less robust respect to wrong selection of elements.

Finally the same strategy, used for the clustering, could be applied for the cluster fusion by means of a cluster of clusters.

3. GRAPH THEORY

Structural relations underlaying the geometry of objects have a topological nature and could be studied by graph theory.

A graph is a set of nodes connected by arcs. In a connected graph a path always exists between two nodes; the shortest path is called distance and the largest distance is called diameter. Each node of a connected graph is reached by at least one arc; the number of arcs of a node is called degree.

A connection matrix could be associated to each graph, being its main diagonal elements coincident with the nodes and its off diagonal elements coincident with the arcs. A corresponding design matrix supplies in form of chart the connections between arcs and nodes.

Since connection and design matrices are generally very sparse ones, they should be stored avoiding their zero elements. Design matrices tabulated by rows or columns and banded, profile or sparse connection matrices are a very suitable way to store them.

Connection matrices may have non zero elements near their main diagonal or anyway located. The arrangement of non zero elements could be improved by changing the order of elements. The arrangement of the elements inside design matrices could be also improved according to the order of the elements achieved for the corrisponding connection matrices.

The most important algorithms for numeration, ordering and dissection of matrix elements use graph theory.

A level structure is a partition of a graph in different levels, whose nodes are connected to the ones of preceding and/or following level only. A level structure, whose first level has one node only, said root, is a tree. The number of nodes belonging to a level is called width; the width of a level structure is the maximum of the level widths, while the number of levels is called depth.

Relationship among matrices, graphs and level structures show that the bandwidth of connection matrices depends upon the numeration of the nodes of the corresponding graphs, an adequate numeration requires a small width of the level structures and this last property is easily achieved as their depth is large. Furthermore since the minimization of bandwidth and profile could not always be performed at the same time, the preliminary dissection of a graph and the subsequent reordering of its dissected parts supply the expected results.

1st algorithm- Search of a diameter of a graph.

The reiteration of the construction of level structures allows to find the diameter of a graph. An arbitrary starting root initiates the first level structure, while the nodes in the last level initiate new level structures, to be substituted to the first one, if their depth is greater. The same procedure has to be repeated until two level structures of equal depths are found.

2nd algorithm- Minimization of the width of a level structure.

When two level structures of equal depths are found, it is possible to merge them suitably, in order to get a new level structure of minimum width. To this aim, the nodes belonging to the same level in the two level structures have already a final destination. On the contrary, the nodes, whose levels are different, should be definetively assigned to the level of one structure (between the two ones).

The selected level is chosen taking into account, level by level, the structure whose local width is smaller. Notice that the new level structure could be unrooted.

3rd algorithm- Numeration.

Node numeration goes on, level by level, starting from a root. In the following it assigns, in a set of nodes, a smaller number to the nodes connected to the nodes, whose number is the smallest one, in a previous set of nodes. In case of ambiguity, the degree of the nodes is taken into account.

4th algorithm- Dissection.

Nodes of a very high degree, or very long connections in a spatial graph should be removed from the graph, so that the previous algorithms supply the expected results. In these cases a preliminary dissection of the graph may be performed, if necessary dividing the graph in dissected parts. The reordering of these parts takes place thereafter, applying the above defined numeration and ordering algorithms.

4. TESTING PROCEDURES

A system of programs able to organize elements of objects in clusters and recognize structural relations, to be formalized as graphs inside the clusters, has been implemented. It plays an important role in a package of methodologies of morphological feature extraction and form descriptors, that respectively precede and follow the presented procedures. Notice that the former is used in the raster vector conversion and it starts the texture analysis, while the latter concludes the understanding and/or explanation at a high level of pattern recognition.

A small synthetic example illustrates the above mentioned procedures and shows that they prove to be effective.

The program of cluster analysis called CLUSTER has, as input data, vector elements, obtained from raster data by a suitable threshold on their argument values. Moreover preliminary cluster points and thresholds for the iterations of the algorithm are added in the input.

The output of the program of cluster analysis supplies:

- the vector elements with the label of the related cluster
- the cluster points
- information about internal dispersion of the clusters.

Notice that the vector elements can be easily sorted putting them, cluster by cluster, together.

Furthermore the cluster points provide information about the distance among the clusters, in terms of cluster point coordinate differences.

Figures 1,2 and 3 show respectively raster data, threshold sorted vector elements and the clusters achieved by the cluster analysis algorithm.

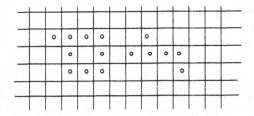

Fig. 1, Raster data

Fig. 2, Threshold sorted vector elements

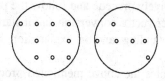

Fig. 3, Clusters

The program of graph theory called TEXTURE has, as input data, the vector elements sorted cluster by cluster. Moreover in order to form graphs, the size of a window should be defined. This additional input is useful to select connections between elements, forming arcs between nodes, when the data are random located. In case of dense fields of points, coming from raster vector conversion, the size of a window is often equal to one pixel.

The output of the program of graph theory supplies design and connection matrices corresponding to some graphs formed by:
- the clustered vector elements flatly
- principal nodes and interest points only
- circuits and open lines.

The three representations of the graphs show different complexity levels. Indeed while the first is quite immediate, after having achieved the clusters, the second recognizes elements where bifurcations or curvatures occur; moreover the third goes further in the interpretation of structural relations. Notice that the second case deals with the enhancement of relevant features (principal nodes and interest points), going over some useless elements.

Figures 4, 5 and 6 show respectively the graphs according to the threee above mentioned cases.

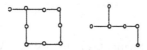

Fig. 4, Graphs of vector elements

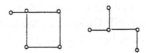

Fig. 5, Principal nodes and interest points of the graphs

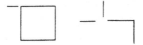

Fig. 6, Circuits and open lines of the graphs

The illustrated procedures are useful not only for the data classification, evaluation and understanding, but also produce data compression that brings some advantages in data archiving.

The procedures should be refined taking into account an improvement of data quality and a preliminary understanding and/or explanation of structural relations underlaying the geometry of objects. To this aim, the thinning of the lines allows that they represent arcs and not chains of arcs. Furthermore the erosion of small areas, obtaining well defined points, allows that they represent nodes, enhancing the lines departing from them. Finally the dilatation along some interrupted lines, going over (if any) gaps, allows that they represent connected arcs.

Anyway the authors would state that the present paper is referred only to a preliminary example and the procedures should be also applied to real examples involving large volumes of spatial data.

REFERENCES

1. Cressie, N.A.C.: Statististics for Spatial Data, John Wiley & Sons, New York 1991.
2. George, A. and J.W.H. Liu: Computer Solution of Large Sparse Positive Definite Systems, Prentice-Hall, Englewood Cliffs (N.J) 1981.
3. Hampel, F.R., Ronchetti, E.M., Rousseeuw, P.J. and W.A. Stahel: Robust Statistics, John Wiley & Sons, New York 1986.
4. Kaufman, L. and P.J. Rousseeuw: Finding Groups in Data: Cluster Analysis with Computer Programs, John Wiley & Sons, New York 1988.
5. Kennedy, W.J. and J.E. Gentle: Statistical Computing, Marcel Dekker, New York 1980.
6. Melhorn, K.: Graph Algorithms and NP-Completeness, Springer-Verlag, Berlin 1984.
7. Rousseeuw, P.J. and A.M. Leroy: Robust Regression and Outlier Detection, John Wiley & Sons, New York 1987.
8. Sachs, L.: Applied Statistics-Handbook of Techniques, Springer-Verlag, New York 1982.

DIGITAL IMAGES IN SUPPORT OF HISTORICAL BUILDING DATA BASES

M. Barbarella

University of Rome "La Sapienza", Rome, Italy

M. Fiani

University of Ferrara, Ferrara, Italy

ABSTRACT

The study of complex monuments is highly interdisciplinary, requiring input from surveyors, architects, archaeologists, historians, geotechnical and structural engineers, etc.
A specialised Data Base is required in order to organise and preserve all of this information. In particular, rectification and orthoprojection techniques allow for geometrically corrected raster images to be inserted into a Data Base; particular software applications are required to preserve this characteristic.
This paper presents the results collected during the study of an extremely complex monument, both in terms of single photograph treatment as well as mosaicking of multiple images. Particular attention was given to defining specific criteria which could be applied for the manual creation of a Digital Elevation Model (DEM), which is necessary for the orthoprojection of a monument image.

1. INTRODUCTION

Monument investigations have a strongly interdisciplinary character since the goal of studying and restoring these structures requires various techniques, including structural, geotechnical, historical and architectural approaches.

Such studies create, in the case of complex situations, a large mass of data; it is imperative to organise this data properly to avoid it being dispersed and to guarantee its complete exploitation by researchers in different disciplines. The necessity for creating a Data Base that is devoted to monuments is clearly evident.

The creation of an integrated Data Base in which it is possible to insert, in a classified way, all information related to the studied monument (including location, restorations and any technical, architectural or iconographic information related to its historic and present condition) is of particular importance. The managed information must be available in different ways, considering the various levels of user access or the different goals of the data applications.

The development of digital photogrammetry has allowed large amounts of highly precise information to be inserted into the Data Base. In particular rectification and orthoprojection allows spatially correct digital images to be introduced. The management of these images must be accomplished using software that can preserve the accuracy from the rectified images.

1.1. The Data Base

The diverse nature of the existing documents necessitates the management of various types of elements in the system, such as:
- data originating from historical sources, i.e. documentation (written sources; bibliographies; archive, photograph and iconograph documents) and metric data (old surveys);
- data originating from new research, i.e. documented, numerical, and graphical data.

All available data should be inserted in the Data Base according to four principle classification formats:
- (i) structural fields, through the use of programs that have guided and controlled data input;
- (ii) text, through the direct digitisation of the data or through text interpretation programs;
- (iii) image-raster, acquired through the scanning of paper documents or, if they are from previously processed raster images from other hardware environments, inputed directly;
- (iv) graphic-vector, through the construction of a "numerical model" obtained from field data, graphic digitalisation and photogrammetric plotting.

The surveyor must supply, among other things, the geometric support of the data; in other words the "three-dimensional reference grid" in which all the available data are inserted, despite their various origins.

1.2. Case study

In the case of a complex monument, the restoration project can be organised into the following main disciplines:
- Historical-Archaeology and Historical-Architecture;
- Geology-Hydrogeology and Geotechnics;
- Structure;
- Surveying.

The historical archaeology / architecture section should supply the body of knowledge that is essential for the correct preservation of the monument.

Investigations and studies concerning the geology, hydrogeology and geotechnics of the site are proposed in order to obtain detailed knowledge of the subsurface on which the monument is constructed, with the goal of understanding the relationship between the ancient structure and its foundation.

Structural research is proposed for a rigorous interpretation of the structural behaviour of the monument from its construction to present, with the objective of defining criteria and specific techniques and technologies to improve the global and local stability of the structure, particularly with reference to seismic events.

Surveying will address two fundamental goals: the first consists of the creation of metric documentation concerning the present state of the monument and its surroundings, as well as the creation of a control system that is active over time; the second consists of the numerical and graphical representation of the items of specific interest for each other discipline. Within this framework the survey section will not only contribute to the knowledge of the monument but will also provide unifying support for all results obtained in the future.

A particularly interesting case is that of the Colosseo, which represents an example of a particularly complex monument.

Presently only the surveying of the studied monument is addressed; this preliminary phase must be completed, because it supplies study material for the successive investigations.

2. SURVEYING

The primary objective of the surveying is to supply support to the study and restoration of the monument. Being the foundation of the restoration project, the topographic and photogrammetric surveying becomes support for the other study areas connected to the restoration. These areas range from the simulation of static, dynamic and physio-chemical behaviour to the Data Base, whose creation could facilitate consultation of the enormous quantity of existing information, as well as the updating of data with results from research in progress.

For the Colosseo survey the expected activities are as follows:

General monument network

This general network will consist of groups of chosen points (both inside and outside the Colosseo) that will be surveyed using classical topographic techniques and calculated three-dimensionally within a single reference system. The network points could later be used to insert all the various surveys conducted on the monument into a single reference system; every survey should be linked to this framework so that information is not lost and it is in a form compatible for input into the Data Base. The network will have, in short, an analogous function for the Colosseo as the network of the trigonometric vertices has for the Country.

Control of relative and absolute altimetric movement
The goal of the altimetric network is to monitor potential vertical movements, whether differentially or throughout the entire monument. In addition if significant movements are not expected over the short term, it is advisable to arrange for monitoring and repetition of the altimetric survey over a long period of time, either for the safeguarding of the monument or for the control of the foundation behaviour during the study (i.e. conservation and restoration).

Insertion of the surveys into a city- or country-wide network
It is the intention of the authors to link the Colosseo to other surrounding monuments or, in a more general context, to a Rome-wide network which could eventually be inserted into a national network.

Photogrammetric survey of the external wall
Such activity will consist of various phases: topographic survey, photographic exposure, definition of the plotting criteria and codes, numerical and graphical archiving and analytical plotting at a scale of 1:50.

Aerial photogrammetric survey and numerical cartography
Surveying of horizontal and vertical sections
Digital rectification and orthoprojection of the external wall
This is an area that has been focused upon with greater detail, and as such it forms a large part of the subsequent discussion.

3. DIGITAL RECTIFICATION AND ORTHOPROJECTON

The application of the digital rectification and orthoprojection techniques is of great interest for the study of architecture; the experimental application of these techniques to the survey of the Colosseo's external wall is in progress.

The rectified digital image has numerous advantages with respect to traditional plotting, including:
- photogrammetric plotting requires a choice of the information to be represented, whereas a rectified photograph contains all the information on the represented object; the user can thus select what interests him without the "filter" of the operator who compiled the map;
- the management of the raster image (in terms of measurements and plotting) can easily be performed using video, even by a person who is not an expert in photogrammetry;
- there is no degradation of quality over time;

- it is also possible to store an archive of any related documentation, together with images, in the same computer (with the possible integration with Geographical Information System).
Disadvantages of the method include:
- the amount of information can appear excessive and the presence of "disturbances" can cause problems for users who are not accustomed to interpreting images; as a result they may prefer the already interpreted traditional plottings (a less significant problem in architecture);
- the high cost of input and output image devices;
- the enormous quantity of developed data;
- the technical creation problems;
- the "integration" problems.
The execution of a rectified image, limited to essentially flat objects, does not create particular problems; qualitatively the results are always very good, however from a metric point of view the results depend mainly on the effective planar nature of the object (which is not the case for an orthophoto).
The question that we posed was the following: considering the present "state of the art", is it possible to apply the digital rectification and orthoprojection techniques to architectural surveying? If the response is yes, with what (eventual) limits and with what modality?
The experimental method used can be summarised in the following points:

• acquisition of photographs;

• discretisation of the images using a scanner;

• orientation of the images;

• acquisition of the DEM using an analytical plotter (for the differential rectification);

• rectification of the oriented image;

• comparison of the plotter and rectified image measurements;

• statistical transformation and processing of the acquired data;

• printing of the rectified image.

The software Orthomap (Galileo Siscam) was used for the rectification.
The choice of a suitable reference system for the Colosseo survey was particularly important; the following principle co-ordinate systems were used:
- external systems, for the general framework of the monument (the National co-ordinate system, the cadastral system or WGS84, if the survey uses GPS);
- a local system, used for plotting and rectification, with the goal of defining local rectification planes for the monument every certain number of arches;
- an elliptical system, which was defined on the Colosseo by Birardi et.al. [3] with the goal of surveying the external wall, considered as a elliptical surface, in a single-surface reference system.

4. RECTIFICATION ON A PLANE

The experiment was performed on the most complex sector (western sector), having various projections and the presence of curvature, and on the least articulate sector (southern sector), where the curvature also has an influence and the relief can be considered to be almost the same as the "physiological" relieves of a manufactured wall.
For the western sector some tests were carried out on a complete photograph and on a single arch within the same picture. For the southern sector only portions within photographs were studied; some mosaicking tests were also performed on the rectified images belonging to this sector.
The digitally obtained rectified images were evaluated, both from a aesthetic and a metrically-correct point of view, to verify the application of the rectification method to the Colosseo.
The aesthetic evaluation allows the use of the product to be assessed, whereas metric evaluation indicates its applicability.
Assuming that the evaluation is positive, the extension of the rectified surface (relative to the scale of the drawing) and the possibility of creating an image mosaic should also be evaluated.
In order to carry out an evaluation of the rectified image from the metric point of view, the co-ordinates of nearly 100 sample control points (natural, not marked details of the wall) were measured both on the rectified image with the computer and on photographs using an analytical plotter; these results were then compared.
The mean co-ordinate difference and associated r.m.s. values for the X and Y components are shown in Tab. 1.

ΔX	r.m.s.	ΔY	r.m.s.
-3.4	21.1	-10.7	23.6

Table 1, Co-ordinate differences on sample control points. Mean and r.m.s. values for each measurement (cm)

Even though the mean values are not extremely high, note that the associated r.m.s. values indicate a large variability in the observed differences and, as a consequence, poor accuracy of the point positions on the rectified photograph.
On the less articulate sector (southern sector) a similar metrical evaluation was performed; nearly 80 control points, delimited by a ground control point, were measured (just a couple of arches were considered).
The results of this comparison are shown in Tab. 2.

Δx	r.m.s.	Δy	r.m.s.
-0.2	3.5	4.2	8.5

Table 2, Co-ordinate differences on sample control points. Mean and
          r.m.s. values for each measurement (cm)

Subsequently further image portions, situated in the outer ground-
control-points area and characterised by much more relief, were
considered and the co-ordinate differences on nearly 130 control points
were calculated.
The observed differences are shown in Tab. 3.

Δx	r.m.s.	Δy	r.m.s.
-3.5	10.3	-9.7	23.8

Table 3, Co-ordinate differences on sample control points. Mean and
          r.m.s. values for each measurement (cm)

The observed co-ordinate differences are caused by:
- measurement error due to the analytical plotter;
- measurement error due to the Orthomap;
- points not belonging to the rectification plane.
In figure 1a and 1b height-displacement vectors obtained on the control
points are shown for western and southern sectors, respectively: the
circle represents the true point position (co-ordinates measured using
the analytical plotter) while the other end of the line represents the
point position on the rectified image; the scale of the displacement
vectors is four times the drawing scale.
Looking at the graph it is possible to draw some conclusions related to
the rectified images: the height displacements are generally too high to
be within an acceptable tolerance, even if the image is quite good from
an aesthetic point of view (the perspective present in the original
photograph are only partially eliminated in the rectified image).
Theoretically distance measurements performed between two point on the
same plane will differ from the true distance, however it should be
noted that this difference is much smaller than the absolute position
error of either two points.
The relative measurement between points at the same elevation on the
rectification plane can therefore be accomplished on the rectified image
with a greater precision than that obtained with the absolute
measurement conducted on the same two points.

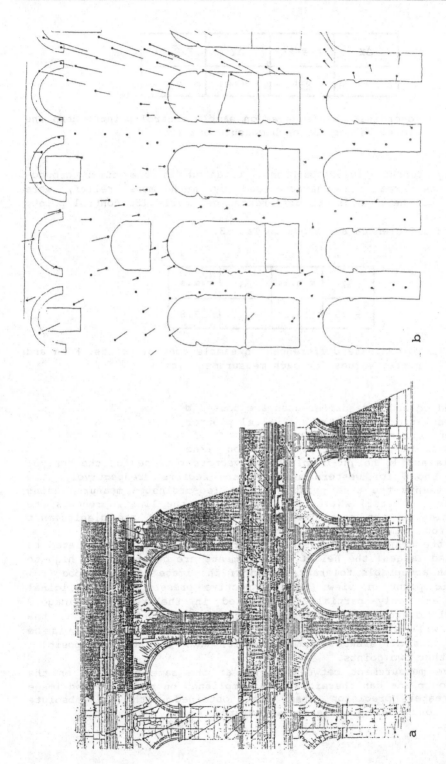

Fig.1, Height-displacements vectors on ground control points.
     a) analogical plotting of western sector
     b) southern sector

4.1. Rectification and height-error calculation of a single arch

It is possible to locate three different planes on the western sector arch, as indicated in figure 2 by the letters A, B and C, and represented by:
A) pillars, arch extrados, balustrade;
B) half-column basements, half-column highest-point, architrave, frieze;
C) trabeation.

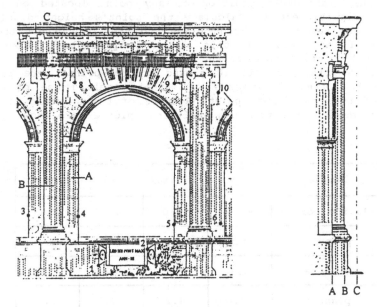

Fig.2, A western sector arch. Ground control points and different planes

The results of the metric evaluation made on nearly 60 control points located on the different planes are shown in table 4.

Sample	Δx	r.m.s.	Δy	r.m.s.
all the points	7.7	8.6	18.9	27.7
plane A	2.7	1.7	2.1	3.6
plane B	13.7	6.5	35.0	10.0
plane C	25.3	10.6	92.2	9.9

Table 4, Co-ordinate differences on sample control points. Mean and r.m.s. values for each measurement (cm)

The results are undoubtedly:
- very good for the control points located on plane A,
- mediocre for the points located on plane B,
- very poor for the points on plane C.
The occurrence of small characteristic relief on the rectification plane
produces very negligible errors, suggesting that different rectification
planes could be used for different portions of the object.
To reach this goal it is not necessary that a rectification plane has
all the required ground control-points because it is possible to
calculate the planimetric position of any point located on a given
plane: it is sufficient to calculate the point's height displacement
relative to the new rectification plane and then to correct the point
co-ordinate of that value. In this way it is possible to make use of
ground control-points located on any plane.
Following this principle, the 10 ground control-points were projected
onto plane B; a rectified image was created making use of the coordinates of the transformed point. Results of the standard comparison
of co-ordinates obtained from the analytical plotter are shown in table
5.

Sampie	Δx	r.m.s.	Δy	r.m.s.
all the points	-3.0	10.0	-13.5	22.3
plane A	-8.0	8.2	-26.8	7.3
plane B	3.3	2.5	-0.4	3.6
plane C	14.0	4.2	43.1	8.1

Table 5, Co-ordinate differences on sample control points. Mean and
r.m.s. values for each measurements (cm)

Note the good results obtained for the points located on plane B; these
are of the same magnitude for the points on plane A measured during the
previous test.
The resultant displacement vectors are shown in figure 3; note the
difference in displacement caused by changing the rectification plane.
From a metrical point of view, these experimental results show that it
is also possible to use rectification for highly non-planar objects by
simply changing the rectification plane according to the points of
interest belonging to that plane.
The advantage of rectification versus differential rectification is that
the former is much easier, since DEM is not required and the rectified
image does not show visible deformations that would reduce its possible
use (in contrast to orthoprojections). The prudent use of rectification
is therefore feasible.

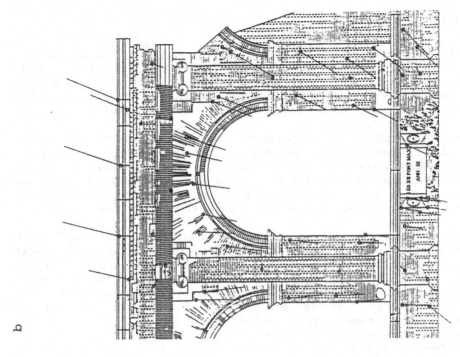

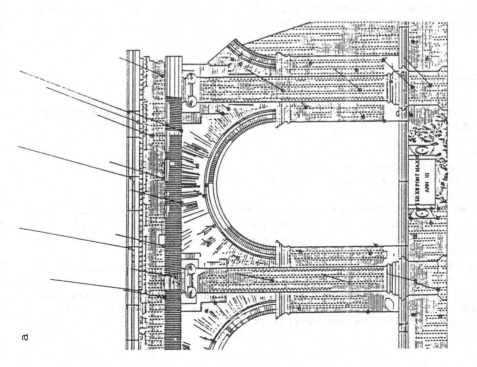

Fig.3, Analogical plotting of a western sector arch. Height displacements vectors on ground control points.
a) rectification plane parallel to plane A
b) rectification plane parallel to plane b

5. RECTIFIED IMAGE MOSAIC

Since the metric evaluation on the rectified images of the Colosseo's southern sector are quite good (when rectification of image surface extension is limited to the area between two nearby arches) a mosaic test was performed on two images belonging to that sector. The goal of the experiment was to test if mosaicking is a viable alternative to differential rectification.

The goal of mosaicking is to preserve the geometric precision of the single images, removing the radiometric differences of the original images caused by different illumination of the object during exposure or by scanning errors. From a geometric point of view the original images must be in the same absolute reference system, while from a radiometric point of view the more similar the image tonalities the better the results.

Since the two rectified images chosen for mosaicking had two different local reference systems (rectification planes were parallel to the local co-ordinate plane XY) a false origin was added to the ground control-points of the right image: the X co-ordinate of the most extreme right ground control-point of the left image, common to both systems. The two systems are rotated with respect to each other, but for software purposes they are mathematically treated as one system.

The procedure used is the following:
- new reference system chosen for the single image portions to be rectified,
- rectification of the two images,
- mosaicking.

From an aesthetic point of view the experimental results are satisfactory; from a geometrical point of view only a "step" is visible in correspondence with the contact between images.

6. DIFFERENTIAL RECTIFICATION

The most important problems to discuss relative to differential rectification are:
- image quality,
- metric accuracy;

Both fundamentally depend on two factors:
- good orientation parameters,
- exact object description;

In architectural applications Digital Elevation Model (DEM) modalities, the acquisition density and the applied interpolation technique are fundamentally important; in our case the Delaunay algorithms were used for DEM triangulation.

Clean surface discontinuities resulted in major description difficulties, and consequently breaklines were necessary.

The goal of experiment was to define the DEM and breakline modalities, as well as the acquisition density; this was done in order to:
- minimise the calculation and plotting time,

- increase the aesthetic quality,
- increase the metric quality.

The experiments were carried out on both sectors which were first rectified on a plane. Orthoprojection, mosaicking and digital image correlation were performed.

6.1. Orthophoto deformations

Information regarding standard DEM and breakline acquisition methods was collected by means of the experiments described below. One should note that aesthetic image problems always hide metric problems; as a result it was necessary to first obtain a qualitatively acceptable image and then verify the method validity from a metric point of view.

Major entity deformations occur on the orthophotos in correspondence with:
- areas in shadow;
- half-column edges;
- half-column horizontal returns;
- capitals, mouldings, cornices and others.

<u>Areas in shadow</u>

Owing to the exposure geometry, projected elements (such as the trabeation and the half-column basement with respect to the balauster or the half-column with respect to the pilaster) may hide some portions of the background elements which don't appear on the original photographs.

Clearly such zones cannot appear on the rectified image. The resulting effect on the rectified image is that of a "shade", located in those zones which lack information. Note that this problem is also present in traditional analytical plotting.

The only possible solution to this problem is to make exposures as nadiral as possible, only making use of the central part of the photograph and mosaicking the images.

Nevertheless the image shade effect is not disfiguring from an aesthetic point of view; the image is therefore satisfactory and metrically correct.

<u>Half-column edges</u>

Another image deformation problem appears in correspondence with half-column borders. Owing to the monument's curvature, the column-pilaster seam is never visible and as a consequence the traced breakline corresponds to the visible column height and will not be in right position.

On the orthophoto the result is a half-column with an incorrect size and a blurred border line. The above-mentioned shade effect is also observed in areas lacking information.

As a result one metric and two aesthetic problems exist.

With the aim of improving orthophoto appearance, some experiments were performed in order to artificially reconstruct the missing part of the half-column. Note, however, that the true half-column dimension has to be known, for example by measuring in the field.

From an aesthetic point of view, no attempt is fully satisfactory; the solution still requires exposures to be more nadiral and to make use of

only the central part of the photograph, followed by the mosaicking of
the images.

Horizontal half-column returns

On the original photograph the repeating horizontal ashlars appear as
curved rather than horizontal lines since the half-column plan is a
semicircle. This effect is also observed on the orthophoto due to a poor
altimetric description related to the points located on the half-column
surface. The obvious solution is to locally acquire an irregular DEM in
the X direction, with an increasingly thicker step from the centre to
the half-column borders.

Capitals, mouldings, cornices and others

Other critical aesthetic elements are the capitals as well as all
mouldings and cornices. An improper DEM combination, together with
breaklines, cause some deformation; in order to avoid these it is
usually sufficient to eliminate elevation points located very close to
breakline points.

6.2. General standards

The following general DEM and breakline standards were defined:

• breaklines have to follow the course of all principal surfaces and
must not be too numerous;

• elevation points must be traced in correspondence with all sharp
discontinuities, in particular corresponding to the border line between
the more hollow plane and the horizontal plane, if it is visible, or on
the hollow plane at the border of the shadow areas if it isn't visible;

• the regular grid of elevation points must be traced:
- within the area defined by wide surfaces and delimited by breaklines,
- on the half-columns, with a much thicker step in X than Y direction;

• elevation points must be eliminated on the following element surfaces:
- capitals,
- columns basements,
- arches cornices,
- mouldings,
- small surfaces delimited by breaklines,
- in any close point-breakline situation.

Regarding the less articulate sector, which is more representative of
classic architectural situations, it was possible to utilise wider
photograph portions than those of the western sector without introducing
image deformation; this was due to the smaller surface relieves (and
consequently less areas in shadow) and the generally smaller surface
articulation.

Considering the general criteria defined in other sector tests, it was
possible to substantially simplify the DEM, resulting in a considerable
decrease in data acquisition time.

General speaking it is possible to state that:
- many aesthetic problems may be resolved by the use of nadiral
exposures and by using the central portion of the photograph;
- DEM acquisition density and modality are extremely important in order
to obtain a good orthoprojection: the density obviously depends on the

surface articulation, the presence of different planes and the desired
accuracy;
- the surfaces must be delimited by the breaklines while regular
point grid must describe the altimetry (if it is an undulating surface)
or speeds up triangulation (if it is planar).
In contrast sparse elevation points are necessary to describe some
particular situations and must also support breaklines. DEM acquisition
and elaboration times were as follows (using a 486/66 PC):
- DEM acquisition - 6 hours for the western sector and 4 hours for the
southern sector (nearly 2000 elevation points and 2000 breakline points
in both cases);
- DEM triangulation - 15 minutes for the western sector and 40 minutes
for the southern sector; this difference was due to the different size
of the two surfaces, as the first one is much smaller than the second;
- orthophoto elaboration - 6 minutes for the western sector and 10
minutes for the southern sector.
Note that DEM triangulation and elaboration times were much higher in
the first experiment and that triangulation never converges.
In figures 4 and 5 the final DEM and corresponding orthophoto are shown
for both sectors. Aesthetically the final product is good for the
western sector and quite positive for the southern sector; it is
possible, therefore, to make full use of both rectified images.

6.3. Final orthophotos metrical evaluation

To metrically evaluate the accuracy of the differential
rectification procedure, and consequently the effectiveness of digital
orthoprojection as a full alternative method for traditional analytical
plotting of architectural object images, two sample control point coordinates were measured for nearly 120 points on the two final
orthophotos.
The orthoprojection accuracy was determined by comparing the above-
mentioned co-ordinate with the corresponding co-ordinates measured with
the analytical plotter. Table 6 shows the compared results; for both
orthophotos the observed mean differences are nearly 1-2 pixels, with
r.m.s. values for each measurement equal to nearly 2 pixels.

observed displacement mean						
	Δx	r.m.s.	Δy	r.m.s.	Δz	r.m.s.
western sector	1.5	1.8	0.7	2.9	-1.7	3.3
southern sector	1.0	2.2	-1.2	1.5	-2.0	1.9

Table 6, Observed displacements. Mean and r.m.s. values for each
measurement (cm)

Fig. 4, Western sector. Final DEM and corresponding orthophoto.

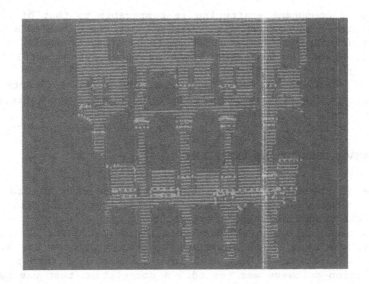

Fig. 5, Southern sector. Final DEM and corresponding orthophoto.

The fact that Z component scattering is comparable to that of the other two co-ordinates indicates the quality of the interpolation and differential orthoprojection results: in fact, it is possible to measure point elevations on the orthophoto with an accuracy comparable to that obtained using traditional analytical plotting.

The accuracy obtained is more or less the same for both sector orthophotos, in spite of the different complexities of the surfaces.

7. DIGITAL IMAGE CORRELATION EXPERIMENT

The use of the digital image correlation technique should be very useful for obtaining DEM's considering the possible time savings relative to the traditional laborious methods.

Even if such techniques are not theoretically suitable for architectural applications, the idea of a completely automated process for elevation acquisition is inviting; it was with this goal in mind that a digital image correlation software was tested. A correlation test was performed on two of the Colosseo's southern sector images.

The software used was a Galileo Siscam prototype program; the algorithm used in this program is a "Least Square (template) Matching" (LSC) with the so called "image pyramid". The DEM acquisition step, as in the manual test, was fixed at 40 by 40 centimetres, and nearly 2600 elevation points were acquired automatically, covering a surface of more than 700 square metres; time required to correlate homologous points was approximately 3.5 hours using a 486/66 PC.

The correlation results are:
- 70% of the correlated points have a high cross-correlation coefficient, indicating good points;
- 20% of the points have middle values;
- 10% of the points have low coefficients.

The correlated DEM is shown in figure 6. Note that the greater part of DEM points located on the upper side of the wall are incorrect; the reason for this is that the first approximation elevation values were so far from reality that it was not possible to reach convergence.

The software begins with an automatic object description procedure in order to create an initial wide step DEM, starting from less resolved image levels; during this phase it is not possible to manually interfere in order to exclude some areas. In the above-mentioned areas - sky and lacking arches - the grey level is homogeneous and the algorithm is unable to correlate.

Rectified images have, as expected, large deformations relative to the top of the wall and other mentioned areas, such as near border lines between full and empty areas owing to the lack of breaklines.

A subsequent experiment was performed by adding the breaklines already acquired by plotter and manually correcting point elevations with very low cross-correlation coefficients; points belonging to the upper side of the wall were eliminated since their correction takes too long. Aesthetically the resulting orthophoto, shown in figure 7, is still quite good even though it still has some residual deformation.

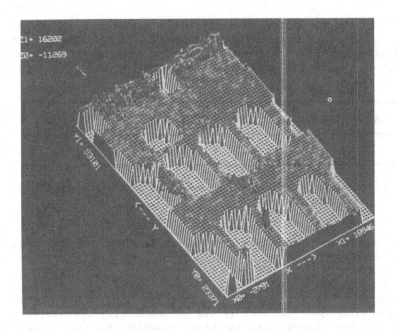

Fig. 6, Correlated DEM.

Fig. 7, Orthofoto elaborated with correlated and then corrected DEM.

Note that the time necessary to correct the elevations was higher than
the time spent for the corresponding DEM acquisition using the
analytical plotter, but it did not require the presence of an operator.
The results of the metric evaluation performed on the final orthophoto
are shown in table 7.

Δx	r.m.s.	Δy	r.m.s.	Δz	r.m.s.
1.2	1.6	-0.2	3.1	-5.3	4.6

Table 7, Observed displacements. Mean and r.m.s. values for each
measurement (cm)

In order to evaluate performance results it is necessary to consider at
least two parameters:
- image quality;
- method productivity.
Orthophoto quality is not as high as that obtained using the manually
acquired DEM; time spent to rectify the automatically correlated DEM and
to produce the breaklines was, as said, superior to that required for
ex-novo altimetry generation. At present the method is inappropriate;
nevertheless the experiments performed make it possible to define
potential changes in order to improve the software in terms of higher
general productivity. Metric quality is, on the whole, acceptable.

REFERENCES:

1. Ackermann F.: Digital image correlation: performance and potential
   application in photogrammetry, Photogrammetric record, 11(64), (1984).

2. Barbarella M. and Fiani M.: Raddrizzamento e ortoproiezione di
   immagini del Colosseo, Boll. SIFET n.2 (1993).

3. Birardi G., Carlucci R., Ferrara E., Giannoni U. and Maruffi G.
   (1988): The photogrammetric survey of the Coliseum in Rome. In:
   International Archives of Photogrammetry and Remote Sensing, XVI
   Congress of the ISPRS, Kyoto.

4. Fiani M.: Problemi di applicazione delle tecniche di ortoproiezione
   digitale all'architettura, Boll. SIFET n.1 (1994).

5. Fiani M.: Esperienze di raddrizzamento e ortoproiezione digitale, in:
   La fotogrammetria per il restauro e la storia, tecniche   analitiche
   e digitali, Atti del 1° Colloquio Internazionale, Bari, 1994.

6. Lee L.T. and Lin A.K.: Generalized Delaunay triangulation for planar
   graph, in: Discrete Computational Geometry, Springer-Verlag, New York,
   1986.

GIS GEOREFERENCING BY GPS

G. Bitelli and A. Pellegrinelli
University of Bologna, Bologna, Italy

ABSTRACT

With the recent growth in the development of GPS surveying techniques and Geographical Information Systems an increasing number of applications of GPS in the GIS environment has been proposed.

In the paper some aspects of this integration are discussed, in particular the problems related to the use of GPS in georeferencing GIS objects and in connection with software designated for generic field data entry. An example is presented.

1. INTRODUCTION

The recent few years have seen an expansion in the use of Geographical Information Systems (GIS), aided by the availability of more powerful software and hardware tools. As well as the application fields, the kind of data and the opportunities of analysis and manipulation by users have become larger.

Among the problems that still exist for an optimal use of the GIS systems, the greatest are perhaps tied to the availability of geographical data of adequate quality and quantity, acquired with highly productive and reliable techniques at low cost.

The diffusion of the satellite Global Positioning System (GPS), in its multifarious techniques of acquisition and processing, could make an interesting contribution to this framework.

A primary use of GPS for GIS data capture is to aid classical airborne or space data acquisition techniques, such as photogrammetry and remote sensing. Using GPS measurements in kinematic mode, frequently combined with inertial systems, coordinates of the projection center and attitude of an aerial photogrammetric camera can be determined [1]. Moreover, by static or fast-static surveys ground control points coordinates for absolute orientation of images are calculated.

Another important field of GPS application could be the quality control of large scale digital geographical databases, whose elements very often come from different sources, characterized by different degrees of accuracy, different nominal scales, different original cartographic systems. In its static or fast-static mode, GPS can in fact provide high precision measurements, with guarantees of homogeneity generally at lower costs than classical surveying techniques [2]: this kind of survey can then constitute a good way to audit the quality of digital cartography.

A very interesting use of GPS, which constitutes the main topic of this paper and will be discussed in the following paragraphs, is finally related to the direct georeferencing of objects during a GIS data acquisition phase or in updating existing geographical databases. By this kind of techniques it is possible to acquire the position of an object together with its geometric and alphanumeric attributes. As will be seen, a certain similarity among these techniques and the methods of stereo plotting in photogrammetry or manual digitizing can be stated. It has been found that with these tools it is as if the GPS antenna on the face of the earth emulate the action of the cursor of a digitizer table: obviously GPS always provides three-dimensional data. It should be emphasised that the methods that will be treated here do not have to be seen as an alternative to the traditional survey methods for data collection, like photogrammetry, which in general remain more productive. GPS could complement them; in some cases, if it is not possible to apply classical methods, if the quantity of data to be entered or updated is limited or in continual evolution, or if a direct field survey of the object is required, this type of solution can be very useful.

2. GPS FOR GEOREFERENCING GIS OBJECTS

It is fairly difficult to point out all the possible hardware and software solutions, that could be adopted in using GPS for GIS data capture, and even more to present a complete list of GIS applications that can make use of the GPS system. We will therefore give an overview, also because these techniques are still in evolution.

2.1 Instruments and applications

The georeferencing of objects in a region requires that their position is provided in a certain reference system with a precision compatible with the purposes and the characteristics of the GIS in which such objects will be handled. As is well known, using the GPS system the position (in the absolute or relative mode) of the phase center of the antenna is obtained in the WGS84 reference system and with a precision that changes greatly depending on instrumentation and survey techniques [2]. In order to properly locate the objects in a GIS, a decimeter or sub-meter accuracy is required for applications at large scale (e.g. GIS at urban level) whereas one meter to ten meters accuracy is sufficient for medium scale applications (e.g. GIS at regional level). In the first case, fast-static or kinematic techniques in the relative mode will be required while in the second case, relative code measurements (differential GPS, DGPS) or code-smoothing will normally be chosen. GPS measurements in absolute positioning, with a single receiver, are today subject, as is known, to uncertainty in positioning that could be in the order of a one or two hundred of meters; in some cases this uncertainty could be acceptable (e.g. GIS at national or international level or phenomena with large positional uncertainity). The greater demand and the more frequent applications belong to the intermediary level, mainly by using DGPS techniques, and this will be discussed in the paper.

As is well known, DGPS requires code receivers and correction to be made from the data acquired at a master fixed station (situated at a known position) to the data acquired by rover stations moved in succession at the objects to be georeferenced. The correction could be realized in real time, using radio-modem systems, or in a post-processing phase using appropriate software modules.

The minimum hardware required for a rover station consists normally of the GPS receiver (L1, C/A code) fitted with a practical, compact antenna and a data storage system such as a field computer or a standard portable computer (e.g. with PCMCIA interface); for efficiency in the field, it could be useful to have devices for facilitating the interface for a rapid and reliable data entry (pentop, barcode, voice input). The availability of a radio-modem system, that might receive the corrections from the master station through the RTCM protocol, enables the recording of coordinates already corrected (with a accuracy varying from a meter to several meters according to the receiver technology) and then cancels the need for a post-processing work. The systems with real-time correction are useful in particular in those operational situations where it could be necessary to know the position of the operator with a high degree of accuracy in order to reach an object of known coordinates or to retrieve the operator position on existing cartography. If the survey is carried out using a vehicle, simple dead-reckoning systems integrated with the GPS receiver can prevent the loss of positional data when satellites are not visible.

The software can be of various kinds:
- products supplied by GPS receiver manufacturers, running on PCs or dataloggers: these are able to monitor and then automatically process GPS acquisition, and associate the

derived position with description of what is mapped or in some cases measurements from other kinds of sensor. The results of the survey can then be exported into a GIS environment;
- products specifically designed for "field data entry" and provided with optional interfaces for GPS receivers or OEM GPS modules, and sometimes support for displaying vector- or raster-based cartography as background for the GPS position visualization;
- add-on modules or customizable software libraries for commercial GIS, which permit communication from the GIS kernel routines to the GPS receiver, the display of the current position in the GIS environment, as well as the normal data entry procedures provided by the GIS for alphanumeric data entry.

The range of potential applications that can benefit from these instruments is very large; we mention here only the following:
- inventories and mapping of distributed assets and objects (utilities management, cadastral applications, archaeology, forestry, roads or trails, ...) or finding of objects of known coordinates;
- environmental planning and assessment (data sampling of land use, agriculture, vegetation, geology, geophysical surveys, monitoring of air or water pollution sources and extent, ...);
- vehicle tracking and dispatching;
- georeferencing data produced by different sensor devices (e.g. meteo, echosounders, gas or chemical analyzers, Geiger counters, video or infrared equipment, ...), with integration of these sensors into complex systems; multimedia geographic data production.

2.2 Execution of a DGPS survey for GIS purposes

A DGPS survey designed for GIS data collection consists of the following stages: survey planning (GPS and database), data acquisition in the field, GPS differential post-processing (if real time processing is not available), data export to a GIS product.

2.2.1 Survey planning

Designing the survey requires the definition of both GPS reception and database characteristics.

For GPS acquisition, a first priority is to locate the master base station at a point of known WGS84 coordinates. As already shown by many experiences, e.g. [3], the reliability of these coordinates is very important, because this effects the quality of the overall mapping process; the distance between the master and rover stations should be maintained in a range of about two hundred kilometers in order to provide acceptable differential corrections. Obviously, a preliminary investigation could be performed in order to establish if a commercial or governmental base station is already operating in the area.

A second topic concerns the choice between real-time or post-processing differential data correction. Some businesses require the real-time option and in any case this certainly simplifies the overall process: the data are stored in the field already corrected, that is to within a meter accuracy if code smoothing technique is applied. Post-processing could enable even better solutions but the difference is normally negligible, and furthermore some packages do not easily permit rebuilding of the link between location and descriptive data after changing the coordinates by means of a differential data correction performed by an external program. In considering real-time acquisition it is very important to verify that the data capture software can specifically mark the records corresponding with a loss of the radio communication link (no correction in real time), because these data only have to be corrected in post-processing.

The GPS survey planning includes obviously the classical receiver setup, by choosing appropriate parameters like minimum number of tracked satellites, maximum allowable PDOP, elevation angle, time sync and specification of the kind of solution required (2D if only three satellites are visibles, 3D by using four satellites, an overdetermined solution if more than the sufficient number of satellites is used). The reception parameters must be related to the survey environment; for some urban surveys, satellite sky plots could be examined in order to choose the best time to carry out the measurements in order to minimize the effects of the obstructions along the specific route.

Regarding the GIS database, different properties have to be defined for the geometrical and alphanumerical attributes of the features to be surveyed. The features themselves could be defined as three-dimensional points, lines or areas; the basic elements stored are points with 3D coordinates and reconstruction of linear or area elements is realized by appropriate data management (fig.1). In certain systems, some graphic specifications can be described for the entity display representation in a GIS or CAD, depending on the entity type or also its attribute values.

It should be emphasised that, unlike a manual digitizing session, it is normally not possible to handle complex objects, i.e. entities composed of geometrically or semantically different elementary components (in layer-based GIS, objects which belong to different layers and which have different dimensionality); their reconstruction, by aggregation of elementary primitives, must be delegated to successive batch processing developed and operating in the GIS environment. Another issue, very important for building a coherent geographical database, is to respect the topological properties of the entities. Also this question cannot be normally automatically achieved by using the kind of systems here described, at least for practical reasons; this could be crucial in developing databases strongly based on topological properties, like network structures (e.g. registre of roads) and it will require subsequent specialized processing.

For each entity some attributes can be defined, together with their properties and validation rules. Moreover their characteristics related to the field recordings can be defined: that is, if the attribute is mandatory, optional or not enterable and which interface is available for data entry, for instance realized by menu displayed in a datalogger. As in other data acquisition techniques, it could be decided to omit some attributes if their values could be implicitly derived from existing GIS layers (e.g. soil use along a surveyed path).

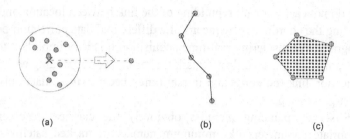

Fig. 1, Entity dimensions: (a) point entity: if some GPS positions are acquired, the point location is normally assigned at the baricentre of these coordinates, (b) line entity, described by a chain of points, (c) area entity: the coordinates of the first and last points coincide.

The entities and their attributes reflect the database structure of the GIS (if a relational database is used, each entity is in general associated with a table), but obviously it is generally not possible to manage complex relationships in the field like those established among the entities in a database environment; in the project it is therefore very important to ensure that attributes which support GIS database relations (such as univocal feature identification codes) are entered and maintained.

Finally, an important point is to choose the reference system and the coordinate type (geographical or projected) in which the results are required. Datum different from the WGS84 require a transformation of the GPS solutions; in some products this task could be performed automatically at the data collection stage, using standard approximate parameters or using appropriate parameters entered by the user and valid for the area in question (e.g. Helmert parameters or Molodensky constants). Some systems require on the other hand that this operation is performed in batch mode at the end of the survey. In some applications it could also be useful to refer the survey to a local system by recording some points whose coordinates are known in such a system.

2.2.2 Data acquisition

In a GPS-GIS system, the data acquisition phase consists of obtaining coordinates by GPS and entering entity attributes in the field.

In GPS data acquisition the antenna set-up is especially important (especially if the antenna is mobile) and the antenna must be raised to prevent interference from obstructions such as trees and buildings [4]. Non visibility of a sufficient number of satellites in some parts of a journey could require a successive passage with different satellite constellation and appropriate algorithms to derive the best final solution. The problem of data acquisition in a difficult environment, like urban areas or forests, requires further studies both in the techniques of data gathering and in the choice of better suited hardware equipment [5]. These questions are further complicated by using real-time systems, where radio communication maintenance becomes another critical factor; the experiences of the authors in historic city centers showed radio and GPS signals losses ranging from 30% to

60% of the total route. For better results in positioning, an integrated dead-reckoning system is at least required.

The user can furthermore choose to remain stationary on each point for only one GPS measurement or for a set-up number of measurements, in order to improve the position accuracy.

Attribute data entry is performed using the database designed in the setup phase, and using tools like a keyboard of a datalogger or a laptop computer, a barcode reader, a pentop, a voice input device. If a computer is used, current location ca be dinamically displayed in a background raster or vector map. It is important that the receiver continues to achieve positions, at the predefined sync interval, during alphanumeric data entry, in this way increasing the number of positions determined. It must be stressed here that a well designed data structure and user interface in the setup phase allows quick staying at the features, increases productivity of the method and reliability of the descriptive data.

In the better designed systems various methods are available for overcoming obstacles. For example, pausing and restarting the data acquisition at a barrier (fig. 2a); or stopping a linear object temporarily so as to enable data acquisition from a point entity to the side of it (fig. 2b). Rapid acquisition of points along linear features, without attribute entry, can be performed by means of quickmark tools for instantly recording the points; the point positions, if the mark time does not coincide with a GPS fix, can be derived by subsequent interpolation between two GPS measurements, the preceding and the next one (fig. 2c). Finally, line objects could be acquired by dynamic segmentation or automatically updating p-lines as an increment of time or of space.

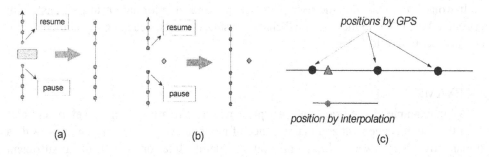

Figure 2, (a) obstacle along a way; (b) point entity acquired during line entity survey; (c) quickmarks for fast entry of positional point data.

An improvement in data gathering performance can be obtained by developing integrated systems that combine GPS with other surveying instruments, like laser rangefinders or video cameras, with tagging by GPS time. Digital image systems can, for example, constitute a means to build a multimedia and multisource GIS, where still images or video sequences can be georeferenced [6].

2.2.3 Data processing

Post-processing GPS differential data correction is required when real-time instruments are not available or when real-time data effected by radio communication failures require correction; this procedure can be normally performed using software provided with the GPS instrument.

Despite the relative simplicity of the method, the datasets always need careful inspection in order to detect problems related either to GPS performance, often affected by multi-path or loss of data due to obstructions, and to user errors, like code entity mismatches.

In difficult environments, no more than three satellites could be visible at the same time; in such a case 2D positions can be provided by processing, and the elevation must then be derived by interpolating from neighbouring 3D points, or by other methods.

Finally, obtaining the geoidal height is very important for many applications since it makes it possible to combine data derived from GPS with other three dimensional GIS information; this requires the knowledge of a local WGS84 geoidal undulation, but in some cases rough global models provided by receiver software could be sufficiently accurate.

2.2.4 Data export to GIS

The software provided by firms with GPS receivers must guarantee a correct data export toward the most widely used GIS packages, both for the geometric and the alphanumeric components. This can be accomplished, depending on the commercial standards adopted, by producing single or multiple data files; in the last case, a great data fragmentation could be useful in order to support and recreate more complex GIS data structures.

It would be crucial for the user to select the kind of information to be transferred; filters could be applied to the positional characteristics (for instance rejecting non-corrected or non 3D records) or to feature codes.

3. AN EXAMPLE

As an example, some brief notes are presented here relating to the development of a GIS for the use, management and maintenance of mountain paths in the Apennines. A data dictionary of about twenty kinds of feature (point, line or areas) of a sufficient completeness was developed, together with attributes and interface for data entry. The survey was conduced using Trimble PathFinder Pro XL receiver and real time DGPS corrections from a distance of about 140 km, with a Trimble 4000 SSE as a base station. The location and descriptive data of a lot of objects were acquired by walking along some paths, already mapped or not. As an example, in figure 3 the simple point feature "path sign" is described. The planimetric accuracy was on the average about one-two meters, sufficient for application purposes.

During the data acquisition phase it is possible to associate different kind of data, such as pictures, to a feature (figure 4) and, after processing, the overall information is available in a GIS environment (figure 5) for selective display or interactive spatial and attribute queries.

In figure 6 a part of a path is superimposed on a Digital Terrain Model derived from contour lines of a 1:5000 map.

The experiment showed the great potential of the technique for this application for georeferencing new entities and for describing entities already mapped. This method, integrated in a wider approach to data management of paths, can permit to maintain and update the related informations as a whole, with good accuracy, high data consistency and low cost.

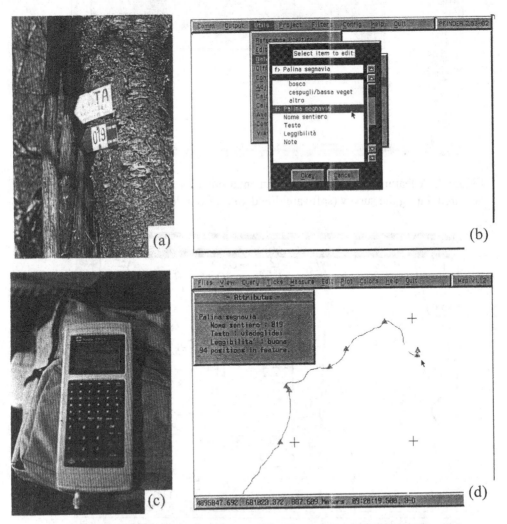

Figure 3, Point feature description. (a) Path sign (b) Definition of some attributes and data entry interface (c) Field data entry of the attributes (d) After post processing, query by feature shows coordinates and attribute values (software: Trimble Pfinder).

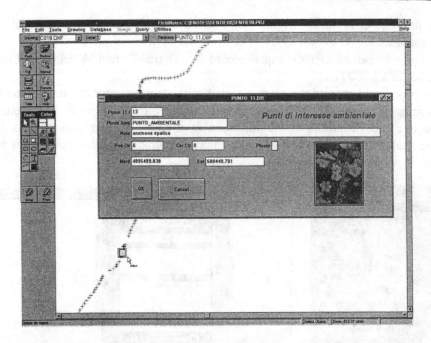

Figure 4, A feature, interesting environmental point, is associated with a picture acquired during the survey (software: FieldNotes, from PenMetrics).

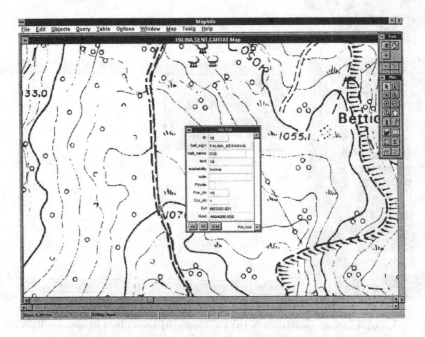

Figure 5, Query on a path sign object located in a path already in cartography; in background a raster image from 1:5000 map (software: MapInfo).

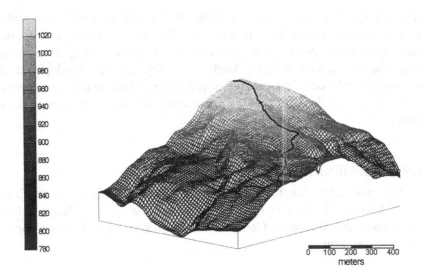

Figure 6, Three-dimensional view of a part of a path depicted on digital terrain model.

4. CONCLUSIONS

GPS surveys for GIS data gathering and updating show very interesting characteristics in terms of productivity and flexibility for a wide range of applications. These techniques could constitute a first method for georeferencig or a method for completing and complementing other kinds of survey; the full development of such integration and the solution of typical GPS acquisition problems in different environments can provide a powerful instrument for extending the scope of geographical data management.

ACKNOWLEDGEMENTS

For the experiment described in paragraph 3, acknowledgements to Codevintec Italiana - Milano for Trimble equipment, to Ash - Pisa for FieldNotes software, to Massimo Pancaldi and Alessandro Geri of the Italian Alpine Club, Bologna, for database definition and general support.

REFERENCES

1. Ackermann, F.: Operational rules and accuracy models for GPS-aerotriangulation, International Archives of Photogrammetry and Remote Sensing, Vol.29, Commission III, Washington D.C., 1992, 691-700

2. Hofmann-Wellenhof B., H. Lichtenegger and J. Collins: Global Positioning System - Theory and practice, Springer-Verlag, Wien, 1992

3. Bitelli, G., A. Capra and L. Vittuari: DGPS on long distances, Proceedings IV Italy-Poland Geodetic Meeting, Warsaw, 1994

4. Santerre, R. and M. Boulianne: New tools for Urban GPS Surveyors, GPS World, February 1995, 49-54

5. Lachapelle G. and J. Henriksen: GPS under cover: the effect of foliage on vehicular navigation, GPS World, March 1995, 26-35

6. El-Sheimy N. and K.P. Schwarz: Integrating GPS receivers, inertial navigation systems (INS) and CCD cameras for a mobile GIS data collection system, Proceedings ISPRS94, Ottawa, 1994, 241-248

TEST AND CALIBRATION OF A DTP SCANNER
FOR GIS DATA ACQUISITION

E.P. Baltsavias and M. Crosetto
ETH Zurich, Zurich, Switzerland

ABSTRACT

DTP scanners have undergone rapid developments the last few years and have characteristics that make them attractive for use in photogrammetric, cartographic and GIS applications, as for example for scanning of aerial and satellite film imagery, plans and small format maps. This data or other derived products like DTMs and digital orthoimages are important data layers in GIS. However, the DTP scanners also have some deficiencies like format limited in most cases to A3 or A4, a geometric resolution that in best case is 1200 dpi and most importantly exhibit geometric inaccuracies that usually exceed 0.1 mm. While the first two deficiencies can be corrected only by the scanner manufacturers, the last one maybe be accounted for by calibration procedures that can be applied by the user. An improvement of the geometric accuracy of DTP scanners by calibration would make their use possible in much more applications. In this paper test patterns and a method for the geometric calibration of the Agfa Horizon DTP scanner are presented. The method, at least conceptually, is applicable to other scanners with similar characteristics. First results show a reduction of the geometric errors after calibration to 7 µm in CCD-line direction and 15 µm in scanning direction.

1. INTRODUCTION

1.1. Motivation

Scanners have been used as input devices in GIS applications for digitisation and eventually subsequent vectorisation of existing hardcopy maps. Other applications that can increasingly be found, particularly in relation to a GIS, make use of scanned images, especially aerial imagery, which are usually transformed into orthoimages and are subsequently used for generation or update of databases, and creation of different visual products in digital or analogue form like orthoimage maps and 3D perspective views. Orthoimages seem to become a central layer within GIS, and have been used in local, regional and national projects by mapping organisations, especially for the update of topographic maps and the establishment of a geometrically accurate basemap. A prerequisite for orthoimage production is the scanning of mostly analogue aerial images, which however is generally accomplished by using expensive so-called photogrammetric scanners. Thus, the motivation of the paper is the investigation as to what extent lower-priced DTP scanners, which are rapidly improving during the few last years, can be used for GIS data acquisition. For certain applications, e.g. production of analogue orthoimages or orthoimage maps, the geometric accuracy of DTP scanners may suffice, while for other applications like updating of digital databases it is generally insufficient. In order to extend the usage of DTP scanners their geometric inaccuracies must be reduced by means of calibration. The paper presents a general method for geometric calibration and first results using the Agfa Horizon.

1.2. Scanning Requirements for GIS Data Acquisition

The scanning requirements for scanning film imagery and maps differ. Maps/plans are black and white or colour, can be transparent or opaque, require a large scanning format (e.g. A1), a geometric resolution of 400 - 1000 dpi, a geometric accuracy that is below the map accuracy (usually 0.2 - 0.3 mm), and a radiometric resolution usually of 1 - 4 bit (2 - 16 grey values). Aerial images are scanned in grey levels or colour, require a format of 25 x 25 cm, a geometric resolution of at least 600 - 1200 dpi, a geometric accuracy of 2 - 5 μm (for high accuracy applications), a radiometric resolution of 10 - 12 bit and a density range of 2.5 D (panchromatic images) to 3.5 D (colour images). Satellite images have the same scanning requirements as aerial images with the exception of the scan format (up to 30 x 45 cm). There is no single scanner, as far as the authors know, that can fulfil all these requirements.

1.3. Overview of DTP Scanners

DTP scanners can be roughly divided into two groups: (a) flatbed scanners that are generally low-cost and up to A3 format, and (b) drum scanners that are generally more expensive and geometrically less accurate than flatbed ones, but with better radiometric performance and larger scan size. The paper will mainly concentrate on flatbed scanners with aim the scanning of aerial images. Flatbed scanners typically employ one or more linear CCDs, and move in direction vertical to the CCD to scan a document. They can scan binary, halftone, grey level and colour data (with one or three passes), may have good and cheap software for setting the scanner parameters, image processing and editing, and can be connected to many computer platforms (mainly Macs and PCs, but also Unix workstations) via standard interfaces. They can usually scan A4 format, but some can scan up to A3 or even more. Some do not scan transparencies, others do so but only of smaller format (typically with a 8″- 8.5″ width). There exist a handful of scanners which can scan aerial images (characteristic representatives are the 1200 dpi Agfa Horizon Plus and Horizon, and the 600 x 1200 dpi Sharp JX 610).

Flatbed scanners have a resolution of up to 1200 dpi (21 μm pixel size) over the whole scan width. Few scanners offer the option to increase the resolution (e.g. up to 2400 dpi) by projecting a document portion (smaller than the full width) on the CCD. Their price range, with few exceptions, is 1,000 - 30,000 $. The big price jump occurs when going from A4 to A3 format. The transition from 600 dpi to 1200 dpi costs less. A3 scanners with 600 x 1200 dpi start at ca. 12,000 $. A4 scanners with 600 x 1200 dpi and transparency options cost much less (2,000 - 4,000 $). Their radiometric resolution and quality, and scanning speed can be comparable to or even exceed that of the more expensive photogrammetric film scanners. DTP scanners with automatic density control and user definable tone curves that can be applied during scanning need for the setting of the scan parameters a few minutes as compared to more time (even one hour) required by some photogrammetric scanners. In particular, the sensor chip and the electronics of DTP scanners are updated faster and are in most cases more modern that the respective parts of photogrammetric scanners. New generation DTP scanners employ 10 - 12 bit digitisation and have a density range of up to 3.4D. Some employ modern 3-colour linear CCDs (like the 2,000 - 8,000 pixels KODAK linear CCDs) and scan colour documents in one pass. Functions that can be encountered in DTP scanners include sharpening, noise removal, automatic brightness and contrast adjustment, manual and automatic thresholding, white and colour balancing, black/white point setting, negative scanning, automatic colour calibration, self-defined screens for scanning halftone documents and printing images, multiple self-defined thresholding for each colour channel to scan multi-colour documents, preview (sometimes with variable zoom) and scan area selection, CMYK scanning, colour correction, integrated JPEG compression, and batch processing. The scanners can be bundled with other packages for image processing, editing, and retouching, colour management and calibration, image management etc. Their quality is rising while their price drops (especially for the A4 format scanners). The main disadvantage of DTP scanners are the small format and the insufficient geometric accuracy and stability, caused mainly by mechanical positioning errors and instabilities, large lens distortions, and lack of geometric calibration software.

1.4. Basic Technical Characteristics of Agfa Horizon

The technical characteristics of the Agfa Horizon and tests of its geometric and radiometric performance are explained in detail in [1]. Table 1 summarises its main characteristics.

Mechanical movement	Sensor type	Scanning format (mm)	Geometric resolution (μm)	Radiometric resolution (bits) (internal/output)	Colour passes
flatbed, stationary stage	3 linear CCDs, 3 x 5,000 pixels	A3 (reflective) 240 x 340 (transparent)	21.2 - 1270	12/10 or 8	3

Table 1: Basic characteristics of Agfa Horizon

2. ERROR TYPES AND TEST PATTERNS

Details on different errors that can occur with scanners and with the Agfa Horizon in particular can be found in [2] and [1] respectively. Following the proposal in [2] the errors are classified in slowly and frequently varying. In the following investigations only the major geometric errors will be considered. These errors include the lens distortions, mechanical positioning errors and vibrations. They are common to one or the other extent in all DTP scanners. Other errors can occur depending on the design, construction, and parts of each individual scanner. All flatbed DTP scanners use one or multiple (optically butted) linear CCDs. In the rest of the paper the following convention will be used. Horizontal direction is the direction of the linear CCD (denoted as x-direction), vertical the direction of the scanning movement (denoted as y-direction). Errors due to lens distortion are slowly varying and have an influence mainly in the horizontal direction. Errors due to mechanical positioning and vibrations are frequently varying and occur in the vertical direction, whereby vibrations occur also in the horizontal direction (e.g. for the Horizon horizontal vibrations up to 0.5 pixels have been observed).

For the quantitative evaluation of the geometric errors and the calibration two plates have been used. For more details on different test patterns and their desired properties see [2].

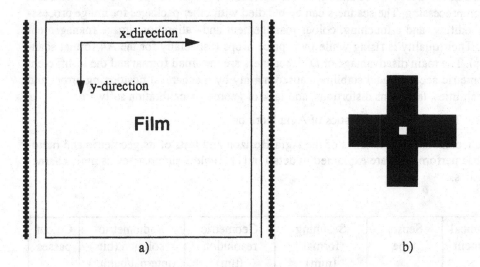

Figure 1.   a) Calibration plate to be scanned together with the film in on-line tests. b) Enlargement of a cross (line intersection).

The first plate includes a 25 x 25 grid with 1 cm spacing and straight lines at the four border sides. The grid nodes are used for evaluation of the geometric accuracy of the scanner and for correcting the slowly varying errors (off-line tests). Since these comprise mainly of the lens distortion, a grid spacing of 1 cm completely suffices. The straight lines are used for checking other geometric errors, e.g. the lines in y-direction can be used to detect horizontal vibrations. However, in this work the straight lines have not been used. The second plate (see Figure 1a) is used for correcting the frequently varying errors. It must be scanned together with the document to be scanned (on-line tests). The distance of the crosses along the left and right vertical line was ca. 1 mm. The high density of the crosses is necessary because errors due to mechanical positioning can occur abruptly. The total number of crosses is 474. The straight vertical lines are for checking the horizontal vibrations but again they were not used in this test.

The dimensions of both plates were 25.4 cm x 25.4 cm x 4 mm. Ideally, they should cover the whole scan area but this would increase the cost of the plates and the errors due to unflatness of the plates. By scanning the plate at different positions the whole scan area can be covered. The plates were fabricated as follows. The plates were created as binary images using a raster editor and other programs. The images were transferred to an Intergraph station and after some transformations were plotted at an Optronics 5040 Laser Scanner/Plotter on film with thick Estar base. The films were delivered to a company producing high precision patterns on glass and other material where they were copied on high quality glass. By using this procedure the costs can be kept low (ca. 800 SFr. per plate) and additionally any custom-made patterns can be produced. The cross centres of both plates were measured at a Wild AC3 analytical plotter using repetitive measurements. For the measurement the small white square at the center of each cross (see Figure 1b) was used. Its size was 37.5 µm and fitted the size of the measuring mark of the plotter. The estimated accuracy of the measurements is 2 - 3 µm.

3. TESTS AND GEOMETRIC PERFORMANCE OF THE HORIZON

The first plate was scanned with 600 and 1200 dpi, at two different positions in the y-direction (in x-direction the whole scan width was covered by the plate) and at different time epochs. The cross positions were measured by Least Squares Template Matching [3], whereby different noise-free synthetic templates were created for the 600 and the 1200 dpi scan resolutions. The accuracy of the matching results is at least 0.1 pixels, i.e. 4 and 2 µm for 600 and 1200 dpi respectively. Using these measurements and the measurements of the analytical plotter (reference values) an affine transformation between these two sets was computed. Firstly, the affine transformation was computed using only the 4 corner points as control and the remaining 621 points as check points. This is a simulation of the case of aerial imagery where only the four corner fiducials are used in an affine transformation to determine the interior orientation. The RMS error of the check points was ca. 150 µm in x- and 60 µm in y-direction. Although the scanner is geometrically much more stable in x-direction, the errors were larger than in y-direction (in these tests), indicating large lens distortions. The second affine transformation was computed using all 625 points as control points. The residuals in x- and y-direction were analysed separately because the expectation was that those in x-direction will be stable but those in y-direction not. These expectations were verified. The x-residuals were very similar independently of scan resolution, plate position and time epoch.

Small variations exist and are partly due to horizontal vibrations. The y-residuals on the other hand were different for each scan.

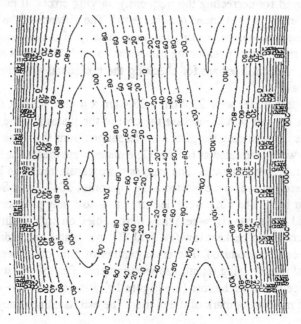

Figure 2. Isolines of x-residuals after affine transformation using 625 control points.

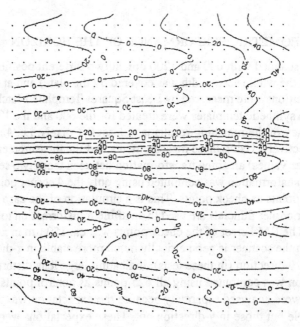

Figure 3. Isolines of y-residuals after affine transformation using 625 control points.

Based on these findings, it was decided to use the first plate and the x-residuals of the affine transformation using 625 control points in order to model and correct the errors in x-direction. These errors are slowly varying and in our experience they stay stable over several months. The x-corrections are applied to each scanned document. For the y-corrections the second plate is used. Again all crosses along the two border lines are measured by Least Squares Template Matching and an affine transformation is computed using all crosses as control points. The y-residuals of this transformation are used to model and correct the errors in y-direction. This calibration must be done for each document to be scanned.

4. CALIBRATION METHOD AND RESULTS

Using the residuals (errors) as observations, a model of the errors and the estimation of the model parameters should be performed. The models usually employed are either physical or polynomials or combinations thereof. The type of the errors in y-direction (abrupt errors at varying positions and with varying magnitude and form) and their temporal instability make the use of a physical or a polynomial model very difficult. In x-direction a physical model for lens distortion could be used. However, the estimation of the distortion parameters would be influenced by temporally varying errors like horizontal vibrations. In addition, as Figure 2 shows, the x-error varies slightly in y-direction, so it would not be appropriate to use the same corrections based on lens distortion parameters independently of the y-position. Thus, we preferred a fast and simpler solution. The residuals themselves were used as corrections. For the x-corrections the residuals of several scans with different scan resolutions and y-position of the grid plate on the scanner stage were combined and averaged. The averaging smooths out the noise and reduces the effect of temporally varying errors, like vibrations, on the residuals. The combination permits computation of x-corrections over the whole scan area and was performed as follows. After measuring the crosses in the digital images, their pixel coordinates were transferred to the scanner coordinate system by using a transformation with two shifts and two scales. The two shifts are given by the scanner coordinates of the upper left pixel which are shown in the scanner software. The two scales are taken from the affine transformation between the pixel and the analytical plotter coordinates. Thus, for each scanned plate the 625 crosses are known in the scanner coordinate system. For each plate, an affine transformation between these values and the reference values, and the respective x-corrections are computed. Then, a regular grid with 1 mm grid spacing is defined in the scanner coordinate system and corrections are interpolated from the x-corrections of each plate. At grid nodes with interpolated values from more than one scanned plate, the respective corrections are averaged.

To test the above procedure an x-correction grid was computed at a time epoch. The first plate was scanned with 600 dpi four and six months after the computation of the correction grid and an affine transformation between pixel and reference coordinates using the 4 corner points as control points was computed. Without correcting the pixel coordinates of the 621 check points the RMS error in x was ca. 140 μm. After making corrections based on the correction grid the RMS dropped to 6 - 7 μm, i.e. 1/7 - 1/6 of a pixel.

In the y-direction the corrections are computed directly from the residuals of an affine transformation between pixel and reference coordinates of all crosses of the second plate. A regular grid, in pixel space, with a spacing of ca. 1 mm is defined and corrections are interpolated from the corrections at the cross positions.

To check the procedure for y-correction we scanned with 600 dpi the second plate whereby the first plate was placed on top of it. The aim was to use the 625 points of the first plate as check points. This causes a small defocusing of the crosses of the first plate as they do not lie on the scanner stage but this does not pose a problem for matching because the contrast of the crosses is very good. In addition, it introduces displacements of the imaged crosses but we expected that this effect would be accounted for by the affine parameters. Of course, it would be better to design the first plate such that it includes the left and right vertical border lines with the crosses. An affine transformation of the 625 points using the 4 corner points as control points was performed with and without applying the x- and y-corrections. The RMS error of the 621 check points without corrections was 138 μm in x and 268 μm in y, and after corrections 7 μm in x and 15 μm in y (see Figure 4 and Figure 5).

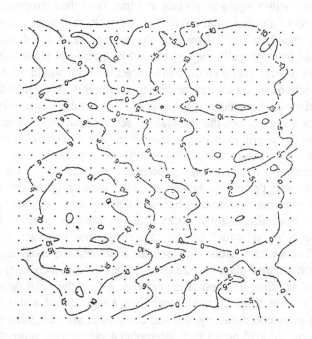

Figure 4. Isolines of x-errors after calibration.

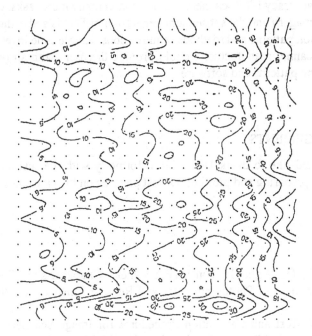

Figure 5. Isolines of y-errors after calibration.

The x-errors without calibration were very similar in form and magnitude to the errors shown in Figure 2. However, the y-errors were now much larger and differed in form that the y-errors shown in Figure 3. After calibration there is still a remaining systematic part in the y-errors as shown on the right part of Figure 5. It is obvious that the results in y-direction are not yet as good as in x but these are just first investigations. Initially, it was thought that the y-errors should be almost equal along a scan line, i.e. due to a faster or slower mechanical movement. However, Figure 3 and Figure 5 show that this is not the case. Other effects like abrupt slight rotation of the scan line, nonparallelism of the scan line to the scanner stage, lens distortion influencing the y-direction can not be excluded and will be further investigated. Nevertheless, the current results constitute a tremendous improvent over the geometric errors without calibration.

5. CONCLUSIONS

Test patterns and a calibration method for a DTP scanner were presented. They should be applicable for other DTP scanners as well, since they all have similar characteristics (large lens distortion, mechanical positioning inaccuracies, scanning in one swath using linear CCDs). This assumption will be tested by using other DTP scanners. The proposed approach is simple and leads to very good results in the x-direction. The modelling in y-direction will be further investigated with aim to reduce the error to ca. 10 μm. By improving their geometric accuracy DTP scanners can be used in many more tasks, especially for digitisation of film imagery and the subsequent generation of DTMs and orthoimages. Development of calibration patterns and methods requires considerable time investment, so it would be ideal if the scanner manufacturers could provide the users with an optional calibration package including patterns and software.

ACKNOWLEDGEMENTS

The authors would like to thank Heinz Stoll, formerly with the Institute of Cartography, ETH Zurich, for preparing and plotting our test patterns on the Optronics.

REFERENCES

1. Baltsavias E.: The Agfa Horizon DTP Scanner - Characteristics, Testing and Evaluation, Int'l Archives of Photogrammetry and Remote Sensing, Vol. 30 - 1(1994), 171 - 179.
2. Baltsavias E.: Test and Calibration Procedures for Image Scanners, Int'l Archives of Photogrammetry and Remote Sensing, Vol. 30 - 1(1994), 163 - 170.
3. Gruen A.: Adaptive Least Squares Correlation: A Powerful Image Matching Technique, South African Journal of Photogrammetry, Remote Sensing and Cartography, 14(3), 1985, 175 - 187.

GEOMETRIC AND RADIOMETRIC CALIBRATION FOR LOW COST FLATBED SCANNERS FOR PHOTOGRAMMETRIC APPLICATIONS

P. Boccardo, A. Lingua and F. Rinaudo
Polytechnic of Turin, Turin, Italy

ABSTRACT

In recent years, low cost scanners have been widely used in different applications such as, desk top publishing, graphics and rendering techniques.

For photogrammetric applications, because of the geometric (resolution and accuracy) and radiometric (dynamic range) constraints, only few high performance scanners (such as the Zeiss, Helava, Rollei and Vexcel ones) have been used.

The aim of this paper is to investigate the possibility to utilise low cost flatbed scanner for photogrammetric application, paying particular attention to the correction of the geometric distortions of the digital image acquired.

Using image processing techniques (such as filtering, enhancement and edge detection algorithms) the geometrically distorted digital image scanned by these devices, can be corrected by the mean of a calibration procedure, using a high precision reseau that gives the acquired image distortion function.

In addition, using feature extraction and matching procedures with subpixel precision, the fiducial marks position could be determined, in order to perform an interior orientation.

The image, geometrically and radiometrically corrected by the mean of high order interpolation algorithms, can be directly utilised for photogrammetric applications (digital stereoplotting devices), and geographical information systems (used as a raster layer) multidata integration (raster and vector merging).

1. INTRODUCTION

Affordable desktop scanners are no longer merely devices to get images inside a computer. As standards have improved, superior units are being touted as serious alternatives to expensive drum scanners.

Particularly noteworthy is the radiometric performances of the new off shelf flatbed scanners, ranging from 30 bits up to 36 bits radiometric resolution. This massively expanded range is said to be different from the old breed, and closer to the flatbed drum scanners necessary to produce professional quality (and thus, available for photogrammetric usage) images.

From the geometric (accuracy and resolution) point of view, the maximum resolution allowed, ranges from 400 dpi up to 1600 dpi, but software and/or hardware interpolations, currently available for most of the desktop scanners, permits geometric resolution up to 2400-3200 dpi. The geometric accuracy (scanning repetability and maximum distortion allowed), is not being tested yet on a systematyic basis; this is due to the desktop scanners usage, not really involved in metric usage.

In the following sections, a preliminary analysis of the geometric and radiometric assessment have been carried out.

1.1 How flatbed scanners work

All scanners work on the same principle of reflection or transmission. The analogue image is placed before the scanning head, consisting of a light source of sensor. The amount of light reflected or transmitted through the image is picked up by the sensor, then converted to a voltage proportional to the light intensity; the brighter the part of the image, the more light is reflected or transmitted, resulting in a higher voltage. This is finally converted by an analogue to digital converter (A/D converter) into information the computer can understand.

The sensor in many scanners is a Charge Coupled Device (CCD). A CCD consists of many photo-senssitive elements, arranged in a grid in the case of a digital camera, or in a long, thin line in the case of DTP scanners.

1.2 Geometric resolution

A DTP scanner claiming a horizontal optical resolution of 600 dpi and a maximum width of 8 inches will have 8×600, or 4800 usable elements on the CCD.

The vertical resolution of a DTP scanner is dictated by the degree of fineness with which the head can be physically directed over the image. In the case of flatbed scanner, the head is driven by a step motor. It is common that the maximum vertical resolution may exceed the horizontal resolution thanks to the step-motor being highly geared: an optical resolution of 600×1200 dpi is not unusual.

The optical resolution represents the maximum resolution of the CCD and the step-motor as described above. It is, however, possible for the apparent resolution to be increased using a technique known as interpolation, which under software or hardware control, guesses intermediate values and inserts them between real ones. Some scanners do this much more effectively than others.

1.3 Colour acquisition

Colour scanners have three light sources; one each for red, green and blue primary. Some scanning heads contain a single fluorescent tube with three filtered CCDs, while others have three coloured tubes and a single CCD. The former produce the entire colour image in a single pass, while the latter will have to go back and forth three times. Obviously, three-pass scanners suffer from geometric inconsistencies due to losses of the origin of the coordinates.

The range of colour shades captured by a scanner is down to the dynamic range of the A/D converter, along with the purity of the illuminating light and any system noise. In theory, a 24 bits scanner offers an 8 bits range of 256 levels for each primary colour, a 30 bits scanner offers a 10 bits range (1024 levels), and a 36 bits scanner stretches to a 12 bits range of 4096 levels.

1.4 Software drivers

The major part of a DTP scanners are provided with Twain compliant drivers for Windows. Twain is not an acronym; it is a very important standard in image acquisition, developed by HP, Kodak, Aldus, Logitech and Caere.

Better twain drivers offer a high degree of overall image adjustment such as brightness, contrast and colour; in addition, several offer gamma correction.

Not all twain drivers are the same. It is up to the device manufacturer to write a driver and decide what options it should offer. All scanner drivers offer a preview which quickly displays a small representation of the image to be digitised. From here the scanning area may be adjusted along with the resolution and pixel depth.

1.5 Other type of scanners

DTP scanners are not the only available type of devices. Other type of instruments are currently accessible, and in particular:

- Hand scanners: unexpensive with low geometric and radiometric resolution, are used dragging the whole device manually.

- Sheet feeding scanners. equipped with a fixed scanning head over wich the image is fed. High scanning speed (up to 40 pages per minute), acquire 1 bit (black and white) images suitable for Optical Character Recognition (OCR) work.

- Drum scanners: equipped with a cilinder into which the image is placed and curved to fit. Very expensive, but suitable for photogrammetric applications.

- Film scanners: designed to digitise transparency film (usually small format), but suitable for photogrammetric applications.

- Digital cameras: like other cameras except for a matrix array CCD where the film usually go.

- Barcode scanners, used for scanning barcodes.

2. DTP SCANNER GEOMETRIC AND RADIOMETRIC QUALITY TESTS

2.1 State of the art of investigation tests

In recent years, the massive usage of raster images suitable for digital photogrammetric systems, have pointed out the researchers' attention to the geometric and radiometric quality tests. An OEEPE working group, coordinated by Prof. Kölbl in Lausanne, is considering the radiometric performances of photogrammetric scanners, while several scientists in different Universities (ETH in Zurich, Graz, Ohio State, Munich, etc.) are evaluating others problem such as geometric tests, accuracy assessment, repetability, etc.

In general, the different test procedures performed (Seywald, Leberl, 1993; Baltsavias, 1994), point out some radiometric problem as:

- noise related to the scanner electronics (thermal variations, single sensor gain, dimensional CCD sensors differences, etc.);

- illumination system non homogeneity and instability (both for reflective and transparency modes);

- tones' saturation and dynamic range variations;

- optical system inconsistency that causes shadowing and discrepancies;

- spectral noise (with scattering anomalies in the blue band)

- non-perfect sampling along scanning samples due to unprecise mechanical movements.

Taking into account geometric discrepancies:

- CCD sensors misalignement (in the case of three passes scanners), that causes three bands shift;

- vibrations along the horizontal axes;

- optical distortions, that can be corrected only using calibration reseau;

- acquisition discontinuity due to the buffer capacity storage;

- non-constant geometric resolution along the two axis;

- deformations on the scan area, that can be corrected if they are systematic.

UMAX PS2400 TECHNICAL SPECIFICATION	
Scanner type	Flatbed
Scan speed	66 s/LT (600 dpi, Colour Mode)
Preview speed	25 s/LT
Maximum scannable area	212×297 mm
Optical resolution	600×1200 dpi
Output (HW) resolution	600×1200 dpi
Maximum resolution	2400×2400 dpi
Scanning density	1 dpi to 2400 dpi
Colour scanning method	one pass with colour CCD
Sample depth	
Colour mode	30 bits/pixel internal
Grayscale mode	10 bits/pixel internal
Halftone mode	1 bit/pixel
Line-art mode	1 bit/pixel
Scan mode	
Colour mode	quality/speed
Grayscale mode	quality/speed
Interface	built-in SCSI II
Environmental ranges	
Operating temperature	10°C-40°C
Relative humidity	20%-80%
Noise	under 50 Db (operating)
Dimensions	539×334×138 mm
Net weight	9.2 kg
Systems supported	PS and Macintosh
Options	Transparency adapter UTA-II

Tab. 1, Umax PS2400 technical specifications

3. SCANNING PROCEDURES IN A PHOTOGRAMMETRIC SYSTEM

In order to precisely define a digital photogrammetric stereoplotter, it is necessary to configure it as a single unit of a broader digital photogrammetric system; this new system has to integrate both a stereoplotting unit and a completely independent acquisition unit. In particular, the acquisition unit have to produce a digital image, where each pixel have to be associated to a coordinate pair (in the fiducial system); in addiction, the image must be corrected *from* all the systematic errors (optical distortion, scanner miscalibration, etc.).

The separation between the two units is a very innovative concept among the digital photogrammetric stereoplotters on the market; infact, it is possible to distinguish, also theoretically, the acquisition and the restitution operations, being the first related to calibration and internal orientation algorithms, and passing to the second unit metrically corrected images sampled according to an epipolar geometry.

The ultimate problem is the correction of the geometric and radiometric distortions induced by the scanner used, that have to be corrected performing an automatic recognition on a digital acquired reseau (with known coordinates) and evaluating and correcting the position of the misplaced crosses centres.

In fig. 1 a geometric and radiometric calibration flow chart is shown.

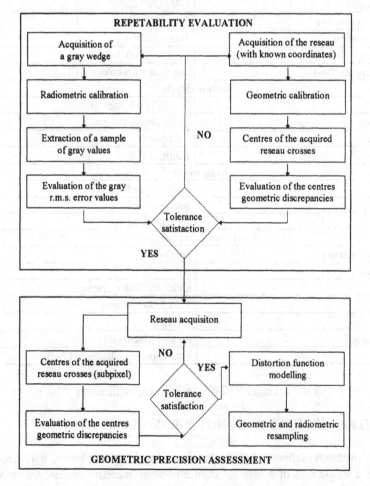

Fig. 1, Scanner geometric and radiometric calibration procedure

4. SCANNER USED

For the radiometric and geometric assessment an off-shelf DTP scanner has been used; manufactured by Umax, the trade name is PS2400. In the table 1 the principal specification are shown.

5. REPETABILITY TESTS

5.1 Radiometric repetability test

In order to assess the radiometric repeatability a calibrated gray wedge (fig. 2) has been acquired 10 times (600 dpi, 8 bits per pixel), and a sample of 10.000 pixel per bar has been extracted. For the different acquisition the r.m.s. error has been calculated.

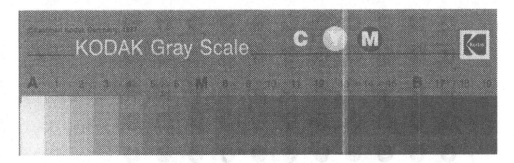

Fig. 2, The grayscale wedge

Depending on the bar considered (the darker is the bar the higher in the noise, as shown in fig. 3), the maximum radiometric noise has never been higher than 2 radiometric tones. This stability, certainly fulfills the radiometric repeatability standards.

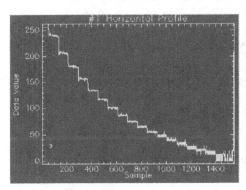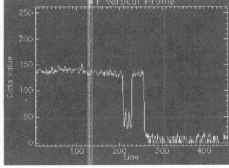

Fig. 3, Horizontal and vertical (in corrispondence of the central bar) contrast profile of the wedge

5.2 Geometric repetability test

In order to evaluate the geometrical repetability, a calibration reseau (with known coordinates) has been acquired 10 times; the position of the image crosses centres has been calculated by means of an automatic procedure based on a cross-correlation algorithm (in this case a search area extracted from the original image has been used).

In fig. 4 the average discrepancies of a couple of successive acquired reseau are shown.

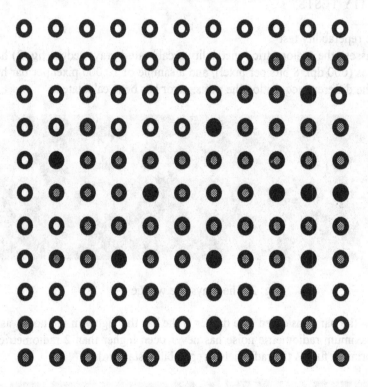

● Discrepancy ≥ 0.90 pixel

◉ 0.90 pixel <Discrepancy ≥ 0.10 pixel

○ Discrepancy< 0.10 pixel

Fig. 4, Acquired reseau repetability considerd on two successive reseau acquisitons

6. GEOMETRIC PRECISION ASSESSMENT

In order to assess the geometric precision of the acquisition the same 10 reseau acquisitions have been taken into account. By means of a subpixel cross-correlation technique, the acquired crosses centres have been determined and thus compared with the real ones. In fig. 5 the single cross discrepancies are shown.

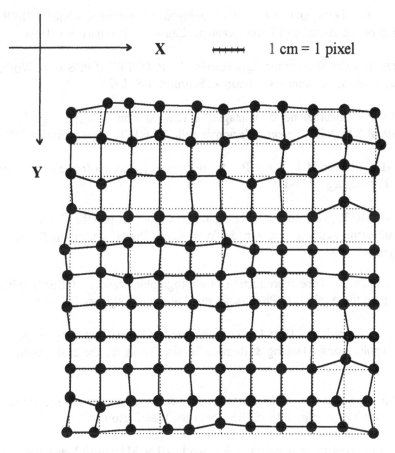

Fig. 5, Acquired reseau discrepancies calculated for every single cross

The results shown that the maximum discrepancy is never higher than ¼ pixel (in this case the pixel dimension is approximately 42 μm in both directions). This precision seems enough for the some photogrammetric applications (orthophoto generation, medium scale mapping updating and DEM extraction, etc.)

7. CONCLUSIONS AND FURTHER DEVELOPMENTS

In conclusion, the different tests performed demonstrate the affordableness of the scanner used, for some photogrammetric applications. The future efforts have now to be directed over a comprehensive scanners' tests (evaluating the most part of the off-shelf scanners), and over an automatic homogeneization procedure usable also by non-experts users.

REFERENCES:

1. Baltsavias E., Tests and performance analysis of scanners, OEEPE-ISPRS Joint Workshop on the Analysis of Photo-scanners, Lausanne, February 7-8, 1994

2. Baltsavias E., Evaluation of the Agfa Horizon DTP, OEEPE-ISPRS Joint Workshop on the Analysis of Photo-scanners, Lausanne, February 7-8, 1994

3. Boccardo P., Calibrazione di immagini a reticolo con tecniche digitali, relazione presentata al XXXVII Convegno Nazionale della S.I.F.E.T., Stresa, ottobre 1992

4. Bosma M., Drummond J., Raidt B., A preliminary report on low-cost scanners, ITC Journal, 1989-2,pagg. 115-120

5. Cramer M., Bill R., Glemser M., Investigations of low-cost peripheral devices for digital photogrammertic systems, , in Atti del Workshop "Digital sensors and Applications", Trento, giugno 1993

6. Kölbl O., Bach U., Tone reproduction of photographic scanners, OEEPE-ISPRS Joint Workshop on the Analysis of Photo-scanners, Lausanne, February 7-8, 1994

7. Philbrick R., Erhardt H., Titus H., The efficiency of linear solid state imagers for film scanning applications, Mapping & Remote Sensing Tools for the 21st Century, August 26th, 1994, Washington DC

8. Sarjakoski T., Suitability of the Sharp JX-600 desktop scanner for the digitization of aerial colour photographs, 17th ISPRS Congress, Washington DC, 1992

9. Seiter C., Low-cost color scanners, in Macworld - The Macintosh Magazine, November 1993

10. Seitz P., Reflections on the efficiency of solid-state photo sensors, OEEPE-ISPRS Joint Workshop on the Analysis of Photo-scanners, Lausanne, February 7-8, 1994

11. Seywald N., Leberl F., Requirements of a system to analyze film scanners, OEEPE-ISPRS Joint Workshop on the Analysis of Photo-scanners, Lausanne, February 7-8, 1994

RESTORATION AND VALIDATION OF IMAGE DATA

T. Bellone
Polytechnic of Turin, Turin, Italy

B. Crippa and L. Mussio
Polytechnic of Milan, Milan, Italy

Abstract
Images supply large amounts of data, that need appropriate statistical and numerical techniques, in order to achieve their restoration and validation. In this work some procedures of data processing are presented; they combine suitably optimality from the statistical point of view and practicability from the numerical one. Furthermore they have been applied, since a relatively long time, to surface reconstruction and deformation monitoring, but they are now specialized and applied to preprocessing of image data (i.e. data assessment, image quality control). A pilot experiment using a SPOT image has been done and its results are reported.

1   Introduction

Images supply large amounts of data, that need appropriate statistical and numerical techniques, in order to achieve their restoration and validation.

Covariance estimation, collocation filtering, prediction and covariance propagation are powerful tools, useful in data processing, and allow for:

- separation of the signal from the noise;

- enhancement of characteristic features;

- identification and elimination of blunders, leverages and outliers;

- assessment of accuracy and reliability.

The fields of application of the presented procedure are very broad. Indeed they range from preprocessing techniques, to achieve image quality assessment and to generate datasets suitable for analysis and measurements, to algorithms for geometric determination and analysis of image data (e.g. feature extraction, image matching), including the semantic aspects of image understanding.

A pilot experiment using SPOT image has been done and its results are reported.

2   Stochastic processes

A stochastic process is a family X of functions of one- or multi-dimensional random variables ξ depending of a group of field parameters θ:

$$X = \{x(\xi, \theta)\}$$

being the probability distribution of the stochastic process assumed by induction from the probability distributions of random variables (e.g. the normal function):

$$P(x(\xi, \theta)) = P(\xi, \theta) \qquad \forall \theta.$$

Notice that, if in each value of the field parameters only an argument of the random variables is sampled, the sample of all these arguments furnish a realization of the stochastic process.

Concerning the applications of interest, images are supposed 2D realizations of scalar stochastic processes, where the grey values represent the sampled arguments, while multispectral images are 2D realizations of vectorial stochastic processes.

According to the random variable statistics, it is possible to define a stochastic process statistics, whose principal elements are the average, the variances and covariances:

$$\bar{x}(\theta) = E(x(\xi,\theta)p(x(\xi,\theta))) \qquad \forall \theta$$

$$C(\theta) = D(x(\xi,\theta)p(x(\xi,\theta))) \qquad \forall \theta$$

furthermore, considering two random variables at two different values of the field parameters, the auto- and cross-covariances are defined as follows:

$$C(\theta_1,\theta_2) = D(x(\xi,\theta_1,\theta_2), p(x(\xi,\theta_1,\theta_2))) \qquad \forall \theta_1, \theta_2.$$

A stationary stochastic process is a stochastic process in which all moments maintain some invariance properties according to invariant transformations of the group of field parameters. In case the normal function is the probability distribution of a stationary stochastic process, its average, its variances and its covariances are constant:

$$\bar{x}(\theta) = const. \qquad \forall \theta$$

$$C(\theta) = const. \qquad \forall \theta$$

moreover the auto- and cross-covariances depending on the chosen norm only:

$$C(\theta_1,\theta_2) = C(||\theta_1 - \theta_2||) \qquad \forall \theta$$

the L_2 euclidean one (i.e. isotropic stochastic process) being the most common norm; otherwise these properties are only preliminary conditions for the definition of stationary stochastic process.

Notice that an isotropic stochastic process depends on an invariant similarity transformation of the group of field parameters, this means that a more general stochastic process depends on an invariant affine transformation. Furthermore the degree of the norm could be changed; in this case particular interest is assumed by the L_1 norm (Manhattan distance) and the L_∞ norm.

Up till now the moments of a stochastic process are been defined inside the random variables; on the contrary moments along the realizations are to be defined too. A necessary and sufficient condition proves that the two classes of moments are coincident when the stochastic process is ergotic and, in case the normal function is probability distribution of the stochastic process, it satisfies the above mentioned stationariness conditions only. Therefore averages, variances and covariances, auto- and cross-covariances along a realization are defined as follows:

$$\bar{x} = E(\xi(\theta))$$
$$C = D(\xi(\theta))$$
$$C(||\Delta\theta||) = D(\xi(\theta)\xi(\theta + \Delta\theta)) \qquad \forall \Delta\theta.$$

Out of normality the stationeriness conditions must be broadly investigated, in order to prove the ergotic property of the stochastic process.

3   Optimal filtering and prediction [1]

The separation of the signal from the noise can be performed in an optimal way, taking into account the results of the covariance estimation; then the prediction of the signal and its functionals follows using the results of the filtering. In the most simple case signal and noise are two addenda, whose sum is the observables:

$$\alpha = s + n.$$

More general functional models, which link signal and noise by means of design matrices, including also some deterministic unknown parameters,

---

[1]This paragraph and the three paragraphs in the following are an extended and reviewed version of Mussio L. et al. (1995): A Recommended Procedure for Image Quality Assessment, presented at ISPRS WG I/1 Workshop in Dessau (Germany).

don't change the principle, but give some management problems, that are here omitted for sake of brevity.

These hypotheses define the stochastic model (notice that it could be complicated a little, taking into account the weights of observables):

$$E(s) = 0$$
$$D(s) = C_{ss}$$
$$E(n) = 0$$
$$D(n) = \sigma_n^2 I$$

$$D(s,n) = 0 \longrightarrow C_{\alpha\alpha} = C_{ss} + \sigma_n^2 I.$$

The Wiener-Kolmogorov principle, introducing the estimation error of the signal: $e = s - \hat{s}$, leads to the minimization of its mean square error (mse), assuming the signal to be a linear combination of the observations:

$$\hat{s} = \lambda^t \alpha^0$$

$$E^2 = \sigma_e^2 = \sigma_{\hat{s}}^2 - 2\sigma_{\hat{s}s} + \sigma_s^2 = \lambda^t C_{\alpha\alpha}^{-1}\lambda - 2\lambda^t c_{\alpha s} + \sigma_s^2 = min.$$

Minimizing this quadratic form, one gets:

$$\lambda = C_{\alpha\alpha}^{-1} c_{\alpha s}$$

that provides the estimates of the signal and its mse:

$$\hat{s} = c_{s\alpha} C_{\alpha\alpha}^{-1} \alpha^0$$
$$\sigma_e^2 = \sigma_s^2 - c_{s\alpha} C_{\alpha\alpha}^{-1} c_{\alpha s}.$$

From the computational point of view, these formulae can be substituted by the following ones, respectively, suitable for filtering:

$$\hat{n} = \sigma_n^2 C_{\alpha\alpha}^{-1} \alpha^0$$
$$\hat{s} = \alpha^0 - \hat{n}$$
$$\sigma_e^2 = \sigma_n^2 I - \sigma_n^4 C_{\alpha\alpha}^{-1}$$

and the prediction:

$$\hat{s}_p = c_{s_p s} z$$

where:

$$z = C_{\alpha\alpha}^{-1} \alpha^0.$$

In such a way, computation go rather easily in the filtering and very fast in the prediction. The optimal way of separation of the signal from the noise answers to a minimum variance principle, i.e. supplies the most efficient, unbiased estimates; however robustness is missing among the properties of the estimator. On the other hand, it is possible to improve the robustness of the estimator, introducing into the weights of the observables a positive inverse function of the residual noise and iterating the solution until a reproduction point is reached.

Furthermore because collocation filtering and prediction require covariance estimation, blunders, leverages and outliers should be identificated and eliminated already in this step. To this aim it is easy to compute the mid point of the median absolute value of a set of data neighbouring a given point. Therefore by repeating this operation for each step point in the data set, the data in the tails will be automatically rejected, because their discrepancies with the estimated mid points are larger than a small multiple of median absolute values.

4   The Taylor-Karman structure

Prediction is an operation that, as already said, can be performed in a lot of applications; therefore not only the signal, but also its functionals can be estimated and, among these, differential operators to the signal are very important, because they are linear and supply information about shape and pattern.

In the following, the covariance propagation of a 2D isotropic stochastic process is investigated, showing the Taylor-Karman structure, when differential operators are applied.

The matrices of the two first derivatives of a function are respectively the so called Jacobian and Hessian matrices:

$$J = \begin{vmatrix} \partial\xi/\partial x \\ \partial\xi/\partial y \end{vmatrix} = \nabla\xi$$

$$H = \begin{vmatrix} \partial^2\xi/\partial x^2 & \partial^2\xi/\partial x\partial y \\ simm. & \partial^2\xi/\partial y^2 \end{vmatrix} = \nabla\nabla^t\xi$$

Therefore given the autocovariance of the stochastic process, the cross-covariance between the first derivatives and the stochastic process itself has the following expression, according to covariance propagation law and the commutability of the operators, because of their linearity:

$$\begin{aligned} C(\Delta_Q\xi_Q, \xi_P) &= \nabla_Q C(r_{PQ}) = C'(r_{PQ}\nabla_Q(|r_{PQ}|)) = \\ &= -C'(r_{PQ})r_{PQ}/|r_{PQ}| = -C'(\xi_Q, \xi_P)r_{PQ}/|r_{PQ}| \end{aligned}$$

being: $r_{QP} = -r_{PQ}$, and: $\nabla|r| = r/|r| = |cos\theta \quad sin\theta|^t$, where the direction θ begins from the x axis and goes anticlockwise. An important function of the first derivatives is the modulus of the gradient:

$$|\nabla\xi| = (\partial\xi/\partial x)^2 + (\partial\xi/\partial y)^2.$$

In the same way, the cross-covariance between the second derivatives of a stochastic process and the stochastic process itself has the following expression:

$$\begin{aligned} C(\nabla_Q\nabla_Q^t\xi_Q, \xi_P) &= \nabla_Q\nabla_Q^t C(r_{PQ}) = -\nabla_Q C'(r_{PQ})r_{PQ}^t/|r_{PQ}| = \\ &= C''(r_{PQ})r_{PQ}r_{PQ}^t/|r_{PQ}|^2 - C'(r_{PQ})\nabla_Q(r_{PQ}/|r_{PQ}|) = \\ &= C''(r_{PQ})\Pi r_{PQ} - C'(r_{PQ})(-I/|r_{PQ}| + r_{PQ}r_{PQ}^t/|r_{PQ}|^3) = \\ &= (C''(r_{PQ}) - C'(r_{PQ})/|r_{PQ}|)\Pi r_{PQ} + C'(r_{PQ})I/|r_{PQ}| = \\ &= (C''(\xi_Q, \xi_P) - C'(\xi_Q, \xi_P)/|r_{PQ}|)\Pi r_{PQ} + C'(\xi_Q, \xi_P)I/|r_{PQ}| \end{aligned}$$

being: $\nabla r^t = I$; $\nabla(1/|r|) = -\nabla(|r|)/|r^2|^2 = -r/|r|^3$, and:

$$\Pi r = rr^t/|r|^2 = \begin{vmatrix} cos^2\theta & sin\theta cos\theta \\ simm. & sin^2\theta \end{vmatrix}$$

where: $(\Pi r)^2 = \Pi r$; $Tr(\Pi r) = 1$; $Det(\Pi r) = 0$. An important function of the second derivatives is the Laplacian:

$$\nabla^2 \xi = (\partial^2 \xi / \partial x^2 + \partial^2 \xi / \partial y^2)$$

that represents an areal dilatation or contraction, while the maximum local shear strain is given by:

$$\gamma = \sqrt{(\partial^2 \xi / \partial x^2 - \partial^2 \xi / \partial y^2)^2 + (\partial^2 \xi / (\partial x \partial y))^2 / 4}.$$

Notice that the Taylor-Karman structure belongs also to multidimensional isotropic stochastic process, e.g. a 3D isotropic stochastic process, where the vector $\nabla |r|$ and the projector matrix Πr have the following expression:

$$\nabla |r| = |cos\theta cos\alpha \quad sin\theta cos\alpha \quad sin\alpha|^t$$

$$\Pi r = \begin{vmatrix} cos^2\theta cos^2\alpha & sin\theta cos\theta cos^2\alpha & cos\theta sin\alpha cos\alpha \\ & sin^2\theta cos^2\alpha & sin\theta sin\alpha cos\alpha \\ simm. & & sin^2\alpha \end{vmatrix}$$

Concerning the applications of interest, obsevations in the object space are 3D data of GIS's, that could supply prior information useful in the management of images.

5   The Kronecker separability and the Toeplitz matrix

The isotropic 2D stochastic processes are very important, from the theoretical point of view; however they give some troubles, from the pratical point of view, particularly concerning their computability. Indeed full matrices require large core storage and their processing demands long computing time; therefore it is preferable to lose some geometrical properties (e.g. isotropy), obtaining a big gain in data management and processing.

The already mentioned L_1 norm uses the so called Manhattan distance and implies orthogonal separability among the two coordinates of the 2D stochastic processes. The orthogonal separability of covariance estimation implies a similar separation in the filtering and prediction, because the systems to be solved can be separated by Kronecker decomposition.

The Kronecker product is the generalization of the product of a scalar and a matrix (or vector),

$$U \otimes V = [u_{ij}V] = [u_{ij}[v_{kl}]] \qquad i,j = 1,m \quad k,l = 1,n.$$

The most important property of the Kronecker product is the application of Kronecker decomposition to the inverse matrices:

$$(U \otimes V)^{-1} = U^{-1} \otimes V^{-1}.$$

This means that a system of which the coefficient matrix can be split according to Kronecker decomposition: the system,

$$(U \otimes V)x = \vartheta$$

$$x = (U \otimes V)^{-1}\vartheta = U^{-1} \otimes V^{-1}\vartheta$$

can be solved in two steps:

$$y_i = V^{-1}\vartheta_i \qquad i = 1,m$$

$$x_j = U^{-1}z_j \qquad j = 1,n.$$

The vectors z_j, $j = 1,n$, contain the elements of the intermediate vectors y_i, $i = 1,m$, after a sorting of their elements according to the indices of the vectors themselves:

$$z_j = [y_{ij}] \qquad i = 1,m \quad j = 1,n$$

The matrix form of the procedure is as follows:

$$Y = V^{-1}\Theta$$
$$X = U^{-1}Z$$

where:

$$\Theta = [\vartheta_i] \qquad Y = [y_i] \quad i = 1, m$$

$$Z = [z_j] \qquad X = [x_j] \quad j = 1, n$$

and, obviously, $Z = Y^t$. The Kronecker product is, of course, associative.

The gain in data management and processing can be done much bigger, if the data present regularity in their structure. This means that the systems to be solved contain observations from raster data and unknowns concerning grid parameters. Therefore the corresponding matrices, after the application of Kronecker decomposition, become Toeplitz matrices.

Toeplitz matrices are a particular type of symmetric and positive definite matrices whose elements depend on the absolute difference of row and column indices:

$$W = [w_{ij}] = [\varphi_{|i-j|}] \qquad i, j = 1, n.$$

The most important property of a Toeplitz matrix is the form of its inverse matrix. Indeed the inverse of a Toeplitz matrix is persymmetric, i.e. the elements are symmetric not only with respect to the main diagonal, $w_{ij} = w_{ji}$ (Hermit's symmetry), but also with respect to the secondary diagonal $w_{ij} = w_{n-j+1,n-i+1}$ (persymmetry by Wise).

Toeplitz matrices are useful to represent normal and covariance matrices in one dimensional problems. Therefore when Kronecker decomposition can be applied, the product of two Toeplitz matrices can represent also normal and covariance matrices in 2D problems.

Notice that L_1 norm, Kronecker decomposition and Toeplitz matrices can be applied also to regular sample data of multidimensional stochastic processes, e.g. in a 3D stochastic process management and processing go, in a similar way, applying three times the same rules defined for 2D problems.

6 Trench's algorithms and its performance

Trench's algorithm provides an efficient method to solve a linear system and to invert a symmetric and positive definite matrix, when the coefficient matrix is a Toeplitz matrix.

As a first step, the inverse matrices of bottom-up and top-down Gauss blocking, with innovation of rank one in both cases are considered. Indeed Gauss blocking is a stepwise procedure, which can be repeated, step by step, increasing the dimension of the system at each iteration.

A second and conclusive step provides the rules to compute the elements of the innovation vector f. To avoid useless computations and to save storage, Gauss blocking is used to compute the last element of the innovation vector and then the other elements, taking into account the elements of the innovation vector obtained in the preceding iteration.

The following formulae represent Trench's algorithm:

$$f_l^{(l)} = \left(b_l - \sum_{i=1}^{l-1} f_i^{(l-1)} b_{l-i+1}\right)/e^{(l-1)}$$

$$f_i^{(l)} = f_i^{(l-1)} - f_l^{(l)} \cdot f_{(l-i+1)}^{(l-1)} \quad i = 1, l-1 \; ; \; l = 2, n-1$$

$$e^{(l)} = \left(1 - f_l^{2(l)}\right) \cdot e^{(l-1)}$$

being b a vector containing the elements of the first row of the Toeplitz matrix, but for the first element, always identically equal to one.

These formulae are directly operative and allow for a stepwise solution of a linear system corresponding to a Toeplitz coefficient matrix.

At the first step, the conditions:

$$e^{(1)} = 1 \qquad f_1^{(l)} = b_1$$

initialize the procedure.

The solution of the system is obtained by applying Gauss blocking, bottom-up, taking into account, at each step, the results obtained by Trench's algorithm:

$$\begin{pmatrix} 1 & b^t \\ b & C \end{pmatrix} \begin{pmatrix} y \\ x \end{pmatrix} = \begin{pmatrix} \eta \\ \xi \end{pmatrix}$$

$$\begin{pmatrix} y \\ x \end{pmatrix} = \begin{pmatrix} \alpha(\eta - b^t x_0) \\ x_0 - C^{-1} b y \end{pmatrix} = \begin{pmatrix} (\eta - b^t x_0)/e \\ x_0 - fy \end{pmatrix}.$$

At the first step, an equation with one unknown is solved:

$$y = \eta$$

being the last observation of the data set.

The unknown y initializes the vector x_0. The observations are then processed sequentially, in reverse order, and each new unknown y adds an element to the vector x_0.

If necessary, the inverse matrix can be determined at the end of the algorithm, suitably applying the results of Gauss blocking. Therefore the inverse matrix can be computed as follows:

$$\psi_{11} = \alpha = 1/e$$

$$\psi_{1,i} = \beta_{i-1} = -f_{i-1}/e \qquad i = 2, n$$

$$\psi_{ij} = \gamma_{i-1,j-1} =$$
$$= \psi_{i-1,j-1} + (f_{i-1}f_{j-1} - f_{n-i+1}f_{n-j+1})/e \quad i, j = 2, n.$$

Note that the formulae allow for the computation of the main diagonal elements only:

$$\omega_1 = 1/e$$

$$\omega_i = \omega_{i-1} + (f_{i-1}^2 - f_{n-i+1}^2)/e \quad i = 2, n$$

avoiding the computation of the off-diagonal elements. Furthermore suitably applying the properties of Hermit's symmetry and persymmetry by Wise, the number of elements to be computed can be shortened to one quater for the full inverse matrix and to a half for the main diagonal only.

Trench's algorithm doesn't require heavy computations. Indeed at each step it presents only two sums:

$$\sum_{i=1}^{l-1} f_i^{(l-1)} b_{l-i+1}$$

$$\sum_{i=1}^{l-1} b_i x_{0_{n-l+i}}^{(l-1)}.$$

These sums can be shortened, if the elements of the vector b are all equal zero except for a small number of them put at the beginning. Being λ the number of non-zero elements in the vector b, the above mentioned sums become:

$$\sum_{i=\kappa}^{l-1} f_i^{(l-1)} b_{l-i+1} \qquad \kappa = max(1, l - \lambda + 1)$$

$$\sum_{i=1}^{\kappa} b_i x_{0_{n-l+i}}^{(l-1)} \qquad \kappa = min(\lambda - 1, 1)$$

and automatically omit products involving zero elements.

Trench's algorithm carries out computations of order n^2, instead of order n^3, being n the dimension of the system. Indeed the solution of the system itself and the inverse matrix can be directly computed after having computed the elements of the innovation vector, obtained, at each step, with one sum only. Notice that, if this sum can be shortened because of the presence of (many) zero elements in the normal or covariance matrix, the results can be obtained after computations of order $n\lambda$, being λ the bandwidth of the above mentioned matrix.

A comparison with direct algorithms, which require computations of order n^3 in case of a full matrix, or $n\lambda^2$ in case of a banded matrix, proves the efficiency and power of Trench's algorithm.

Besides, remembering the advantages given by Kronecker decomposition, 2D (or 3D) problems can be solved applying two (or three) times Trench's algorithm. Indeed, when Kronecker decomposition can be applied, the product of two (or three) Toeplitz matrices suitably represent 2D (and 3D) problems. In these cases the solution can be obtained after computations of order $2n^2$ (and $3n^2$) in case of full matrices, and $2n\lambda$ (or $3n\lambda$) in the case of banded matrices.

7   A pilot experiment

The proof of the goodness and the effectiveness of the procedures, explained in the previous paragraphs, has been obtained by a pilot experiment. To this task a portion (500 × 500 pixels) of a SPOT image, belonging to OEEPE test organized in the past years by the SPOT-image OEEPE WG, has been selected, processed and evaluated (see fig. 1 and 2).

The pilot experiment has consisted of the following steps:

- covariance estimation;

- collocation filtering;

- prediction and covariance propagation.

Covariance estimation and covariance function modelling are very important, because collocation filtering and prediction require appropriate models to interpolate the empirical auto- and cross-covariance functions of the signal.[2]

An hypothesis has been made: the data can be seen as realizations of a continuous and normal stochastic process which is stationary of 2nd order with mean zero and covariance function of the kind:

$$C(P_1, P_2) = C(||P_1 - P_2||).$$

Notice that 2D (or 3D) stochastic processes can be assumed isotropic, as well as with orthogonal separability among the two (or three) coordinates. The former uses the euclidean distance, that is invariant on a circle (a sphere); the latter uses the Manhattan distance, that is invariant on a square (a cube). As already said, L_1 norm implies orthogonal separability, that offers a big gain from the numerical point of view.

With $X(P_i)$ the n observations at the different points $P_1, \cdots, P_n$ the estimates of the empirical auto-covariance function at the interval $\Delta^{(l)}$ are calculated from:

---

[2]The description of the procedures about covariance estimation and covariance function modelling is quoted, with minor changes, by Mussio L. et al. (1995): A Recommended Procedure for Image Quality Assessment, presented at ISPRS WG I/1 Workshop in Dessau (Germany).

Original SPOT image (OEEPE WG test)

 a portion (500 x 500 pixels) of the image

Original data

Covariance function
of the rows

Covariance function
of the columns

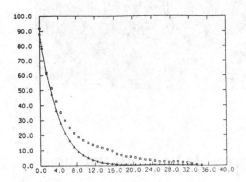

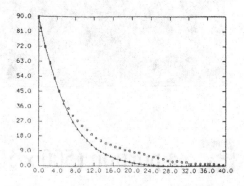

$$\gamma(\Delta^{(l)}) = \frac{1}{n} \sum_{i=1}^{n} v_i \frac{1}{n_i^{(l)}} \sum_{j=1}^{n_i^{(l)}} v_j^{(l)}$$

where:

$$\{v_j^{(l)} : P_j \Rightarrow \Delta^{(l-1)} < ||P_i - P_j|| \le \Delta^{(l)}\}$$

and:

$$v_k = x_k - \bar{x} \quad k = 1, \cdots, n$$

while the estimates of empirical cross-covariance function at the interval $\Delta^{(l)}$ are computed in similar way.

A criterion for the choice of the interval including the first auto-covariance zone is maximizing the first autocovariance estimate as follows:

$$r_{(1)} \;\; : \;\; \gamma(\Delta^{(1)}) = max(\gamma(\Delta^{(l)}))$$

where:

$$\gamma(\Delta^{(l)}) = \frac{1}{n} \sum_{i=1}^{n} v_i \frac{1}{n_i^{(l)}} \sum_{j=1}^{n_i^{(l)}} v_j^{(l)}$$

and:

$$\{v_j^{(l)} : P_j \Rightarrow 0 < ||P_i - P_j|| \le \Delta^{(l)}\}$$

Notice that, when processing raster data, the optimal interval is often equal to one and, otherwise, always an integer number.

Furthermore by using the orthogonal separability among the two co-ordinates, it is easy to obtain covariance estimates for each row and each column. The median of the row covariance estimates and the median of the column covariance estimates give the empirical auto-covariance functions for the rows and the columns.

The covariance estimation continues then by multiplying the row covariance estimates by the column covariance estimates, in order to obtain the covariance function of the data. The same procedure has to be done for the "best fit" covariance function and the noise variance.

In case of 3D stochastic processes, the orthogonal separability among the three coordinates goes to the estimation of covariances for each row, each column and each pile. Successively the medians give the empirical auto-covariance functions and the product among the three covariance components supplies the covariance function of the data.

Because covariance estimation doesn't satisfy positive definite property automatically, this last must be achieved by modelling the covariance estimates with a suitable set of positive definite models.

The "best fit" of the auto-covariance function of the signal is then chosen among some available models, namely, when euclidean distance is used:

$$\gamma(\Delta) = a \ \ exp(-b\Delta)$$
$$\gamma(\Delta) = a \ \ exp(-b\Delta^2)$$
$$\gamma(\Delta) = a \ \ exp(-b\Delta)(1 - c\Delta^2)$$
$$\gamma(\Delta) = a \ \ exp(-b\Delta^2)(1 - c\Delta^2)$$
$$\gamma(\Delta) = a \ \ exp(-b\Delta)J_0(c\Delta)$$
$$\gamma(\Delta) = a \ \ exp(-b\Delta^2)J_0(c\Delta)$$
$$\gamma(\Delta) = 2a \ \ exp(-b\Delta)J_1(c\Delta)/(c\Delta)$$
$$\gamma(\Delta) = 2a \ \ exp(-b\Delta^2)J_1(c\Delta)/(c\Delta)$$

or alternatively, when Manhattan distance is used:

$$\gamma(\Delta) = a \ \ exp(-b\Delta)$$
$$\gamma(\Delta) = a \ \ exp(-b\Delta^2)$$
$$\gamma(\Delta) = a \ \ exp(-b\Delta)(1 - c\Delta^2)$$
$$\gamma(\Delta) = a \ \ exp(-b\Delta^2)(1 - c\Delta^2)$$
$$\gamma(\Delta) = a \ \ exp(-b\Delta)cos(c\Delta)$$
$$\gamma(\Delta) = a \ \ exp(-b\Delta^2)cos(c\Delta)$$
$$\gamma(\Delta) = a \ \ exp(-b\Delta)sin(c\Delta)/(c\Delta)$$
$$\gamma(\Delta) = a \ \ exp(-b\Delta^2)sin(c\Delta)/(c\Delta)$$

where the smoothness given by the coefficient b is in both cases very high.

This list has been built according to the properties of covariance function: positive power spectrum, i.e. positive Fourier transform, and Schwarz inequality for vector processes.

New covariance function can be created from old by applying following fundamental theorems:

- a linear combination with positive coefficients;

- a product;

- a convolution;

The same list is used to interpolate a cross-covariance function: it is not correct in principle, but is acceptable in practice, provided that cross-covariance estimates are low enough.

Besides finite covariance functions are obtained by multiplying the "best fit" by positive definite finite functions, that are given by convolution of a finite function with itself, respectively, in the first (isotropic) case:

$$\gamma(\Delta) = 5\pi - 15\Delta^2\pi/2 + (5\Delta + 20\Delta^3/3 - 5\Delta^5/12)\sqrt{1 - \Delta^2/4} +$$
$$+ (15\Delta^2 - 10)arcsin(\Delta/2) \qquad \text{with} \quad \Delta \leq 2$$

$$\gamma(\Delta) = 0 \qquad \text{with} \quad \Delta \geq 2$$

and in the second (with ortogonal separability) case:

$$\gamma(\Delta) = 16/15 - 4\Delta^2/3 + 2\Delta^3/3 - \Delta^5/30 \qquad \text{with} \quad \Delta \leq 2$$

$$\gamma(\Delta) = 0 \qquad \text{with} \quad \Delta \geq 2.$$

Finally, the noise variance is found as:

$$\sigma_n^2 = \sigma^2 - \sigma_s^2 = \sigma^2 - a$$

and the noise covariance can be found with a similar formula.

In case of 3D stochastic processes, a library of auto-covariance function models, especially when they are finite covariance functions, is difficult to construct. On the other hand, the orthogonal separability among the three coordinates allows to use three times the one-dimensional auto-covariance function models. The same is true for the finite covariance functions and the product among the three covariance components supplies positive definite models.

Figures 3 and 4 show the empirical covariance estimates and the model functions for the rows and the columns.

Collocation filtering and prediction have gone straightforward, after having solved the system, taking into account Kronecker decomposition and suitable algorithms for Toeplitz matrices. Thus the noise has been obtained by multiplying the standard vector for the noise variance and the signal has been derived by the difference between the data and the noise (see fig. 5).

Furthermore five masks of 35×40 elements have been prepared, in order to make the prediction of the Jacobian and Hessian matrices, by using covariance coefficients and their first and second derivatives:

$$
\begin{aligned}
C_{\dot{s}_x s} &= [\dot{\gamma}_x \gamma_y] \\
C_{\dot{s}_y s} &= [\gamma_x \dot{\gamma}_y] \\
C_{\ddot{s}_x s} &= [\ddot{\gamma}_x \gamma_y] \\
C_{\ddot{s}_y s} &= [\gamma_x \ddot{\gamma}_y] \\
C_{\ddot{s}_{xy} s} &= [\dot{\gamma}_x \dot{\gamma}_y]
\end{aligned}
$$

then they have been suitably multiplied by the standard vector, obtained the expected predicted values.

From these values the modulus of the gradient and the Laplacian have been computed (see fig.'s 6 and 7). Both of them are useful for the enhancement of characteristic features, especially the Laplacian; therefore after a definition of a suitable threshold, the zero-crossing points of the Laplacian have been drawn in figure 8.

They show characteristic features, but their manipulation and interpretation need successive investigation by topological operations and form descriptors.

Signal of the data

Modulus of the gradient

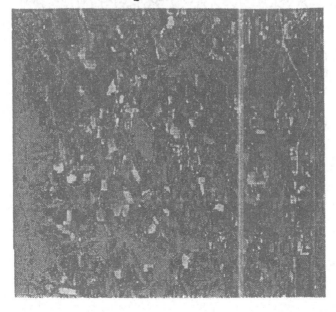

Laplacian operator

Zero crossing of the laplacian operator

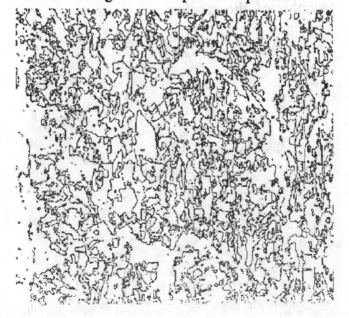

8   Conclusion

Restoration and validation of image data require appropriate statistical and numerical techniques. Indeed because images supply large amounts of data, their processing needs particular care, in order to achieve the expected results.

Therefore as already mentioned, some statistical procedures, like co-variance estimation, collocation filtering, prediction and covariance propagation, assure the optimality, while suitable algorithms for regular structures, taking into account Toeplitz matrices and the property of Kroneker separability, satisfy the numerical feasibility of the procedures.

The results obtained from the pilot experiment encourage to continue and to examine closely the argument and, up till now, allow to formulate the following remarks:

- separation of the signal from the noise and assessment of accuracy and reliability got its goal, because of the statistical and numerical properties of the adopted techniques;

- identification and elimination of blunders, leverages and outliers, especially when contaminated data are a relatively large percentage, require the use of robust estimators;

- enhancement of characteristic features makes an intermediate step, waiting for successive investigation by topological operations and form descriptors;

further on proving the contiguity between preprocessing and processing.

Appendix - Application to finite element interpolation

Finite element interpolation can be performed in different way; one the most promising method uses spline functions (e.g. bicubic spline functions for image data restoration). A bicubic spline function is given by the product of two orthogonal cubic spline functions:

$$S(x,y) = S(x)S(y)$$

The choice for the number of cells and the number of knots depend on the number of observations m and the interpolation step δ.

The number of cells is the product of the number of classes in two directions x and y:

$$\nu = \nu_x \nu_y$$

where: $\nu_x = int(\Delta X/\delta) + 1$, $\nu_y = int(\Delta Y/\delta) + 1$, being ΔX and ΔY the dimensions of the space region in two directions and δ the chosen interpolation step. Consequently the number of knots is:

$$n = n_x n_y = (\nu_x + 3)(\nu_y + 3).$$

The bicubic spline interpolation is performed, as a classical least squares problem, by writing a system of observation equations:

$$\hat{s}_k = s_k^0 + \hat{v}_k = \sum_{i=1}^{4} \sum_{j=1}^{4} \hat{a}_{I+i\ J+j} S_{ij}(\xi_k, \eta_k)$$

and associating it with the least squares norm:

$$\phi = \sum_{k=1}^{m} \hat{v}_k^2 = min.$$

The weights are mostly assumed equal one; however more complex stochastic model should be defined including correlations between the observations, but they are usually omitted for sake of brevity.

The following formulas are the legenda of the functional model; indeed for the x direction, the coordinate of the k-th knot with respect to the initial corner is split in two parts:

$$\Delta x_k = I\delta + \delta x_k$$

where the number of the preceding knots is:

$$I = int(\Delta x_k/\delta)$$

and the position inside the class is:

$$\xi_k = \delta x_k/\delta$$

being: $\delta x_k = \Delta x_k - I\delta$; analogously, for the y direction:

$$\Delta y_k = J\delta + \delta y_k$$

where: $J = int(\Delta y_k/\delta)$, and: $\eta_k = \delta y_k/\delta$, being: $\delta y_k = \Delta y_k - J\delta$.

Notice that suitable constraints for the knots should be introduced at the border and in empty regions, if any.

The functional model, as already said, is given by product of two orthogonal functions. Consequently if there are raster data and grid parameters, like images, the problem can be performed applying Kronecker decomposition and solving two times Toeplitz matrices, e.g. by using Trench's algorithm.

This way is alternative to the stochastic process approach; however it is less powerful, therefore it may be applied as prefiltering, or embedded in advanced multigrid or multilevel techniques.

Notice that observations in the object space, i.e. 3D data of GIS's, when there are regular distributed data and grid parameters, can be processed, in a similar way, applying three times the same rules defined for 2D problems.

References

Ackermann F., (1973). Numerische Photogrammetrie. Herbert Wichmann Verlag, Karlsruhe.

Meissl P.,(1982). Least Squares Adjustment a Modern Approach. Mitteilungen der geodaetischen Inst. der Techn. Univ. Graz, Folge 43, Graz.

Mikhail E.M., with contributions by Ackermann F., (1976). Observation and Least Squares. IEP-A Dun Donnelley Publisher, New York.

Moritz H., (1980). Advanced Physical Geodesy. Herbert Wichmann Verlag, Karlsruhe.

Moritz H., Suenkel H., (1978). Approximation Methods in Geodesy. Herbert Wichmann Verlag, Karlsruhe.

Trench W.F., (1964). An algorithm for inversion of finite Toeplitz matrices. J. Soc. Indust. Appl. Math. vol. 12, n. 3.

Trench W.F., (1974). Inversion of Toeplitz band matrices. Mathematics of Computation, vol. 28, n. 128.

AUTOMATIC DEM GENERATION FROM DISCRETE POINTS IN MULTIPLE IMAGES

H-G. Maas
ETH Zurich, Zurich, Switzerland

Abstract:

One of the major difficulties in automatic DEM generation is the procurement of approximate values. In this presentation a new development based on the extraction of discrete points by an interest-operator and epipolar line intersection techniques in multiple overlapping images will be shown. The method can be considered an extension of the well-known MATCH-T approach from a stereo technique to a multi-image technique based on image quadruplets in a 60/60 block or even 6 images of a 80/60 block, and it solves the problem of provision of approximate values inherently. In contrast to most stereo-based techniques, the approach is not based on image pyramids, but on the consequent exploitation of the geometric strength of multiple images, implemented via the intersection of epipolar lines. Although not outlined for a complete DEM generation yet, the method may be very valuable for a hypothesis-free generation of good approximate values for other (e.g. area-based) DEM generation techniques or for the refinement of DEMs degraded by smoothing effects.

1. Introduction

The automatic generation of digital elevation models (DEM) has gained large attention among photogrammetrists in the past decade. A wide variety of approaches has been developed and presented in the literature, and automatic DEM generation packages are meanwhile commercially available in digital photogrammetric workstations. A rough classification of methods distinguishes between area based techniques and feature based techniques. Techniques based on least-squares-matching (Grün, 1985) are often considered a combination of both.

A critical point in all automatic DEM generation methods is - besides the recognition (and avoidance) of objects above the terrain surface like houses or trees - the procurement of approximate values in regions where no pre-knowledge on the terrain surface is given. Most methods employ image pyramid techniques or interactively selected starting points to solve this problem:

- Image pyramid: in a coarse-to-fine approach DEMs are generated at successively finer resolution levels of the images, starting with a horizontal plane as approximation for the highest level of the pyramid; this approach is e.g. implemented in the commercial DEM generation package MATCH-T (Krzystek, 1991) and was also used by (Baltsavias/ Stallmann, 1992).
- Interactive starting points: an operator sets one (or a few) starting points interactively, and the matching procedure moves into all directions from the starting point(s), assuming a certain maximum terrain slope. This approach is e.g. being used in the INDUSURF system (Schewe, 1987).

Both techniques work well in most cases, but may show problems in regions with extremely steep surface gradients, occlusions or certain frequency patterns in the image texture. The method presented in this paper does not require either one of these techniques, but provides a solution which is independent on the availability of approximate values or pre-knowledge on the maximum terrain slopes; only very rough pre-knowledge (like H_{min}, H_{max} in the observation region) is required. The procedure is similar to the procedure followed by MATCH-T: Discrete points in the images are extracted by an interest operator (Förstner, 1986), and correspondences between points are established using epipolar lines. These epipolar lines may become rather long when the terrain shows large height differences, leading to ambiguities in the establishment of correspondences, which can often not be solved. For that reason MATCH-T uses image pyramids as a coarse-to-fine method, thus limiting the length of the epipolar lines at each level of the pyramid. Furthermore, the 'general overkill philosophy' of MATCH-T (Ackermann, 1994) solves some of the problems connected with DEM data acquisition algorithmicly: By switching from interpolation to adjustment, outliers generated by false matches or by objects above the terrain surface can be removed by robust surface fitting, and it has been shown that even breaklines can often be detected automatically in the dense datasets of typically more than 500'000 surface points per stereo pair.

In the method presented in this paper, the length of the epipolar lines is basically unrestricted, and ambiguities due to multiple candidates on the epipolar lines are solved by epipolar line intersection techniques in multiple images, thus avoiding any smoothing effects introduced by surface fitting techniques or by patch sizes in area based techniques.

2. Epipolar line intersection

If the orientation and camera calibration parameters of images are known, for each point P' in one image an epipolar line in an other image can be defined on which the corresponding point P'' has to be found. The length of this line can be restricted if approximate knowledge about the depth range in object space is available. Adding a certain tolerance width to this epipolar line segment (due to data quality) the search area for the corresponding point location in the other image becomes a narrow two-dimensional window. Depending on the number of detected points per image, the arrangement of points and the complexity and depth extension of the object a problem of ambiguities will occur here, as often more than one candidate will be found in the search area. If point features do not allow for a reliable distinction of candidates, these ambiguities cannot be solved by a system based on only two cameras. It has been shown in (Maas, 1992a) that the probability of such ambiguities grows linearly with the depth extension in object space, linearly with the width of the search area and with the square of the number of points. A solution of this problem based on the inter-section of epipolar lines in an image triplet has been shown in (Maas, 1991), where trinocular correspondences in digital images of some 1000 particles marking a turbulent flow were established. The same technique has been applied in combination with a projected dot raster for surface determination in (Maas, 1992b). Also an extension to a four-camera system is shown in (Maas, 1992).

The method has been extended to be used with an arbitrary number of image coordinate datasets, where the data (= image coordinates) may origin form digital, analog or hybrid systems. The aims of this development were:

- The method should be applicable for the establishment of image correspondences between image coordinate datasets from an arbitrary number of images (with the additional requirement that the images have been taken from at least three camera stations).

- Features which would allow for a distinction of points are not needed.

- Only minimum knowledge about the object space is required. This knowledge can be reduced to rough approximations of the volume boundaries in object space; approximate values for the targets are not required, the targets may be arbitrarily distributed over the object space. A continuous object surface is not required or the object surface may show an arbitrary complexity; this includes objects freely distributed in space (like particles in water), occlusions or targets fixed on wires.

- The method is robust against missing points: Not all points have to be measured in all images, points may be missing due to occlusions, failures in the detection procedure, illumination effects, limitation of the sensor format, etc.

These aims were mainly derived from tasks occuring in close range photogrammetry, and they exceed the problems occuring in DEM generation, but they resulted in an interesting tool, that was also used for deformation measurements on a masonry wall using 18 images (Dold/Maas, 1994) and for the modelling of a human head from 40 images (Ursem, 1994).

As the computational effort connected with the strict solution of ambiguities as published in (Maas, 1992) grows exponentially with the number of images, a reduced version had to be implemented. While the strict solution with three or four images is based on a combinatorics algorithm, the reduced solution is based on probability measures for potential correspondences, which are defined by the number of images the candidate can be traced through. This is implemented in a recursive manner in a way that the longest traces (i.e. the points which can be matched successfully in the largest number of images) are accepted first and in case of ambiguities only a candidate with a trace, which is significantly longer than the traces of the other candidates, is accepted. The flow chart scheme in Figure 1 elucidates this principle.

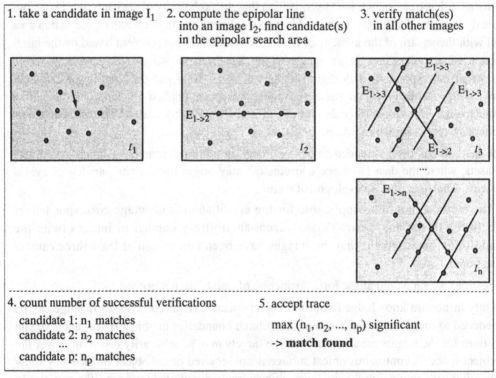

Figure 1: Computation scheme for the automatic establishment of
          correspondences via epipolar line intersection

In this manner, all candidates in the epipolar search area are either confirmed as valid candidates or rejected as false or spurious matches. It is not mandatory that the epipolar lines do really intersect: With a parametrization on the epipolar line also data from multiple images with all projection centers lying on a straight line can be processed.

3. Results

The epipolar line intersection technique described above has been applied to scanned imagery from a flight over the Simplon area in the Swiss Alps. Images of the area, that shows elevations between 1400m and 2700m, were taken from a height of 4600m, using a 150mm lens. The overlap in flight direction was 80% and the overlap between stripes 60%, with some deviations due to terrain height. The black-and-white images were scanned at

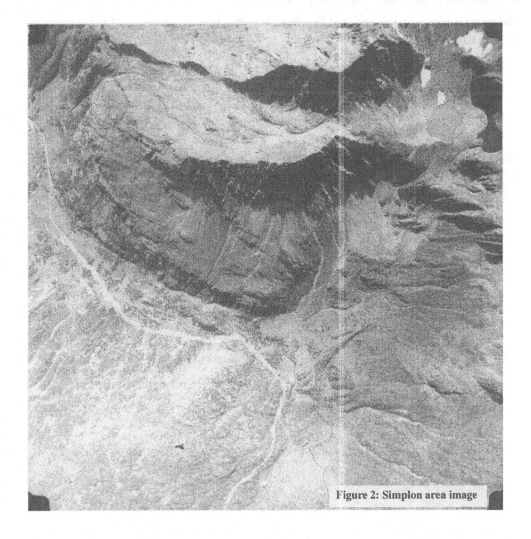

Figure 2: Simplon area image

600dpi resolution (42 micron pixel spacing, image size ~ 5200 x 5200 pixel) on a Agfa Horizon desktop scanner. The fiducial marks in the scanned images were measured with least-squares-matching. The σ_0 of 60-70 micron of the affine transformation for inner orientation indicates the limited accuracy of the scanner.

For the DEM generation with the epipolar line intersection technique a sub-block of 2 x 3 images was taken out of this block, with 36% of the image format imaged in all 6 images. The orientation data were taken from former processing of the images on an analytical plotter.

The Förstner operator detected between 50'000 and 80'000 image points in the overlapping parts of the 6 images. Feeding these points into the epipolar line intersection routine, a total of about 1500 points could be matched in all images.

The result is shown in Figure 3 and Figure 4. Note that no post-processing has been applied to these data.

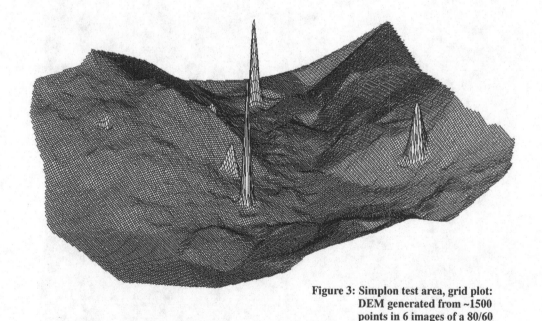

Figure 3: Simplon test area, grid plot: DEM generated from ~1500 points in 6 images of a 80/60 block

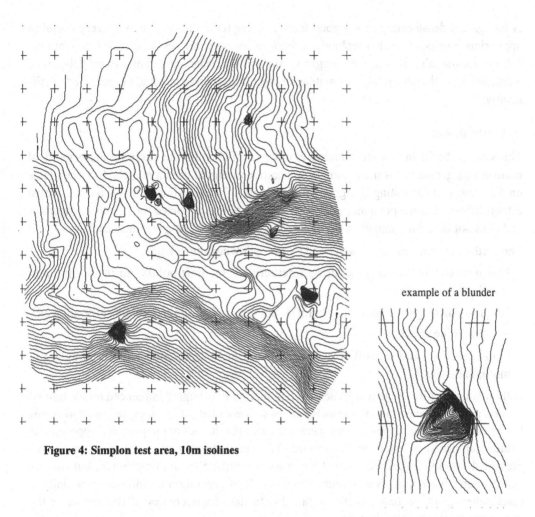

Figure 4: Simplon test area, 10m isolines

example of a blunder

With only about 1500 points matched in all 6 images the yield is relatively sparse and not quite satisfactory. The 5-6 blunders (blunder rate 0.3%) showing up as clear peaks or holes in the visualization can easily be removed by local post-processing methods like median filtering or robust surface fitting (Krzystek, 1991). When in addition all points were accepted which were matched in any 5 of the 6 images, 5500 points could be matched, but with a larger percentage of blunders (~ 2%). The main reason for this low yield is the poor geometric stability of the scanner, which necessitated to work with a rather large tolerance to the epipolar lines (80 micron in this case) thus increasing the number of ambiguities and the number of correct matches rejected as spurious matches significantly. Another reason is the probability of the interest operator detecting identical points in all images, which decreases exponentially with the number of images.

Although not dense enough for a good terrain description, this result may be very useful as approximate values for other techniques; in the same way, the results could be fed into the routine recursively, limiting the length of the epipolar lines and thus the probability of ambiguities, so that the number of matched points in a second run could be increased significantly.

4. Conclusion

The system shown in this presentation is to be considered a pilot study on the benefits of multi-image geometry for the automatic generation of digital terrain models. It concentrates on the aspect of exploiting the geometric strength of a multi-image block for the robust establishment of correspondences between discrete points extracted by an interest operator and does not intend to compete with matured, commercial DEM generation packages.

The method shown can be characterized by the following keywords:

- No approximate values needed, no interactively set start points, no image pyramid involved.
- No pre-knowledge of maximum terrain slopes required, unlimited terrain slope allowed.
- No smoothing effects.
- Good precision and reliability due to subpixel interest-operator and measurement in multiple images.

Although the quality of the results achieved in this study is largely influenced by the limited accuracy of the scanner and the results cannot yet be called satisfactory, the method seems to be quite promising. Its main advantage is the fact that it does not require any approximate values nor pre-knowledge on the terrain. The small number of matched points in the presented example is definitely insufficient for a complete terrain description, but may be very valuable when used as seed points for other DEM generation techniques - especially in cases where pyramid based methods fail due to the characteristics of the terrain or the frequency spectrum of the images.

Using a better scanner or data acquired by high resolution CCD cameras (cameras with 5120 x 5120 pixels will be available soon), one can expect a significant increase of the number of matched points, as the tolerances to the epipolar lines can be set much smaller, which reduces the number of unresolvable ambiguities and will probably allow to use also points that were detected in only 4 of the 6 images. Combining the robustness of multi-image geometry with the advantages of proven techniques a further improvement of automatic DEM generation can be expected.

References

1. Ackermann, F., 1994: Digital elevation models - techniques and application, quality standards, development. IAPRS Vol. 30, Part IV, pp. 421-432

2. Baltsavias, M., Stallmann, D., 1992: Advancement in matching of SPOT images by integration of sensor geometry and treatment of radiometric differences. IAPRS Vol. XXIX, Part B4, pp. 916-924

3. Dold, J., Maas, H.-G., 1994: An application of epipolar line intersection in a hybrid close range photogrammetric system. IAPRS Vol. 30, Part V

4. Förstner W. (1986): A Feature Based Correspondence Algorithm for Image Matching. IAPRS, Vol. 26, Part 3/3, pp. 150-166

5. Grün, A., 1985: Adaptive least squares correlation: A powerful image matching technique. South African Journal of Photogrammetry, Remote Sensing and Cartography 14 (3), pp. 175-187

6. Krzystek, P., 1991: Fully automatic measurement of digital terrain models. Proc. of the 43. Photogrammetric Week, Stuttgart, Germany, pp. 203-214

7. Maas, H.-G., 1991: Digital Photogrammetry for Determination of Tracer Particle Coordinates in Turbulent Flow Research. Photogrammetric Engineering & Remote Sensing Vol. 57, No. 12, pp. 1593-1597

8. Maas, H.-G., 1992a: Complexity analysis for the determination of image correspondences in dense spatial target fields. IAPRS, Vol. XXIX, Part B5, pp. 482-485

9. Maas, H.-G., 1992b: Robust Automatic Surface Reconstruction with Structured Light. IAPRS, Vol. XXIX, Part B5, pp. 102-107

10. Schewe, H., 1987: Automatic Photogrammetric Car-Body Measurement. Proc. of the 41. Photogrammetric Week, Stuttgart, Germany, pp. 47-57

11. Ursem, R., 1994: Accurate Reconstruction of a 3D Wireframe of a Human Head. Diploma Thesis TU Delft, The Netherlands

X-RAY DOSE REDUCTION IN COMPUTERIZED TOMOGRAPHY BY USING WAVELET TRANSFORM

D. Belluzzo, I. Raggio and S. Tarantola
Polytechnic of Milan, Milan, Italy

ABSTRACT

Wavelets has become a very popular matter of investigations, due to their great potential of applications. With respect to the classic windowed Fourier analysis, whose basis functions have the same width, the wavelet bases functions have time–widths adapted to their frequency. Therefore wavelet transform is better able to zoom–in on very short–lived high frequency phenomena.

In diagnostic medicine, image reconstruction techniques such as Computerized Tomography can be used to view internal organs with great precision. However these techniques are characterised by a high sensitivity to noise usually overcome by giving a suitably high dose to the patient. In this paper a new method has been developed, using the properties of the wavelet transform to lower the noise and then to reduce the radiation exposure in X–ray tomography.

1.   INTRODUCTION

In recent years the subject of the wavelet analysis has generated a tremendous interest in many areas of pure and applied science. The importance of the Wavelet Transform (WT) is proved by the useful results attained in a variety of contexts such as computer vision, signal processing and compression, operator analysis etc. The Fourier Transform (FT), which nowadays is the most common technique used in signal processing, yields information over the entire signal support, so that it is not able to extract events localized in time. Actually, a windowed FT may be used if one needs local information in time but the window-length is fixed both in time and in frequency domain. So if one needs high resolution in time, the window must be short, while a large window allows better frequency resolution. A trade-off between time and frequency resolutions can be attained by using WT that enjoys the zoom-in property due to the intrinsic feature of varying support lengths. In this paper we show that it is possible to obtain a consistent X–ray dose reduction in Computerized Tomography (CT) by using the wavelet transform technique for the reduction of the noise due to the poor X–ray statistics involved. In the next section we give a brief summary of the CT principles; in section 3 a WT review is presented; in section 4 we report the result of numerical experiments and finally, in section 5, we draw some conclusions.

2.   COMPUTERIZED TOMOGRAPHY REVIEW

Tomography refers to the evaluation of the density of an object along a planar cut; this is performed using data collected by illuminating the object with X–rays coming from many different directions. As inverse problems in general, tomography is poorly conditioned, that is to say, very sensitive to small variations in the input data [7,8].

In this section we sketch how one can go about recovering the density along the cross section of the object under investigation from the projection data. A projection is formed by combining a set of measurements of the density values integrated along straight lines of the object, here represented by the function $f(x, y)$ (Fig. 1). Actually, in X–ray CT $f(x, y)$ represents the local attenuation coefficient $\mu(x, y)$ strictly related to the local density of the object.

We assume to use a parallel projection experimental setup constituted by a line source of parallel, monoenergetic X–ray photons placed on one side of the object and a segment of detectors on the opposite side. The integral of f along a line univocally identified by the parameters θ and t, is called the Radon Transform of f, defined by the relation

$$R(\theta, t) = \int_{-\infty}^{+\infty} f(s, t)ds.$$

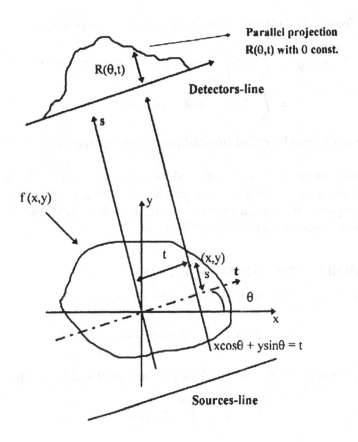

Fig. 1: **An object $f(x,y)$ and its projection $R(\theta,t)$ are shown for an angle θ.**

The connection between the projections and the function to reconstruct is given by the Fourier Slice Theorem (FST) summarized by the relation

$$\hat{R}(\theta, w) = \hat{f}(w\cos\theta, w\sin\theta)$$

(where $\hat{R}(\theta, w)$ is the 1−D Fourier Transform of $R(\theta,t)$ respect to t and $\hat{f}(u,w)$ is the 2–D Fourier Transform of $f(x, y)$), this can be rewritten as

$$f(x, y) = \int_0^\pi d\theta \int_{-\infty}^{+\infty} \hat{R}(\theta, w) |w| e^{i2\pi w(x \cos \theta + y \sin \theta)} dw$$

$$= \int_0^\pi Q(\theta, x \cos \theta + y \sin \theta) d\theta,$$

where the last integral is called the Backprojection Operator applied to the function Q.

In the reality the projections are affected by noise; note that the factor $|w|$, appearing in the above identity is like a high pass filter (called ramp filter) which amplifies the high frequency fluctuations mainly due to the noise component. Therefore, the $Q(\cdot, \cdot)$ function results highly affected by noise.

3. WAVELET TRANSFORM REVIEW

Consider a 1–D function in $\mathbf{L}^2(\mathbf{R})$; the FT of $f(x)$ is

$$\hat{f}(\omega) = \int_{-\infty}^{\infty} f(x) \exp(-i\omega x) dx$$

The convolution between two functions $f(x)$ and $g(x)$ will be written as

$$f * g(x) = \int_{-\infty}^{\infty} f(u) g(x - u) du$$

The dilation of $f(x)$ at scale s ($s \neq 0$, generally $s > 0$) is defined as

$$f_s(x) = \frac{1}{s} f\left(\frac{x}{s}\right)$$

For our purposes we may define a wavelet as a real function which must be oscillatory, with amplitude fast decaying to zero in both directions and whose integral over the real axis is zero. From this original wavelet – called mother wavelet – we construct a set of new wavelets, whose components may be used as elementary blocks for building up functions, as the sines and cosines in the Fourier approach. The wavelets here considered are spline–wavelets with compact support [2,3,4] and belong to the non-orthogonal, redundant class.

The WT maps a function $f(x)$ from the original 1–D space domain to the 2–D scale–space domain: more specifically, the wavelet coefficient of $f(x)$ at location x and scale s is

$$W_s f(x) = f * \psi_s(x)$$

Note that at small scales the wavelet support is narrow and the above coefficient carries local information on $f(x)$; viceversa, at large scales the wavelet coefficient gives a global information on the function, smoothing out local details. Furthermore, the local maxima of $|W_s f(x)|$, called modulus maxima (MMs), are the points of sharp variation of $f(x)$, and may be used for detecting the singularities of the signal [5,6]. An interesting result concerning the white noise is that the average density of the number of MMs due to white noise only is inversely proportional to the scale s. In the present numerical application we discretize the scale $s_j = a^j, j \in \mathbb{Z}$ with a scaling factor a sligthly greater than unity [1], instead of using the most common dyadic one, $a = 2$ [5,6]. In such a way we observe much more in detail the evolution of MMs across scales and we can apply a new denoising technique which allows to separate the MMs due to the signal from those due to the noise [1].

4.   APPLICATIONS

In this section we present a study case and report the numerical values of the parameters. All our tests were performed using a specific phantom whose shape, shown in Fig. 2, is well suited for simulating the human head dimensions. The values involved are related to a reference attenuation coefficient μ_0 of water (H_2O) for 100 KeV photons. Within the phantom, we chose a minimum coefficient difference of

$$\Delta\mu_{min} = 0.02\mu_0$$

embedded in a higher density shield resembling the skull.

The main task of this paper is to enlight the capability of our WT technique to resolve the above small difference in the attenuation coefficient and to compare the results thereby obtained with those of the traditional technique.

To have a first approximation description of the practical implementation we have introduced a model of the noise as due to the randomness of the matter−radiation interaction, assuming Poissonian emission and interaction statistics.

The number of detectors, assumed to be ideal point detectors, for each projection was set to 512. The same value was chosen for the number of projections in 180° and for the resolution of the reconstruction grid (set to 512 × 512).

The wavelet denoising procedure sketched in section 3 has been applied to the problem of CT. Let us consider the noisy dataset $R_i(t)$ related to the ith projection. In a traditional approach, after the canonical ramp operator, these datasets are low pass filtered by a Hamming window, to reduce noise fluctuations. In our experience this filter represents a good choice for these applications but the results are not satisfactory for low exposure levels. By further denoising these datasets with our

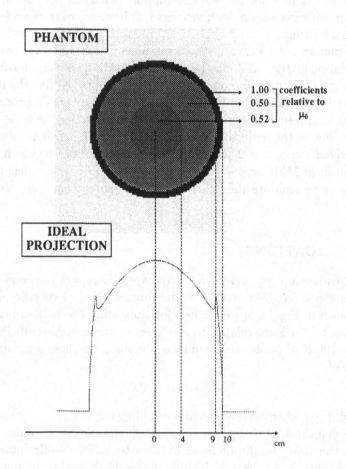

Fig. 2: **Our phantom and one of its ideal projections.**

wavelet procedure (Fig. 3), we have obtained a considerable improvement in the image reconstruction.

A global plan view of the results is given in Fig. 4 which refers to images obtained

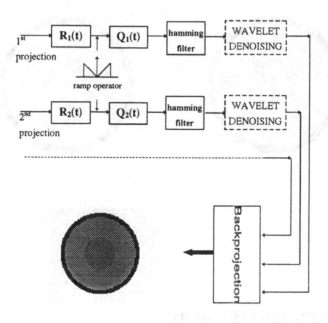

Fig. 3: **Schematic tomographic reconstruction algorithm.**

with an X-ray intensity of $3 \cdot 10^6 \frac{\text{photons}}{\text{cm}^2}$, i.e. ten times lower than the usual intensity used in practice. Making use of the classical approach, oscillations in the gray values inside the skull are present; they disappeared only when the radiation exposure was

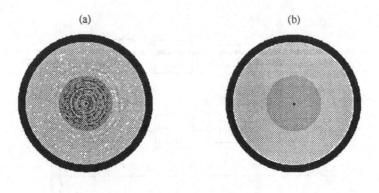

Fig. 4: **Global plan view of the results:**
         (a) classical approach; (b) wavelet approach.

increased by a factor of ten. Instead, with our WT approach no such oscillations appear, even in the low X-ray intensity case.

The difference between the classical and the WT approach is better illustrated in Fig. 5, which reports a detail of the reconstructed phantom cross–section (figures 5.a and 5.b) and the density values obtained with the two methods (figures 5.c, 5.d, 5.e, 5.f).

With the wavelet approach we can see better the differences between the densities of the internal and intermediate disks. In the classical approach, local density spots may be suspected, but they are artifacts. Figures 5.e and 5.f represent the densities with an expanded scale: we can see that oscillations around the true values are much reduced within the WT approach.

5.   CONCLUSIONS

In the present paper we addressed the problem of utilizing the wavelet transform methodology in the tomographic Filtered Backprojection Algorithm, aiming at reducing the noise which corrupts the filtered projection data $Q(\theta, t)$.

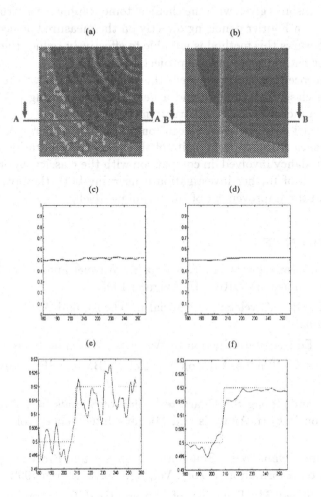

Fig. 5: **Detailed views of the results:**
   (a) classical approach plan view.
   (b) wavelet approach plan view.
   (c) cross–section A–A with full density scale.
   (d) cross–section B–B with full density scale.
   (e) cross–section A–A with expanded density scale.
   (f) cross–section B–B with expanded density scale.

A comparison is given with the classical tomographic reconstruction technique which performs a Fourier denoising directly on the measured projections; with low radiation exposures this method is not able to effectively reduce the noise amplified by ramp filter without spoiling the projection features.

In the procedure here proposed, the wavelet analysis of the Q–functions is added to the classical technique and a better data denoising has been obtained, which allows a more precise reconstruction of the object.

The wavelet coefficients have been computed by considering a non orthogonal, redundant wavelet transform; in spite of the heavier computational efforts due to the high redundancy involved, in comparison with the classical approach, we retain this work worth of further investigation concerning both the optimization of the algorithm and the improvement of the technique itself.

6.   REFERENCES

1. Belluzzo D., Marseguerra M., Tarantola S.: Wavelet analysis of fast, short lived transients with noise, SMORN VII, Avignon 1995.

2. Chui C.K. Ed.: Wavelets: A Tutorial in Theory and Applications, Academic Press, N.Y. 1992

3. Chui C.K. Ed.: An Introduction to Wavelets, Academic Press, N.Y. 1992

4. Daubechies I.: Ten Lectures on Wavelets, CBMS–NSF Series Appl. Math., SIAM, 1992.

5. Mallat S. and Zhong S.: Characterization of Signals from Multiscale Edges, IEEE Trans on Pattern Analysis and Machine Intelligence, Vol. 14, No. 7, July 1992

6. Mallat S. and Hwang W.L.: Singularity Detection and Processing with Wavelets, IEEE Trans. on Information Theory, Vol. 38, No. 2, March 1992

7. Kak A.C., Slaney M.: Principles of Computerized Tomographic Imaging, IEEE Press, New York 1988

8. Herman, G.T.: Image Reconstruction from Projections, Academic Press, Orlando 1980

DEM AND ORTHOPHOTO GENERATION IN
CLOSE RANGE PHOTOGRAMMETRY

G. Forlani and E.S. Malinverni

Polytechnic of Milan, Milan, Italy

ABSTRACT

While digital rectification may be often adequate to document architectural surveys, sometimes the object characteristics prevent such a simple approach. If a truly 3D object model is required, methods for automating object surface reconstruction become necessary, to reduce production time and costs and to avoid the use of highly qualified personnel. Though, at a first sight, just a side issue in the more general context of GIS for architecture, inexpensive and reliable methods for object reconstruction are crucial, since without a data acquisition stage actually affordable, GIS will not gain a foothold in architecture, but for a few, nice looking and expensive projects.

In the following we introduce an automatic approach to DEM generation, which makes use of an initial coarse DEM (to update iteratively) and of the epipolar constraint in order to limit the search space for corresponding points.

The procedure has been applied to a pair of photographs of an architectonic element of the church of S.Pietro in Ciel d'Oro in Pavia (Italy), to generate a digital orthophoto.

1. INTRODUCTION

1.1 Digital photogrammetry and architecture

Those who take an interest in architectonic restoration know how important is to have accurate information about a monument in reasonable time and avoiding the use of highly qualified personnel, to keep costs at an acceptable level. Close-range photogrammetry is therefore ideally suited for the survey of architectonics elements, either buildings or sculptures, statues and so on, which have shapes difficult to survey by classical instruments (such as tapes and so on) and are often inaccessible.

Digital photogrammetry provides a survey either accurate and fast: it is therefore an effective support for planning restoration works, for easy data archiving and interfacing to a database: rectified images and orthoimages (with the underlying three-dimensional object model) merge geometric and pictorial information, overcoming some descriptive limits of the drawings and making more flexible and cost-effective the whole work.

DEMs and orthophotos therefore greatly facilitate the analysis of a monument's state; massive point determination also simplify comparisons in time, allowing to create a data base to follow e.g. the dynamics of the cracks and the changes of the monument's shape in time.

Last, but not least, digital photogrammetry provides with a flexible representation of the architectonic workpiece, since measurements can be taken or densified at any time, if necessary very with high geometric precision. Indeed, what really matters in most cases is the production of a "base-document" to be enclosed in an appropriate information system to allow further analysis, knowledge documentation, easy selection and display of metric characteristics and descriptive data to support the study and/or the restoration of the storic-architectonic heritage.

In this context, inexpensive and reliable methods for object reconstruction are crucial, since without a data acquisition stage actually affordable, GIS will not gain a foothold in architecture, but for a few, nice looking but expensive projects.

1.2 What kind of automation is feasible and affordable?

Given the large variety of objects relevant to architecture, there is not a unique answer to this question.

In most applications, simple rectification may be adequate and therefore digital rectified images can easily be generated with a straightforward interactive procedure, just identifying control points with a pointing device. Significant developments towards automation and/or ground survey reduction have been introduced by searching for straight lines and geometrical constraints to be used to improve the solution [6].

If object surface cannot be approximated by a plane or by a sequence of parallel planes, differential rectification must be used, and therefore object surface reconstruction is required, in order to generate the orthoimage.

Moreover, it is quite often doubtful that parallel projection of the surface elements on a plane may be the best way to represent a generic object: think for instance of a column, or of an approximately cylindrical ceiling; in both cases, projection over a cylinder, which is a developable surface, is by far a better solution.

Still, the surface reconstruction task must be addressed keeping in mind that only automatic or semi-automatic approaches should be used to this aim, if the whole process is to be made sound and economically affordable in architecture. In this respect, unfortunately, architectural objects are among the most difficult to handle, even compare to close-range photogrammetric applications [4]: most of them are in fact truly 3-D, therefore the interpolation methods used in aerial photogrammetry to create DEM's are inadequate. Moreover, they often exhibit breaklines, sharp discontinuities and concave surfaces that generate large perspective distorsions and occlusions: this requires more information (multiple images, careful design of the survey) and a great deal of ingenuity from the surface reconstruction algorithms.

If even orthoimages are not suited to the purpose, due to the lack of any approximate developable surface, the only alternative for the representation is building the solid model of the object; once this is available, its surface may be "illuminated" by reprojecting the original images. As far as the geometric reconstruction is complete and consistent, new images (either perspective or not) of the object may be generated on arbitrary surfaces from arbitrary projection points or directions, while full (3-D) information may be recovered just pointing to an image feature. This is obviously the best possible description, and therefore the most demanding in its automatic implementation.

Contrary to 2.5-D objects, where topology is always explicit, in 3-D object reconstruction data may or may not come with topological information (e.g., by progressive sampling of contour lines...). To represent object surface then, the convex envelope may be recovered by euristics methods (e.g. Delaunay triangulation) and afterwards the so-called sculpture algorithms may be applied, to correctly model the object surface.

2. OBJECT SURFACE RECONSTRUCTION

2.1 DEM generation

Several techniques for automatic DEM generation have been proposed and are now available in software packages ([1], [2]); most of them nevertheless have been designed for mapping application in aerial photogrammetry, that is, to 2.5-D models; none of them can truly claim to be error-free or suited for every kind of application and photoscale.

In the following, we will introduce our approach to this task, which we cannot certainly claim to be fully original in the concept, but perhaps in the procedure. The basic idea is to find corresponding points in pairs of or in multiple images, by improving the object surface reconstruction iteratively; we assume that the exterior orientation of the images is known, therefore we can make use of epipolar constraints to limit the search space.

To effectively ensure reliability and overcome (at least in some cases) problems due to occlusions and large perspective differences, the procedure should be applied to multiple images. In the actual implementation of the procedure, nevertheless, we use a couple of images only: extending it to more images does not change the strategy, but some steps should be modified, making the labelling of the corresponding points more complicated and changing (making it more robust) the way data are checked for outliers.

Our procedure consists of 6 stages, some of which may be repeated, if necessary.

1) *Coarse DEM generation*
   Since the exterior orientation is known, we are likely to have some control points in the images. If they are well distributed and the object shape is very simple, they may provide a preliminary, admittedly very rough, description of its surface. In case this is unsatisfactory, more points, selected by the operator where they are more effective in defining object shape, may be added by just roughly giving their object coordinates; here the goal is only to get an approximate surface description: therefore, in the subsequent steps, all these points but the control ones will be discarded.

2) *Finding optimal candidates for LSM*
   In order to begin the search for conjugate points, it is better to select a set of points which may promise to match with high accuracy, perhaps at the expense of an uneven distribution on the object surface: undersampled areas will be dealt with in later steps. We apply therefore an interest operator (we used Förstner's operator [5]) either to one of the images or to the whole set available, depending on the method used for the search.

3) *Recovering pairs (n-nples) of conjugate points*
   We take one of the images as a reference and we project all its interesting points down to the coarse DEM. Then we go up to each other image along the epipolar line and we may look for candidates in two ways (see fig. 1):
   • (c1) if we apply Förstner's operator to all images, we define a search window around the approximate conjugate point position, looking for the best match of the reference point with all the interesting points falling in that window;
   • (c2) if we use only the set of interesting points of one image, we again define a window around the approximate conjugate point position and we simply move in steps along the epipolar line, looking for the best match point.

4) *Reliability and gross errors check - DEM interpolation*
   In the matching procedure, each point selected in the reference image provides the template, to be matched with all candidates in the search area; the best matching point is retained, if threshold values are satisfied for the correlation coefficient between the matching windows. In addition, the derived ground coordinates are accepted based on the discrepancies of the redundant determination of the Y coordinate and by the smoothing condition of a bicubic spline interpolation in a coarse grid. The refined DEM is then computed, again by splines, in a denser grid.

5) *Getting candidates in undersampled areas*
   Since the point distribution from Förstner's operator may not be optimal or reliable for spline interpolation, we look for new and more candidates in undersampled areas. This may be achieved, thanks to the improved knowledge of the DEM (so that the positional accuracy of the candidates should be better):
   • (c3) in image space, by interactively defining a regular grid in selected areas of the reference image and choosing the grid points as template centers;
   • (c4) in object space, by selecting those grid cells of the spline interpolation including no points or too few points.

6) *If object reconstruction is adequate, then stop; else, go back to 4).*

In principle part (or even the whole) procedure may be applied to an image pyramid; this might be worthy, in order to limit the search space when the object is very complex, so that getting a satisfactory initial approximation would require too many points.

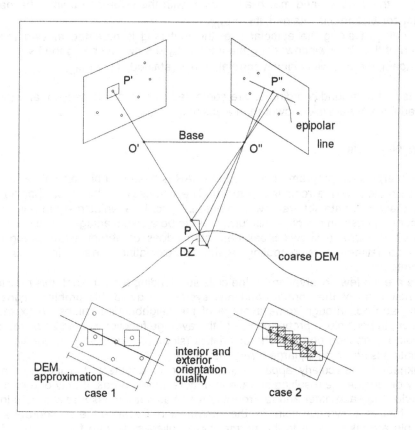

Fig.1 - Finding corresponding points in the images

2.2 Search for conjugate points

As outlined above, we may look for conjugate points in two ways, depending on whether we have already two (or more) sets of points to match or we have got a reference set only. In both cases we define the epipolar line on the slave image(s), then we select a value for a coefficient cw, to scale the error bound e_{DEM} for the DEM at hand; the coefficient cw takes into account the quality of the actual DEM, which is varying along the procedure. We then compute the (approximate) ground coordinates XYZ of the reference

point and intersect its projecting ray with two horizontal planes at $Z \pm cw \, e_{DEM}$: the points so found are projected on the epipolar line to define the search interval (see fig. 1).
At this stage, there is a fork in the procedure:

- if we look in another set of characteristic points, we reduce the number of comparisons by giving the size of the window search perpendicular to the epipolar line; then, through a binary search in the array of the candidates, we select the points falling in the window, and match all of them with the reference point. The pair with highest correlation ρ is retained, if $\rho > \rho_{min}$;
- if we look just along the epipolar line, the matching is repeated moving the initial position of the slave window along the interval by 0.8 nw, nw being the l.s.m. window size. Again, the pair with highest correlation ρ is retained, if $\rho > \rho_{min}$.

In both cases, the ground coordinates are computed by spatial intersection, and the point is accepted if the Y-parallax is lower than a specified Y_{min}.

2.3 Spline interpolation

The DEM interpolation program developed at DIIAR is based on bicubic spline functions, with knots located over a rectangular mesh. The user selects the knot spacing which applies to the whole data set area, with no possibility for local adjustment to point density and object roughness. In our case this turned out to be a disadvantage in two ways:

- if the object surface shows breaklines, step edges or abrupt slope changes, this results sometimes in overshooting of the approximating surface, in areas lacking observations;
- if there are too few observations in the cells surrounding a given knot, this may lead to malconditioning of the normal equations system; to avoid this problem, constraints may be applied, through some average of the neighbouring values, to fix the knot elevation. In principle, a proper choice of the average function should be based on the available information on local characteristics (slope changes, etc); in practice, if this information is not enough, large errors may occur.

This weakness is particularly apparent in presence of undetected blunders, that may completely determine the resulting surface in undersampled areas: if only one or a group of points (close to each other and far from the rest) has been measured in a certain area, they act in fact as a leverage point for the interpolation. Because the photographs in our test example are taken close to the normal case, reliability is zero for X, Z coordinates and anyway poor for Y coordinate. Therefore, in order to at least make up for part of the possible outliers, we run a first interpolation with a relatively coarse grid: so doing, the smoothing of the resulting surface is higher, and therefore, whenever a point would drive the surface far from the trend defined by rest of the points in the cell, the interpolation will (we hope) show a large residual (possibly only) at that specific observation point. By testing the standardized residuals, possible gross errors are then removed and a new l.s. adjustment with a smaller (more suited) mesh size is repeated, yielding a closer approximation to the surface.

2.4 Densification

To increase both the distribution and the number of points, we set up two different options:
- from image space, by choosing grid points as template centers;
- from object space, by selecting the grid cells of the spline interpolation including no points or too few points.

In the first case we interactively select a window in the reference image; then we choose the size nw of the l.s. matching window and move it in steps of 0.8 nw rowwise and columnwise. At each position a window is defined, whichact as the template for the l.s. matching. To find the candidate point, we move along the epipolar line.

In the second approach, we begin by forming a rectangular grid in XY (e.g. the one available from a previous spline interpolation) and we look for the cells with less than n_{min} points inside, where n_{min} may be 0, 1, 2. Then, for each cell to be densified, we generate a set of (five) points, evenly distributed in the cell, and, after computing their Z coordinate, we project all of them onto the reference image. From now on, the procedure runs as if we had started from points selected in the reference image

The densification may not be successful wherever applied, due either to DEM inaccuracies or to the lack of acceptable matching points (the candidates are selected independent of the signal content of their window); therefore this stage may be repeated to increase the number and to extend the distribution of the measured points, after an improved version of the DEM has been derived.

3. OUR TEST EXAMPLE

3.1 The earlier photogrammetric analogue survey

The above procedure has been applied to a pair of photographs taken to monitor the degradation of a portion of the left side column of the portal of S. Pietro in Ciel d'Oro, an ancient medioeval church in Pavia (Italy). Here two surveys (the first in 1988, the second in 1992) were executed to evaluate the progressive erosion of the column [9]: to test our DEM reconstruction procedure, we used the 1992 pair only. After DEM generation, we were interested also in producing a digital orthophoto ([8],[10]) at the scale 1:5.

The pictures have been taken with a Wild P31 camera with principal distance c=106.86 mm, at a maximum distance of about 1.5 m (photoscale approximately 1:14) and with a base of about 40 cm; the range in depth is of about 20 cm. The total model area in object space is about 70 cm wide and 1 m high (63% overlap), while the area of interest is about 40 x 70 cm. As can be seen, both sides of the column are compressed in a few pixels, so that measurement in that are are very difficult even for a trained operator, while some areas on the top are poorly textured. To the left of the column there is a wall, images in a very tiny strip in image 2; to the right there are quite elaborated decoration. Nine control points have been determined by forward intersection with a Wild T2 theodolite, and the DEM has been generated, after relative and absolute orientation, by profiling with a spacing of 1 cm, on an analytical plotter Zeiss Planicomp C100; overall, more than 2000 points were acquired. Since the profiles were measured many times, an empirical evaluation of the precision of the Z coordinate (σ_Z for each point in the measured profiles) gave a value of 0.3 mm, though restricted to the upper part of the column: in this area the analogue DEM will therefore be assumed as a reference for our test results.

3.2 Digitization and orientation procedures

For the digitization of the pictures we used an *HP IIp* scanner, with 8 bit radiometric resolution and maximum geometric resolution of 300 dpi. The geometric resolution was set to 200 dpi (128 μm), corresponding to about 1.8 mm in the farthest area of the object. We used contact prints obtained from the original glass plates, since our scanner cannot be used in trasparency mode. The chosen resolution is clearly not enough for the semi-automatic procedure to match the results of the analogue measurements: to get an accuracy in object space of 0.3 mm would require an accuracy of 6.5 μm in image space, i.e. a matching accuracy of 1/20 of pixelsize, which cannot be obtained without use of targets and correction of all effects coming from scanning and uncertainties in interior and exterior orientation. Given the object characteristics we may rather aim at accuracies in the order of 1/3 - 1/5 of pixelsize.

In order to estimate the distortions due to scanning, a very simple test has been carried out. In a stable polyester film a 1x1 cm reseau was obtained by using the cutter of a plotting table. The coordinates of the centres of the crosses were measured at a Zeiss PK1 monocomparator, providing the nominal reference values. After scanning the film, the centres of the crosses were automatically measured by l.s. matching, and an affine transformation between the two sets of coordinates computed. The residuals after the transformation were rather irregularly distributed, with maximum values of about 100 μm, that is, quite significant but at the same time very hard to compensate for. In the following stages, therefore, no corrections were applied to the image coordinates.

The inner orientation of the digital image (the pixel-to-image coordinates transformation) has been computed by manually measuring the 5 fiducials on the screen, followed by an affine transformation. The maximum residual after the l.s. adjustment was of 0.7 pixel.

To compute the exterior orientation, two bundle adjustment have been executed, including all 9 control points, whose image coordinates were measured manually on the screen. In the first adjustment we fixed only 4 points at the corners, to have some check points: the largest residual (discrepancy) at the control points was less than 1 mm in X,Y and less than 2 mm in Z. In the second adjustment all control points were held fixed; the estimate for σ_0 was 30 μm, while the accuracy for the exterior orientation elements was 3 mm in X,Y, around 1 mm in Z and 0.19^{gon} in attitude.

A lower resolution image (decimation factor=2) was also generated, to check whether an image pyramid approach [1] would be worth doing, after low-pass filtering the image by a gaussian filter box [7] of size 5 x 5.

3.3 DEM generation

We started the procedure from the low resolution pair; the acceptance thresholds for the conjugate points were set as follows:
* correlation coefficient after l.s.m. $\rho \geq 0.9$

- Y-parallaxes ≤ 1 mm.

The interest operator, with a window size of 9 pixels (i.e. about 1.6 cm on the object) selected 546 points in image 1 and 521 points in image 2 (see fig. 2).

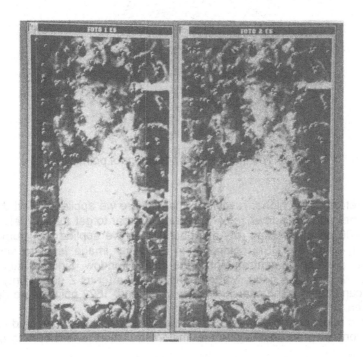

Fig. 2 - The characteristic points selected by the interest operator

To find the corresponding pairs, a coarse DEM (see fig. 3) has been generated by bicubic spline interpolation using just all 9 control points; starting from the points in image 1 (taken as reference) and afterwards from those in image 2, 290 pairs were recognized as conjugate by searching along the epipolar line (case c2) in a band 14 pixels wide (about 25 mm in object space).
We proceeded then with a densification of the DEM from the empty cells of the spline grid (case c4), finding 225 more points.
After each of the preceeding stages, a coarse spline interpolation with a 5 cm grid was performed to reject outliers. The clean data set so obtained was interpolated with a dense grid (3 cm), yielding the final DEM at that stage (see fig. 4).

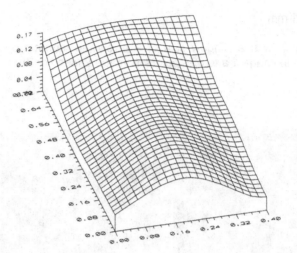

Fig. 3 - The initial DEM from 9 control points

We moved therefore to the higher resolution image. Here we applied at first the interest operator (this time, with a windowsize of 13 pixels), in order to get a set of reliable points (we got about 400 pairs, by using (c1) approach). Then we applied (c3) from a regular grid in image 1, using the DEM computed in the lower image level to sort out the ambiguities or to provide the centre point of the search window. Since at this stage the quality of the DEM should be good, we used a lower value for its Z uncertainty, reducing the number of candidates in the search window. In the end, 1622 points were determined in the model area.

Again, in order to reject possible outliers, a coarse interpolation was executed: about 3 % of the observations were discarded and the final interpolation for orthophoto generation computed.

In the l.s. matching stage, we used at the begininng an affine transformation; this turned out to be rather error prone, since due to object texture and to the relatively large search area along the epipolar line, in several cases we had mismatches with correlation coefficient value as much as 0.95. Therefore we decided to start the l.s.m. with the simpler, but more robust, shift model; if a good correlation is achieved after the first iterations, then the program switches to the affine model.

If we look at the distribution of conjugate points found by the procedure, it is clear that no point could be found on both sides of the column, where the perspective differences are the largest due to the steepness of the surface. In this area of the stereomodel, indeed, also the operator had troubles in measuring the elevations as confirmed by the r.m.s. of the differences of repeated profiles.

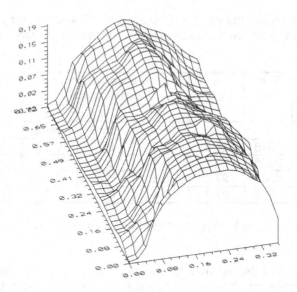

Fig. 4 - Surface reconstruction (lower resolution level)

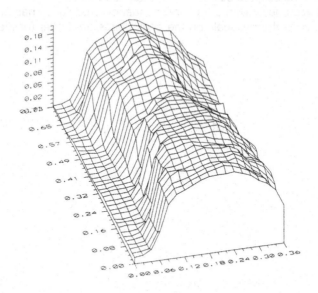

Fig. 5 - Surface reconstruction (original resolution level)

After DEM generation, the image coordinates of all points were put toghether in a bundle adjustment, to estimate the theoretical accuracy of the measurement procedure in image and object space. To this aim, we run two adjustments, one fixing only the control points

at the corners of the model, the other one fixing all c.p. The same has been done for the
image coordinates from the lower resolution level. The results are listed in table 1.

Block adjust.	Pixel size	N. of obs.	σ naught	Theoretical accuracy in object space			Residuals on check points	
				σX	σY	σZ	Z (rms)	Z (max)
	(μm)		(μm)	(mm)	(mm)	(mm)	(mm)	(mm)
4 c. p.	128	3846	25.5	0.5	0.7	2.13	0.93	1.71
All c.p.	128	3846	25.2	0.4	0.5	1.75	-	-
4 c. p.	256	1608	27.3	0.6	0.8	2.29	1.53	2.29
All c.p.	256	1608	27.2	0.4	0.6	1.83	-	-

Table 1 - Results of the bundle adjustments

As it may be seen, since the estimate for σ naught does not change significantly in the
two data sets, so does the theoretical accuracy. This is obviously a bit embarrassing,
because measurement accuracy should be better for the higher resolution images; in a
stereo model, nevertheless, the estimate for σ naught cannot be taken as a reliable figure
for measurement accuracy, given the poor redundancy. Resolution has actually an effect
on the precision, since the residuals on the chek points are larger for the low resolution
images.

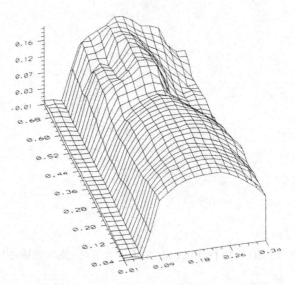

Fig. 6 - Surface reconstruction from Planicomp C100 (DEM from 1 x 1 cm grid)

With the above remark in mind, translating in pixel space the estimates for measurement accuracy, we get about 1/4 ÷1/5 of a pixel (and, for the 256 μm images, 1/8 ÷1/10 of a pixel!), which is an acceptable result, taking into account the object characteristics. The accuracy in object space compares fairly well with the residuals on check points (the residuals for X,Y, not shown, are less than 0.8 mm), though a few points only were available to compute the rms values.

In order to empirically assess the precision of the procedure, we used those profiles measured at the Planicomp C100 analytical plotter (see fig. 6), which, as already mentioned, provided reference values.

In a first stage we predicted by spline interpolation the elevations of the 1992 surface at the X,Y location of the points selected automatically: we expected therefore the interpolation errors be small, since the surface is generated from a dense 1x1 cm sampling of the object; only the errors coming from the photogrammetric point determination were likely to be highlighted. The r.m.s. of the differences turned out to be 6 mm, and the maximum difference of 1.8 cm, showing that several mismatch occurred, though mostly at the left and right border of the area, were the surface of the column is very steep.

In a second attempt to assess the accuracy, we tried to separate further the effect of the exterior orientation on the derived DEM, by measuring by l.s. matching the same points selected by the operator along the profiles. To this aim, we run a modified version of (c3) procedure, taking as starting points from object space the X,Y,Z coordinates of the measured profiles and projecting them back onto the images; since the exterior orientation elements computed in 1992 are not available, we used those coming from our automatic procedure. Out of about 1800 points, only 417 were matched successfully (again with the same threshold for ρ and Y-parallaxes as above). By applying a spatial similarity transformation to the resulting coordinates, which should compensate for the differences in exterior orientation, the residuals were analyzed. The r.m.s. value of the differences in Z (those in X, Y are negligible) turned out to be 4.4 mm; still 12 points were clearly gross errors, with residuals up to 2.5 cm. After rejection of these observations, the final value of the r.m.s. of the differences went down to 3.6 mm, based on 405 object points, mainly located in the upper and lower part of the column; a contour line plot of the Z differences is shown in fig. 7. The differences do not show any clear systematic behaviour; still, they are larger than the expected accuracy and do not match with the figures from the control points, by far better. The discrepancies may be reasonably attributed to the lack of correction for the scanning deformations (which were estimated to be in the order of 100 μm): indeed, given the large image scale, the X,Y locations are not much affected, while Z coordinates may change from 3.5 mm up to 4.7 mm.

By using the DEM, the orthophoto at the scale 1:5 has been generated (see fig. 8), by resampling with bilinear interpolation the right image of the pair. The pixel size in the orthophoto is about 70 dpi, corresponding to a coverage on the object of about 1.8 mm. The maximum error in XY in the orthophoto occur at the corners of the image. Referring to the maximum error in the DEM (1.8 cm) this amounts to 0.18 cm (0.9 cm on the ground), while with respect to the r.m.s. value (0.06 cm) the maximum error would be 0.06 cm (0.3 cm on the ground).

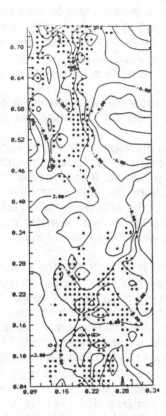

Fig. 7 - Contour plot of the Z differences
between manual and digital measurement

Fig. 8 - The orthophoto

4. CONCLUSIONS

The feasibility of the proposed approach for the case shown has been demonstrated and this was indeed the main aim of this first experiment, where the objective was checking the procedure rather than meeting a specific goal on the accuracy, by appropriate design of project parameters. Taking into account object surface characteristics (roughness, steepness, sharp discontinuities and sometime poor texture) and the poor redundancy (just a stereo pair), the result seems acceptable. Nevertheless, the procedure needs to be tested with more demanding cases: it is not yet clear how critical for the convergence of the method is the initial DEM quality; highly redundant observation schemes are mandatory to tie image coordinate measurement and object coordinate determination, as in the MGCM approach, so that a more robust check against inconsistencies may be provided.

REFERENCES

1. Ackermann F., Haan M. (1991): *Image pyramids for digital photogrammetry*. In: Digital Photogrammetric Systems, Ebner/Fritsch/Heipke (Eds), Wichmann Verlag, 43-58.
2. Baltsavias E. (1991): *Multiphoto geometrically constrained matching*. Institut für Geodasie und Photogrammetrie, Mitteilungen n. 49, ETH, Zurich.
3. Baltsavias E. (1994): *Test and calibration procedures for DTP scanners*. IntArchPhRS vol 30, Part 1/1, 163-170.
4. B. Crippa, G. Forlani, A. de Haan (1993): *Automatic deformation measurement from digital images*. In: Optical 3-D Measurement Techniques II, Gruen/Kahmen (Ed's), Wichman Verlag, 1993, 557-563.
5. Förstner W. (1986): *A feature based corrispondence algorithm for image matching*. IntArchPhRS vol 26, part 3/3, 150-166.
6. Georgopoulos A. (1994): *Digital rectification using a PC*. IntArchPhRS vol 30, Part 5, 102-108.
7. Gonzalez R.C, Woods R:E. (1992): *Digital Image Processing*. Addison-Wesley Publishing Company.
8. Kraus K. (1994): *Photogrammetry*. Dümmler Verlag, Bonn.
9. Malinverni E.S. (1993): *S Pietro in Ciel d'Oro: un'applicazione della fotogrammetria come metodo di controllo non distruttivo dello stato di degrado*. Rivista del Dipartimento del Territorio n° 2-3, 1993.
10. Mayr W., Heipke C. (1988): *A contribution to digital orthophoto generation*. IntArchPhRS, vol. 27, Part B11, 430-438.

MATCHING AND MAPPING OF STARS
IN STAR CATALOGUES

B. Bucciarelli and M.G. Lattanzi
Astronomical Observatory of Turin, Turin, Italy

T. Bellone
Polytechnic of Turin, Turin, Italy

ABSTRACT

The materialization of a reference frame is a fundamental task which is relevant to both astronomy and photogrammetry. A step toward this direction is the construction of a stellar catalogue. After a brief introduction on the main astronomical catalogues presently or soon available and the status of photographic sky surveys, the principles of data reduction of photographic plates are outlined; then, three successful methods for the construction of all-sky photographic catalogues using Schmidt telescopes are presented.

---

* Recent resolutions expressed by the photogrammetry community, relatively to space photogrammetry, have wished a broader connection with the astronomical community. This paper, like the previous "Trento '93", moves in this direction. In particular, B.Bucciarelli and M. Lattanzi are responsible for the formulation and development of this work, while T. Bellone contributed to building the interface with photogrammetry and cartography.

1. INTRODUCTION

The determination of an inertial (or quasi-inertial) reference frame is of utmost importance in astronomy, for it provides a materialization of a coordinate system in which observed motions of celestial bodies can be described by means of Newton's laws. The accuracy with which such a system can be established directly affects studies concerning galactic dynamics, dynamics of the solar system, and geodynamics (Froeschle' and Kovalevsky, 1982). As an example, the geometry and dynamical behaviour of the Earth can be described in the relative sense by using a suitable coordinate system attached to the Earth in some prescribed way. In order to monitor such effects in an absolute sense, the earth based system needs to be referred to an external (inertial) system.

The first step in this direction is the construction of a *star catalogue*, which consists of a list of stellar objects, their position on the celestial sphere at a chosen epoch T_0 and their proper motions, i.e., their angular displacement with time.

The realization of a star catalogue is a very onerous task, for it involves a series of steps which are both time demanding, from an observational point of view, and critical from the point of view of data reduction.

An investigation of star catalogues, the methods of their construction, and their purpose can be found in Eichhorn (1974).

In this context we will be concerned with all-sky (or hemispherical) catalogues, which can be basically subdivided in the following categories:

● Ground based catalogues:

      -meridian circles instruments (fundamental measurements)
      -large field telescopes (relative measurements)

● Space based catalogues:

      -HIPPARCOS: ESA astrometric mission (survey-type astrometric satellite).

Besides their fundamental role in the establishment of an inertial reference system, such catalogues constitute a basic astrophysical tool of immediate use in:

● mission planning: object visibility, simulations of observations;

● precise satellite attitude control and guidance;

● reference stars for adaptive/active optics-controlled ground based telescopes.

The availability of Schmidt cameras (1930-1931) has made possible the construction of all-sky catalogues from photographic sky surveys, which constitute a very precious astronomical resource due to the variety of astrophysical applications they are suitable for.
A typical example is represented by the Guide Star Catalogue (Lasker et al. 1990) - which was constructed for the guiding and pointing of the Hubble Space Telescope - and its future generations (GSC-II, Lasker et al. 1995).
A Schmidt telescope is characterized by a large field of view (tipically 6.5 x 6.5 degrees) and an aperture ratio (ratio between the diameter of the objective and the focal length) ranging tipically from F:3.5 and F:25. This results in a high luminosity efficiency, which, combined with the large field capability, permits to image large portions of the sky down to an interesting magnitude limit with a reasonable effort.
All-sky Schmidt photographic surveys presently (or soon) available in digitized form are given in table 1.

SURVEY	EPOCH	MAGNITUDE LIMIT(VISUAL)
Palomar Observatory Sky Survey (POSS) I	1950	20-21
POSS II	1993	21-22
Palomar "Quick V"	1983	19
UK Schmidt Telescope Survey (SERC-J)	1975-85	22
Anglo Australian Observatory Survey (SES)	1993	22

Table 1.

The present and near-term future situation of all-sky reference catalogues suitable for the astronometric reduction of Schmidt surveys can be summarized by the following - not exhaustive - list:

Catalogue	Coverage	No. of stars	Pos. error at 1995 (arcsec)	Prop. motion error (milliarsec/year)
Available::				
PPM North	North	181731	0.290	4.3
PPM South	South	197179	0.120	3.0
ACRS, I	All sky	250052	0.230	7.5
ACRS, II	Ali sky	70159	0.370	7.5
In progress:				
HIPPARCOS	All sky	118209	0.008	2.0
TYCHO	All sky	1 100000	0.030	–

Table 2

2 IMAGE PROJECTION FOR AN ABERRATION-FREE SCHMIDT TELESCOPE

The geometry of a Schmidt telescope projection is depicted in Fig. 1.
Let's consider a celestial sphere with center O and radius r. At the tangential point T with equatorial coordinates (A,D) a tangent plane touches the sphere. S' is the projection of an arbitrary point S with spherical coordinates (α, δ) onto the plane. We introduce a rectangular system of standard coordinates (ξ, η) in the tangent plane with origin at T, so that the η axis is directed to the projection P' of the celestial pole P.
In the Schmidt telescope, the mirror conformly projects the celestial sphere onto a spherical focal surface on which no distinguished point exists. However, we can define a "neutral point" as the one which remains at rest during the process of bending in the plateholder.

Thus, as first approximation, the neutral point coincides with the intrinsic tangential point T. The geometric projection realized in such way is the so called *azimuthal equidistant projection*.

In this projection, the distance from the projected point S'_E to T is equal to the arc length from S to T, i.e., $TS'_E = \theta$. The mapping equations are:

$$
\begin{cases}
\xi_E = \dfrac{\theta \cos \delta \, sin(\alpha - A)}{sin\theta} \\[4mm]
\eta_E = \dfrac{\theta \left(sin\delta \cos D \cos \delta sinD \cos(\alpha - A) \right)}{sin\theta}
\end{cases}
$$

The inverted mapping equations are:

$$
\begin{cases}
\theta_E^2 = \xi_E^2 + \eta_E^2 \\[3mm]
sin\delta = \eta_E \cos D \, \dfrac{sin\theta}{\theta} + sinD \cos\theta \\[3mm]
sin(\alpha - A) = \dfrac{\xi_E sin\theta}{\theta \cos\delta}
\end{cases}
$$

where θ can be derived from simple spherical trigonometry applied to the spherical triangle $TP'S'_E$.

Most of the problems with modelling the geometry of Schmidt plates lie in the deviation from this projection. In fact, the anelastic stresses induced onto the plate in the bending process are not completely recovered once the plate is brought back to its planar state. Such a process generates *local distortions* of the plate metric. An example of these distortions can be seen in Fig. 2, where residual systematic errors are revealed by using a very precise special astrometric catalogue (Klemala et al., 1984)), realized for the passage of the Halley Comet in 1986.

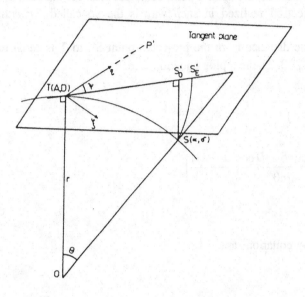

Fig.1, geometry of azimuthal equidistant projection

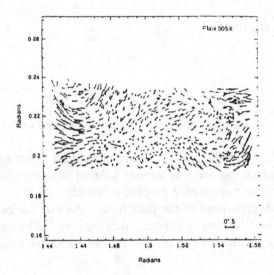

Fig. 2, systematic errors on the GSC

3. PLATE REDUCTION

The first step is the *generation of measurements* (x,y) on a plate-based coordinate system. This process consists of two parts:
• *digitization*, which is done with the aid of a measuring machine,
• *image centroiding*, which is meant to compensate all the effects of image formation effecting the shape of a single star image, as opposed to the location of the image center.
The second step is the *identification of reference objects* (stars or galaxies) on the plate. There exist various algorithms which are suitable to this particular task (see, e.g., Groth, 1986, Murtag, 1992), which essentially produce a match of star lists, given by their two-dimensional coordinates.
The star lists can be of different cardinalities, and the approaches should be unaffected by translation, rotation, rescaling, and random perturbations. Moreover, magnitude information can be used as weight associated to the points to be matched.
The third, and final step is the mapping of (x,y) onto standard coordinates (ξ, η)
The ideal geometrical mapping equation of (x,y) onto standard coordinates (ξ, η) is merely an orthogonal rotation of the coordinate axis plus a magnification factor depending on the actual focal length of the telescope. In reality, certain corrections need to be taken into account because of the following.
The Schmidt plate in its planar state represents a plane pangent to the sphere at the intrinsic tangential point T_I with equatorial coordinates (A_I, D_I). The coordinates (A_A, D_A) of the assumed tangential point T_A differ by the small vaules p and q from (A_I, D_I)

$$p = \left(A_I - A_A\right)\cos D_I \quad , \quad q = D_I - D_A.$$

This is called *plate tilt* (or centering error).
Moreover, when the zero point of the measured coordinates does not coincide with the intrinsic tangential point more quadratic terms may appear as significant in the general reduction polynomial.
Another important correction is the one that accounts for the effect of atmospheric refraction on the area of the sky portrayed on a Schmidt plate. What one wants to estimate in this case is the differential refraction, i.e., the change of the effect as function of position on the plate.
The correction can be made by applying the following formula, which is derived by using radial symmetric atmospheric model

$$R \equiv z - z_0 \cong A \tan z_0 + B \tan^3 z_{0f.}$$

where z is the zenith distance of the light ray outside the atmosphere, z_0 is the observed zenith distance, and A, B are the so-called refractions constraints, which depend on meteoreological conditions, and also vary with wave length. Since the causes of variation in refraction are so complicated, is has been the practice for astronomers to make use of specialized Refraction Tables compiled at several major observations, e.g., the Refraction Tables of the Pulkova Observatory.

Additional residual terms can then be easily taken into account by a third order degree polynomial.

These kind of corrections, however, do not account for all the effects which characterize the metric properties of a Schmidt plate. It has been proven, in fact, that a low-degree polynomial fitting is not adequate to map (x, y) onto (α, δ), its performance degrading quickly as the plate edges are approached.

4. REDUCTION STRATEGIES

In the astronomical literature of the last few years three basically different approaches have been proposed and successfully applied as alternative methods for the astrometric reduction of Schmidt plates (Fresneau, TLB, LB).

For all three approaches it is crucial to have at disposal a dense, accurate reference catalogue.

Several investigations (Andersen 1971, Dodd 1972, Fresneau 1978, de Vegt 1979, Taff et al. 1990) have pointed out the serious difficulties that arise in Schmidt plate reductions and their interpretations. Two of the authors have being studying alternative solutions to this problem in the past few years, and we outline here the methods which have shown to be very effective in curing the presence of residual systematic errors from standard polynomial fitting.

4.1 Least Squares Collocation Method

Figure 2 shows a typical residual pattern left on a Scmidt plate after a classical polynomial fitting has been applied.

Such a set of residuals ($\Delta\xi$, $\Delta\eta$) can be interpreted as a discrete sample of a realization of a bi-dimensional *stochastic process* S ($\Delta\xi$, $\Delta\eta$, Θ), where Θ is a generic point on the plate.

Assuming that the process is stationary in a wide sense and isotropic, then the auto-covariance of the residuals is computed as (Moritz, 1972):

$$c_\xi(r) = \sigma_{\Delta\xi\Delta\xi}(r) = \frac{1}{\left[8\pi\left(\frac{a}{2}-r\right)^2\right]} \sum_{i=1}^{N_{SAMP}} \left(\sum_{k=1}^{N_i} \Delta\xi(\Theta_i)\Delta\xi(\Theta_k)\Delta\alpha_k\right)\Delta\Omega_i$$

where:
a is the actual plate size,
r the distance at which the covariance is being estimated,
$\Delta\alpha_\kappa$ the angle computed from a reference direction to the line connecting stars i and k,
$\Delta\Omega_i$ an element area on the plate,
N_{SAMP} is the number of stars in a square of side $(a-r)$ connected to the plate,
and N_i is the number of stars at the distance between r and $r+dr$ from the star located at Θ_i.
After the empirical covariance function has been computed, a fit to an analytical model suitable to represent a theoretical covariance function is performed (Barzaghi and Sanso', 1983).
Finally, the "signal" representing local distortions left by the model is given by

$$s_\xi(\theta) = \left[c_\xi(r_1), \ldots, c_\xi(r_m)\right]^T \left(C_\xi + C_n\right)^{-1} \Delta\xi$$

where C_n is the covariance matrix of the random part of the residual errors, assumed uncorrelated.
An example of how this method works in practice can be seen in figures 3a and 3b.

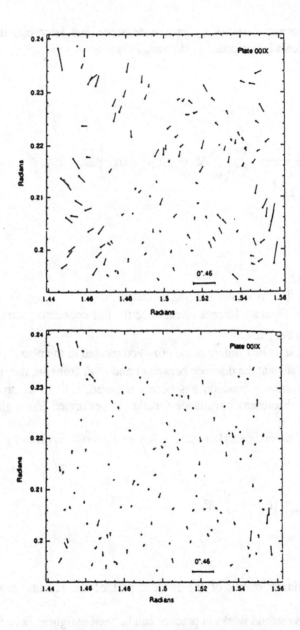

Figs. 3, plate residuals (a) before signal removal via LS collocation; (b) after signal removal, for a Palomar "Quick V" Schmidt plate

4.2 Sub-plate method

This method has been developped in (Taff et al., 1990). Briefly, it consists of an artificial division of the entire (6.5x6.5 deg) Schmidt plate into smaller, more tractable, pieces. To break the embrace of systematic effects, the size of each sub-plate must be below one degree.

Because of the small area covered, a sub-plate can be successfully modeled by a linear relation between measured and spherical coordinates.

The actual size of the artificial sub-plates is a trade-off between the number of model parameters to be determined and the reference star catalogue density. The purpose of overlapping the artifical sub-plates is to reduce the systematic errors in the sub-plate constants induced by the errors in the reference star positions. An example of sub-plate division with two-fold overlapping is shown in Fig. 4.

The number of sub-plates per plate in the case three-fold overlapping is given by the formula (3*NS-2)**2, where NS is the number of subdivisions. In this case, there are tipically ~ 360 sub-plates per plate, which results in 360x6=2160 plate constants to be estimated. Once all these unknowns have been calculated, the position of an object in the plate is computed as:

$$\xi = \frac{\left(\sum\limits_{i=1}^{NSP} \xi_i\right)}{NSP} \quad , \quad \eta = \frac{\left(\sum\limits_{i=1}^{NSP} \eta_i\right)}{NSP}$$

where NSP is the number of sub-plates where the object is found.

4.3 Mask Reduction Method

If the systematic effects present in the plate residuals left after a low-order polynomial solution are stable for a given Schmidt instrument, then they can in principle be discovered a posteriori and recovered. By superimposing residuals form many different plates, the *random noise* present in each single vector is beaten, and the *signal* brought out by integration.

The "mask" obtained in such way, see Fig.5, can then be used to post-correct the residuals on a single-plate basis (see Taff et al., 1990).

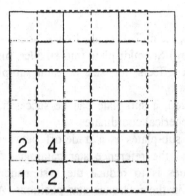

Fig. 4, "mask" obtained superimposing astrometric residuals from the reduction of ~1500 plates of the SER-J Schmidt Survey

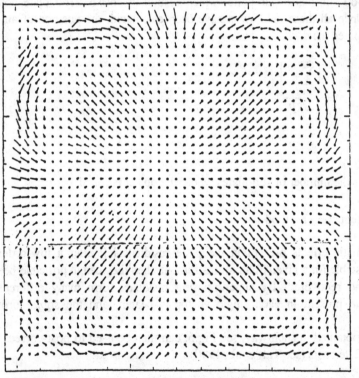

Fig. 5, example of "mask" obtained from residuals of the SERC-J Survey

5. AN ASTRONOMICAL APPLICATION

After having demonstrated the success of the above techniques in performing astrometric reductions of Schmidt plates, the first astronomical application has been the construction of an updated version of the AGK3 catalogue (Heckman, Dieckvoss et al. 1975).
In an effort to improve positions and absolute proper motions of the AGK3, the Palomar "Quick V" photographic survey has been used as source of new observations. To this purpose, the x,y measurements coming from the survey plates have been astrometrically reduced and integrated with the AGK3 catalogue to produce the AGK3U (Bucciarelli et Al., 1992).
The reduction method used was mainly the sub-plate one, with the mask technique as backup solution. The final mean positional error is ~ 0.167 arcsec at an verage epoch of 1950.6, while the total proper motion error is ~ 0.82 arcsec/century, a significant improvement over the former catalogue.
In conclusion, the astrometric potentials of large-scale Schmidt plate can be realized via the application of the methods reviewed above. In such ways, all-sky astronomical catalogues are successfully constructed.

REFERENCES

1. Barzaghi, R. and Sanso', F.:Bollettino di Geodesia e Scienze Affini, 4, 389, 1989.

2. Bucciarelli, B. et Al.: AJ, 103, 1689, 1991.

3. Dick, W. R.: Astron. Nachr., 312, 113, 1991.

4. Dodd, R. J. : AJ, 77, 306, 1972.

5. Eichhorn, H.: Astronomy of Star Positions, Frederick Ungar Pub. Co.,New York, 1974.

6. Fresneau, A.: AJ, 83, 406, 1978.

7. Froeschle, M. and Kovalevsky, J.: A&A, 116, 89, 1982.

8 Green, R. M. : Spherical Astronomy, Cambridge University Press, 1985

9. Groth, E. J.:A.J., 91,1244, 1986.

10. Heckmann, O. et Al.: AGK3. Star Catalogue of Positions and Proper Motions North of 2.5deg Declination, 1975

11. Klemola, A.R. et Al.: Comet Halley Faint References Star Catalogue, IHV Newsletter, 5, 38, 1984.

12. Lasker, B. M. et Al.: AJ, 99, 2019, 1990.

13. Lasker, B. M. et Al.: ESA SP-379, 1995.

14. Lattanzi, M. G. and Bucciarelli, B.: A&A, 250, 565, 1991.

15. Moritz, H.: Reports of the Ohio State University, 175, 1972.

16. Murtagh, F.:PASP ,194, 301, 1992. Groth, E. J. 1986, A.J., 91,1244.

17. Taff, L. G., Lattanzi M.G. and Bucciarelli B. 1990, ApJ, 358, 359.

A QUICK LOOK ON THE SEA SURFACE TOPOGRAPHY OF THE MEDITERRANEAN FROM THE GEOMED GEOID AND THE ERS1 GEODETIC MISSION

B. Crippa and F. Sansò
Polytechnic of Milan, Milan, Italy

ABSTRACT

The purpose of this work is to verify the potential tools offer by GIS (Geographic Information Systems) for storage, analysis and representation of geodetic data. For this experiment two basic data set have been used: the first one is the Geoid on the Mediterranean sea, the second is a new altimetric datum from the ERS1 geodetic mission.

The results obtained are interesting from a geodetic point of view, moreover the use of GIS for the treatment of geodetic data has showed only a little, but meaningful, part of the GIS potential use in the field of geodesy.

Introduction

The study of the gravity field and of the precise point positioning on the earth surface are two apparently different aspects joined by the theory of integrated geodesy. The classical geodetic measurements are performed on the surface of earth; they concern the gravity modulus, the close range variation of gravity (Gradiometry), the direction of gravity (deflections of the vertical) and the gravity vector (inertial systems). Other measurements as azimuth and zenith angles, height differences, are intimately tied to the gravity field. Other sources of geodetic data are supplied by satellites, the most important of which are: laser tracking (Lageos, Startlette, Ajisai, Stella, etc...), doppler (Doris), radaraltimetry (Seasat, Geosat, ERS1/2, TopexPoseidon, etc...), global position system (GPS).

All this measurements are used in two main modes: global and local. Global problems mainly regard: the definition of reference systems (ellipsoid parameters, earth rotation, geodynamic control), the gravity field models (coefficients of spherical harmonics), the time variations of the gravity field. Local problems mainly regard: local networks for cartographic applications, control networks for deformation monitoring, gravimetric networks for local gravity measurements.

The principal target of integrated geodesy is to obtain the precise coordinates of points on a global or local scale together with an improved knowledge of the gravity field. The study of the anomalous gravity field (geoid) is necessary to link the physical heights (orthometric i.e. above the sea level) to the geometric heights.

The different types and sources of geodetic data suggest to explore the potential tools offer by GIS to archive, retrive, analyse and represent the geodetic data.

Mediterranean geoid and ERS1 data sets

For this experiment two basic data sets have been used: the first one is the geoid on the Mediterranean sea, the second one is the new altimetric data from the ERS1 geodetic mission. The geoid on the Mediterranean sea comes from the GEOMED ARCHIVE. The data gathered to build GEOMED ARCHIVE come from different sources, and can be divided into data for geoid and data for the sea surface topography.

The data for geoid come from:
- marine gravimetry;
- land gravimetry;
- global geopotential models;
- digital terrain models;
- bathymetry.

The data for sea surface topography result from the observations done by some altimetric satellites. In addition a new altimetric data set is now available from the last period of ERS1 geodetic mission.

The marine gravimetry data have been obtained by digitizing the Morelli's gravity maps. This task has supplied about 40000 data. Another data set has been obtained form the USA Defence Mapping Agency.

The land gravimetry data set is composed by many data gathered by different National Gravimetric Archives and International Agencies like the Bureau Gravimetrique.

As for the global geopotential models, among the many available, in this experiment the OSU91A has been chosen. This geopotential model is complete up to order and degree 360 and in generally considered as the best available up to date.

Another important data set, in GEOMED ARCHIVE, is the digital terrain model ETOPO5U. This terrain model provides the heights and depths in the mediterranean area with a resolution of 5' by 5'.

The last data set is the bathymetry which is obtained by digitizing the bathymetric Morelli's map with 200 metres of equidistance.

The second component of the GEOMED ARCHIVE is the altimetric data set. From the SEASAT mission 18780 altimeric observations on the Mediterranean sea have been obtained.

The GEOSAT ERM mission has supplied about 55862 altimetric data in the same area. These two data sets have been validated at the Ohio State University.

The data from earlier ERS1 mission have been analysed to detect gross errors. For this task the collinear track analysis, after subtraction of the OSU91A geopotential models, has been used.

All data relative to the SEASAT, GEOSAT and ERS1 missions have been processed with the crossover–adjustment method on a regional scale in order to minimize the radial orbital error. Simultaneously a linear regression has been applied to the altimetric signal against the geoid derived from OSU91A.

All these data sets have been used to build the geoid in Mediterranean area. In particular the marine geoid has been computed in five big patches which are strongly overlapping (fig. 1).

Without entering into details the method used has been the following:

a) remove from data the gravity anomaly contribution from global models (geopotential models) and the terrain correction calculated by DTM;

b) estimate the residual geoid by collocation. At this step only the marine geoid has been preserved;

c) finally restore the contribution to the geoid by the global model and the terrain correction.

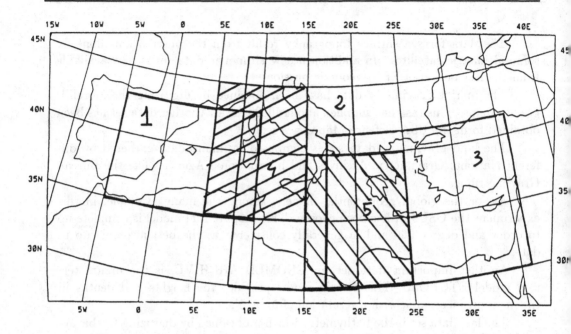

Fig.1 - The five computation areas.

The sea surface topography (SST) has been computed after tracks were adjusted in order to eliminate cross over residuals and to reduce the orbital error and the influence of tidal deformation of the sea surface. The SST is attached to OSU91A geoid in the average by the mentioned linear regression. The SST has been derived by estimating the geoid N at footprints of the altimetric observations and subtracting it from the later; it has been deemed that this procedure would mostly maintain the accuracy of the measurements. In particular this operation has required an interpolation of N from a large gridded data set to points irregularly scattered.

The relations existing between the altimetric observations and the other data are shown in figure 2. The ellipsoidal height h is known from satellite tracking, the orbital error can be modelled by a linear function. The orthometric height H is given by the radaraltimetric observation; N is the geoid undulation (height of the geoid over the ellipsoid); τ represents the time–variable sea surface; ζ represents the SST or the stationary sea surface.

Figure 3 shows the stationary sea surface topography in mediterranean area; the equidistance of the contour level is 0.2 meters.

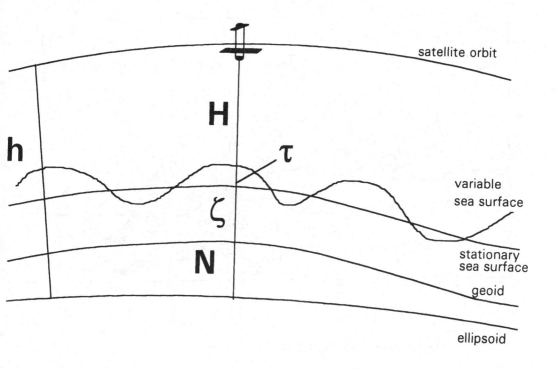

Fig.2 - The observational scheme of altimetry.

The new radaraltimetric data set

In the last year the ERS1 satellite has been put in an orbit with 167 days cycle. This gives a high spread of the data and a major controllability of the data itself. Thus we got a new very dense altimetric data set, particularly on the Mediterranean area (fig. 4).

The actual data set used for this experiment has about 70000 data supplied by KMS Copenhagen after geophysical and instrumental corrections have been applied. Such geophysical corrections are related to the physical effect of the ionosphere, troposphere and so on. The instrumental corrections are due to imperfection in the manufacture of the satellite, to the position of antenna, electronic circuitery etc. This altimetric data set does not contain repetitions. In particular this data set was not cleaned of the outliers so that a quick look processing was necessary to get rid of statistically undesiderable features. For this purpose each new track has been adjusted to Geomed geoid.

The single track adjustment is performed by a simple L^1 robust linear regression on the difference between geoid and altimetric data; as linear parameter the longitude can be coveniently taken instead of time for a small region. For each track

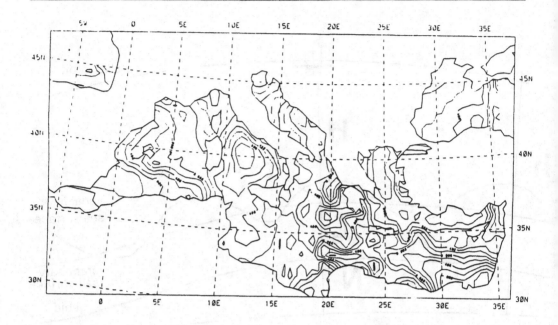

Fig.3 - The estimated stationary SST.

the possible presence of outliers is detected by a robust technique. The method is similar to that of classical linear regression but for the fact that the median is used instead of average. As a dispersion coefficient the median absolute value of the residuals is used instead of the root mean square.

The estimation of the coefficents of regression is perfomed in the following way:

a) compute the two median on the differece data (geoid–altimetry) and linear parameter longitude ($m(y) = median\ of\ y\ data$; $m(\lambda) = median\ of\ \lambda\ data$);

b) estimate the slope coefficent as the median of individual slope ratios ($\hat{a} = median\left[\frac{y_i - m(y)}{\lambda_i - m(\lambda)}\right]$);

c) the shift coefficent b is simply the difference between the median of y and median of longitude times the slope coefficent a ($\hat{b} = m(y) - \hat{a}m(\lambda)$).

The Geomed data set was given on a grid 5' by 5' and to calculate the difference between altimetric data and geomed data it is necessary to interpolate the Geomed geoid at the altimetric measurement points. To this purpose the GIS Grass has been used.

The rejection criterium is quite similar to the classical statistical criterium used in linear regression. For each track we calculated the absolute difference between the observations and their estimated value; after we calculated the median absolute value of these differences:

Fig.4 - Altimetric track pattern (ERS1 geodetic mission).

$$\nu = y_{obs} - \hat{y} \qquad mav = median(|\nu|).$$

When the absolute difference between observed and estimated value is greater than five times the median absolute value, the observation is rejected.

$$|y_{obs} - \hat{y}| \geq 5 \times mav \Longrightarrow Reject.$$

By applying this criterium, we detected about 2000 outliers over 70000 data. Had $y_{obs} - \hat{y}$ been normally distribuited, this criterium would have rejected about $1\%_{00}$ of good observations, so that the 2000 actually rejected values are most likely outliers. Probably the outliers in the Adriatic sea are due to the interruption of data stream on land. The figure 5 shows the outliers detected. Note that many outliers are near the coasts.

From the cleaned altimetric data we derive the new SST (fig. 6). The SST still contains noise and further investigations are necessary.

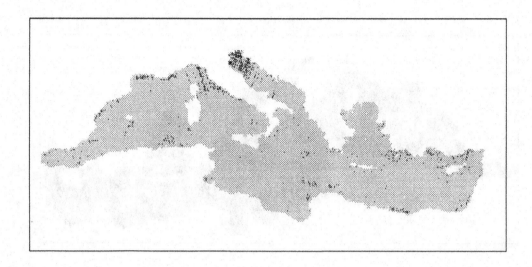

Fig.5 - Outliers detected after robust linear regression.

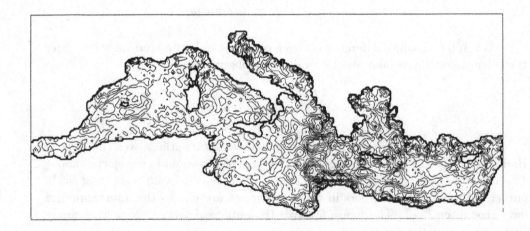

Fig.6 - The new stationary SST after a quick look.

The public domain GIS GRASS

Grass is a GIS public domain software developed by US ARMY Construction Engineering Research Laboratory and US Geological Survey. This GIS software is open for development and it is written in C language and runnnig under UNIX operating system. The source and some data sets are available in internet by ftp anonymous.

It is possible to build specific tasks for particular applications. this is done by following some simple rules when writing the source code. In addition we can read and modify the source for further improvement.

The fundamental capability are:

- general file management;
- window management (under Xwindow);
- data conversion (importing and exporting data between other GIS);
- interface to the other commercial and public domain packages, in particular with DATABASE MANAGEMENT SYSTEMS like Oracle, Postgres and so on, or graphics packages;
- it possible to digitaze maps;
- there are many commands for image processing like rectification. filter. map-projection and so on ...);
- the package for raster structures is very rich in commands which cover a wide range of problems. It is also possible to perform some typical tasks of the terrain analysis.

Discussion

The experiment presented in this paper was to use an ordinary GIS (GRASS) to:

- represent two large data sets, one of gridded (the geoid), the other one sparse (the altimetry);
- to provide a fast interpolation of the geoid on the altimetric points and analyze the residuals, after robust linear regression, to identify and reject the outliers;
- to use the remaining set of residuals as "noisy" sampling of the SST, which was to be reconstructed and visually displayed by smoothing and interpolated back onto the original grid;

in all those functions and others not reported here, like some involving the use of Fourier Transform, GRASS has proved to be a fast and reliable auxilium.

References

Arabelos D., Barzaghi R., Sansò F., Sona G. (1994). The gravimetric geoid and the SST in Eastern Mediterranean. MARE NOSTRUM, Geomed Report n.4, Dept. of Geodesy and Surveying, University of Thessaloniki, Thessaloniki, Greece.

Barzaghi R., Brovelli M.A., Sansò F., Tscherning C. C. (1992). Geoid computation in the Mediterranean area. MARE NOSTRUM, Geomed Report n.1, DIIAR - Politecnico di Milano, Italy.

Barzaghi R., Brovelli M.A., Sansò F. (1992). The gravimetric geoid and the sea surface topography in the central Mediterranean area. MARE NOSTRUM, Geomed Report n.3, DIIAR - Politecnico di Milano, Italy.

Barzaghi R., Sona G. (1994). The geoid and the steady circulation pattern in the Mediterranean sea. International Geoid Service, bulletin n.3, DIIAR - Politecnico di Milano, Italy.

Brovelli M.A., Sansò F. (1993). The GEOMED project: the state of art. Annali di Geofisica, vol. 36, n.5-6.

Shapiro M., Westervelt J., Gerdes D., Larson M., Brownfield K.R. (1993). GRASS4.1 Programmer's Manual. U.S. Army Construction Engineering Research Laboratory.

GEODETIC DATA IN A
MULTIPURPOSE REGIONAL GIS

F. Radicioni
University of Perugia, Perugia, Italy

A. Stoppini
University of Bologna, Bologna, Italy

ABSTRACT

In the present paper, the problems connected with the archiviation of
geodetic data (such as coordinates, heights, benchmark construction
details, images, ...) are examined, referring to the relational database archiving model. A special attention is given to the connection
of the geodetic database to a GIS support.
Some general characteristics and objectives which have to be reached
by a well-designed geodetic database are presented. Then, the different phases in planning and organizing such an archive are discussed.
Finally, as an example, the geodetic database realized by Perugia
University for the Cartographic Service of the italian Region Umbria
is presented.

1. INTRODUCTION

From approximately the mid-nineteen seventies, the Italian Regions have begun producing medium scale technical maps. Moreover, the extensive use of the GPS method in networks for monitoring ground movements by means of repeated measurements poses a series of problems regarding the correct, advantageous management of the vast amount of data received from satellites.

Each Region made choices, and set out priorities. In some cases, ortophotographic maps have been chosen. In other cases, real technical maps were produced: on a scale of 1:5000 in most cases, or 1:10000. But technical maps are also produced on 1:2000 scale, and sometimes cartography on small scales (1:25000, 1:50000, ...). The situations differ, also because of the varying funds made available for cartography by one Region or another.

There has however been some co-ordination, thanks to the Technical Norms established by the Italian Geodetic Commission, and the setting up of the Inter-regional Centre.

In any case, following this extremely relevant cartographic work, all of the Regions now possess a large quantity of geodetic data: all of the densification vertices and levelling benchmarks that are determined as control points for cartography.

On the whole, we are talking about a real wealth of geodetic data, whose usefulness is quite clear, especially if we consider the current state of national and cadastrial geodetic networks.

Sometimes the monumentation of these points leaves something to be desired. But regional vertices are generally easily reachable; their centres are often accessible for our instruments; many can be distinguished in aerial photographs... and they are certainly useful for a lot of applications.

In most cases, this data is not archived in an organised fashion, or it is at least not easy to consult. Often the filecards of the points are confined to the archives with the documentation of single cartographic works...

The usefulness of a system of computerised archives, a database of the regional geodetic elements, is clear to everyone; the Regions could then really make the information available.

Many italian Regions already possess a GIS (Geographic Information System). The levels of information vary from case to case: soil use, viability, the hydrographic network, the digital terrain model,... but, as far as we know, no Region has yet put the geodetic data into the GIS.

However, a geodetic database is just right for a GIS system: the geodetic data are automatically geo-referenced; moreover, the GIS provides on video a clear cartographic interface which is ideal for database consultation, directly visualising the position of the points on the territory.

The aim of this study is to establish some criteria to follow for the creation of a geodetic database, which can be easily integrated into a GIS such as the ones that the italian Regions possess.

2. THE SPECIFIC CHARACTERISTICS OF A GEODETIC DATABASE

Other than the general functions of any archive system (ordering and cataloguing data, conserving data safely for long periods of time, ease of updating and consultation, ...), a geodetic database must have a series of specific characteristics; for example:

- it must be able to easily handle monographic images (drawings, photographs, cartographic extracts,...) together with numerical data and text;
- it must have a clear and well-defined glossary for the various kinds of monumentation;
- it must have within it a series of procedures and rules for data validifying, which prevent double feed-in of information, ambiguity in the definition of points, etc., especially in the not infrequent case of points determined more than once in the course of time;
- it must be able to provide the coordinates of the archived points in diverse geodetic datums according to the case in hand: national system, cadastrial system, WGS 84 system, ... ;
- finally, it must have a security system for a selective access to the data, some of which is the property of institutions which sell such data (Istituto Geografico Militare, Catasto, ...).

Before showing with a concrete example how these requisites can be satisfied, it is advisable to remind ourselves of some of the basic concepts concerning the structure of a modern database system.

3. THE MAIN CHARACTERISTICS OF A DBMS (DATA BASE MANAGEMENT SYSTEM)

A DBMS is a system which can manage a group of data, offering a series of *logical* functions (relations between the data, automatic calculating routines, etc.) and *physical* functions (feed-in of new data, search, etc.).

These functions are carried out at three different levels:

- *External level*: the "view" which is offered to the user, who accesses the data without needing to know how it is structured inside the database;
- *Conceptual level*: the functions which define the logical structure of the database as a whole, and the relations between different data;
- *Internal level*: all the functions which allow one to physically operate on the archives, carry out elaboration or tranfers of data

inside the archive.

On the whole, the DBMS must manage the references between these three different levels. For this reason, there are various techniques, each one different according to the model of data on which the DBMS is based (hierarchical, networked, relational, etc.).

Of them all, the relational model (RDBMS) is the one which has been most successful. Among its main advantages is its extreme versatility and flexibility, which are particularly appreciable for an application like ours, in which the data belongs to so many different types, often variable in the course of time. Hierarchical and networked systems are however more rigid.

Another noteworthy advantage of the relational systems is the relevant economy of their occupation of the memory in the case of repeated data, which are archived once and are then recalled only on a visual level.

4. PLANNING A GEODETIC DATABASE

Today, anyone who has as an objective the setting up of an efficient archive on the database model, does not need to develop specific software "from square one". It is obviously preferable to use pre-existing software which has already been tried and tested. There are currently a number of excellent software packages to manage data, most of which are relational.

These systems already have efficient security procedures, which can safeguard the integrity of the data, limiting the risk of errors.

Operating in a field which at the outset guarantees a high procedural efficiency, the person who creates the archive can concentrate more on the specific characteristics of the data and on the performances to be obtained in order to benefit the future users of the system.

In the specific case of geodetic data, the organisation of the archive can follow these subsequent phases:

a) *Typological analysis of the available data*

First of all, the data to be archived has to be examined, evaluating the real information contents and sorting it out into categories. A scheme of cataloguing is already defined after this first phase.

b) *Construction of a logical scheme for the database*

This is the nucleus of the study. Here the "conceptual level" of the database must be defined.

If the system used is of the relational kind, in this phase it is necessary to define the details of the single archive tables (one for each homogeneous data category), and the relations between these. For each table a "principal key" must be identified so that it is unique. Moreover, the rules for the internal validation of the software would prevent the input of two data items with an equal value in the key field.

c) Definition of the procedures for the input of data

In this phase the external interface of the database is planned for the operators who will be feeding in data.

For numeric and alfanumeric data there are no particular difficulties. At present, even the management of iconographic data is easy, given that most of the sofware systems are able to deal easily with multimedial "objects".

We should not forget that it is strongly advisable to use image compression systems to reduce the memory space occupied. The best technique is to conserve the compressed images in archives outside the database itself, managed by a software package which acts as a "server" to visualize the images, for the "client" application constituted by the database.

In this phase the procedures to check the correctness of the input should also be carried out (for example, a check for double inputs or ambiguous ones). The activation of such procedures should be automatic. If any error condition are found, the operator will be warned by a timely video message.

d) Definition of the internal database routines

Here it is a question of implementing the necessary calculating procedures in the database (in our case, as we have seen, the routines for the transformation of data and coordinates) and the other routines and service macros (for example for the automatic calculation of the cartographic sheet in which a point falls, etc.).

e) Definition of the search and consultation procedures

In this phase the database interface for the external user should be organised, that is to say what we have previously called the "view"of the database.

The interface must be designed so that both at a logical level (implementing into the database all the necessary "queries" to carry out "intelligent" searches for data), and at a physical level (designing appropriate graphic "pages" on the video or in print).

f) Definition of the security procedures

When, as in our case, the archived data have a commercial value, it is necessary to install procedures that will guarantee a controlled access to the data.

Usually, the criterium followed is one of cryptographing the data and then giving access only to those users who have one or more passwords, or with an appropriate hardware protection package.

5. THE UMBRIA REGION GEODETIC DATABASE

As an example, let's look at how a database for the geodetic elements of the Umbria Region is being developed at Perugia University. The work is being carried out in collaboration with the Cartographic Service of the Region, and the University of Perugia.

The Region has got a GIS built up by CRUED (The Umbria Regional Centre for the Elaboration of Data). The system operates in UNIX environment on a local network of HP-A-1630 graphic stations. The background software used is the ITALCAD S-5000 system.

A quantity of varying information has already been archived in this GIS: completed cartography, soil use, viability and so on. Our objective is to add a new information "level" containing the geodetic data.

Cartographic activity in the Umbria Region began in 1977 with the production of an ortophotomap on a scale of 1:10000 covering all the territory. Subsequently, technical maps were begun to be produced to a scale of 5000, proceeding in small areas. Approximately the 50% of the territory has been covered by the technical map until today.

5.1 Typological Analysis of the available data

Cartographic production has "endowed" the Region with a large quantity of geodetic data. Let's look briefly at the typology:

- *IGM trigonometric vertices*

First of all, the Region has acquired the classic "monographs" drawn up by the Military Geographic Institute. Everyone knows their contents, so it is not necessary to insist further. Obviously, this data is the property of the IGM, and the access to it should be strictly controlled.

- *Trigonometrical Vertices of Densification (VTR)*

They are the points determined by the companies that have produced the cartography, densifying the IGM network with geodetic measures.

The companies provide the monographs, which are normally fairly complete. Generally, there is a planimetric sketch and a photo (almost always in colour). Often, only the cartographic plane coordinates are present, while the geographical coordinates are omitted.

- *Photographic Points (PAF)*

     These are points of an inferior order to the preceding ones,
determined through simple connections to the geodetic network. However they do have the characteristic of being visible in aerial photograms. Their monographs are very simple, generally containing only
a planimetric sketch, but the essential data is present.

- *Points determined in the aerotriangulation (TA)*

     Many of these points, whose coordinates are determined during
the aerotriangulation calculus, are "punctured" onto the photograms.
Therefore, they can't be found on the ground, and we are not interested in archiving them.
     But some of them correspond to permanent particulars, and for
these we have simple monographs, similar to those of the preceding
type. It is clear that these are not "geodetic data" in the real
sense, but they can be useful for instance to georeference photograms
or thematic maps, or for other purposes which do do not require high
precision.

- *IGM Levelling Benchmarks*

     Then there are, obviously, all the levelling marks, which are
well known; it is not necessary to go into them further. Let's just
remember that there is also the so called "old" IGM levelling, whose
marks, if still existing, can be useful in zones which are not covered by the "new" levelling.

- *IGM 95 Vertices*

     When the GPS IGM 95 network is completed, the Region will buy
all the relative monographs.
     Other than the traditional Gauss-Boaga coordinates, the mono-
graphs of IGM 95 points also contain the WGS 84 coordinates.

     So, we have quickly gone through the data typologies. Following
the logical scheme that has been already defined, let's now look at
how the database is organised.

5.2 The construction of the logical scheme of the database

     For the moment, we have built only a model of what will be the
definitive database. For this purpose, we have until now worked in
the software environment of Access 2.0, by Microsoft, one of the most
widespread and versatile relational systems.
     In this database model, into which we have put a sample of
data, we are checking the correct functioning of the logical schemes
and of all the routines. The subsequent phase will be to transfer
everything into the Region GIS, producing the definitive archive.

For each one of the categories of data that we have seen, we have predisposed an archive table, which can receive all the contents of the monograph, and any other accessory information.

The archive tables have the classic database structure, in which for each item there is a "field". Most of fields contain numeric or alphanumeric data (ccordinates, text descriptions, ...). In the "monographic sketch" and in the eventual "photographic image" field, there is the name and the path of the corresponding raster file.

The various tables forming the database are in relation to each other according to a simple scheme: a "general catalogue" table contains all the points which have been fed into the archive, with a unique code number for each point, and it is connected with all the singular tables.

Two other tables contain accessory information, like data about the companies involved and on the cartographic studies.

5.3 Definition of the procedures for the input of data

The input of the data has been organised predisposing a series of video filecards, one for each kind of data.

The filecards have the same fields as the tables seen before, and are prepared to receive data which the operator reads on the paper monographs.

As an example the filecard of the VTR (Densification vertices) can be shown (Fig.1). The texts are in Italian, their meaning is still obvious, looking at the values in fields.

The photo and the picture must be prepared before, in the form of raster graphic files; the operator only needs to indicate the name and the path of the files: the images are immediately visualised in their fields on the filecard, and it is therefore possible to check them.

When the input is finished, two routine searches can be activated by the buttons located at the filecard bottom on the right; the first checks if the point has already been fed in (to avoid double inputs); the second checks if there are points near it, within an established range (to avoid ambiguous inputs).

5.4 Definition of the internal database routines

The principle datum and coordinates transformation routines used in Italy have been fed into the database:

- From geographical to cartographical coordinates (using Gauss map projection) and viceversa; for this purpose we have used the direct and inverse formulas of Hirvonen; with this transformation, obviously, we can calculate the plane coordinates in the western zone when we only know the eastern ones. Umbria is in the overlap area between the two zones: the regional convention is to use the east coordinates, but it is a good thing to have them both;

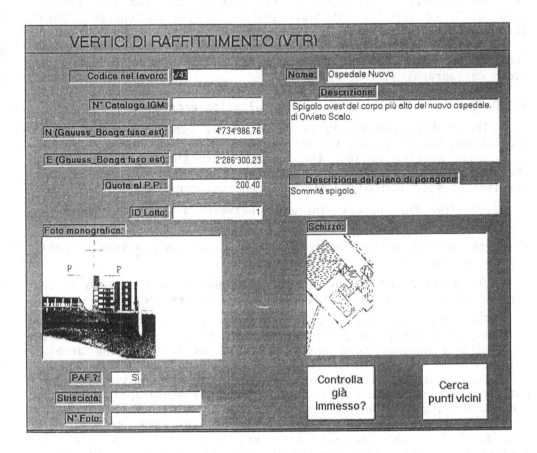

Fig.1 - Densification vertices input filecard

- From the Gauss-Boaga coordinates to the UTM coordinates; on purpose, we have used polynomial formulas of the 4th degree calculated by Bencini and used at IGM [6];

- From the Gauss-Boaga coordinates to the cadastrial coordinates; we have used a procedure described by Antongiovanni [5], based again on polynomial expressions computed at IGM, that transform geographic coordinates on the Bessel/Genua 1902 ellipsoid into geographical coordinates on the International/Rome 1940 ellipsoid;

- From the coordinates in the national system, and orthometric heights, to the WGS 84 coordinates, geocentric or ellipsoidal; we have used Helmert's 7-parameter transformation.

Inside the database there are also other routines, for example for the automatic calculation of the numeric code for the "General Catalogue", or for the automatic search for the cartographic sheet into which the point falls.

This latter operation will no longer be necessary when we have fed the database into the GIS; we will be able to do it using the video cartographical interface.

5.5 Definition of the search and consultation procedures

The archive can easily be consulted by scrolling the video of the tables, but it is not the best way.

While waiting to feed the database into the Region GIS, where the cartographical interface will make everything easier, we have predisposed some automatic search procedures; for example:

- search for the points included between a 1:50.000 sheet, a 1:10.000 element, 1:5.000, etc.;

- search for the points included around a certain planimetric position, within a distance definable by the user.

The point found by the search query appear on the video in the form of tables which can be saved in a file or printed.

For each point the so called "report" can be obtained, either on video or printed. In practice, it is a classic monographic file-card with a graphic form which is the same for all the archive points.

Thanks to the routines implemented into the database, the output reports contain more data than the original monographic filecards from which data have been taken.

For example, in the densification vertices (VTR) report there are, apart from the Gauss-Boaga eastern zone coordinates, also the western zone ones, the geographical ones, the UTM ones, the cadastrial ones, the WGS 84, etc. To distinguish these calculated data from the original ones, we have indicated them on the report in italics, and with a different colour.

The reports in this form can be consulted on the screen or printed. It is a multimedial system which is without doubt very user friendly.

The definitive system will bring notable benefits from the cartographical interface of the GIS. We will be able to have, on the screen, overviews of the points in different scales; and we will be able to zoom in, search in individual Communes, etc. This is what we propose to do in order to complete the work.

REFERENCES

1. Chen, P.S.: The entity relationship model: toward a unified view of data. ACM TODS (1977)
2. Cood, E.F.: A relational mode of data for large shared data banks. Comunication of the ACM 13, n.6 (1970).
3. Date, C.J.: An Introduction to Database System. Vol. 1 e 2, 4th Edition. Addison-Wesley (1985).
4. Bitelli, G., Gatti, M. and Russo, P.: An information system for levelling data in land deformation surveys. Proc. of the 7th FIG Symposium on Deformation Measurements, Banff, Canada (1993).
5. Antongiovanni, R.: Procedimento rigoroso per la trasformazione delle coordinate catastali in coordinate Gauss-Boaga e viceversa, risolto con la calcolatrice TI 59. Bollettino ASIT, n.12 (1979).
6. Stoppini, A. and Surace, L.: Formule approssimate per la trasformazione diretta e inversa tra sistema UTM (ED 50) e sistema nazionale (Gauss-Boaga). Bollettino di Geodesia e Scienze Affini, n.3 (1991).

THE ETNEAN GEOMORPHOLOGICAL AND GEOPHYSICAL DATA BASE

V. Barrile

University of Reggio Calabria, Reggio Calabria, Italy

R. Barzaghi

Polytechnic of Milan, Milan, Italy

ABSTRACT

In this paper an overview of the different data collected in the Mount Etna area is presented.
Clinometric observations, geodetic trilateration, gravity measurements and spirit leveling have
been performed for a long period, so that comparisons and integrated results can be derived.
These classical geodetic data can be also linked to seismic and geological investigations to obtain
a reliable risk evaluation model suitable for preventing from catastrophic events.
Furthermore, GPS observations carried out during the period 1990-1992 have been deeply
analyzed to define their effectiveness in detecting the global displacement features of the volcano.
Provisional conclusions of this investigation are derived and some comments on the method and
the present network configuration are given.

1. INTRODUCTION

The Etna volcano can be considered as a scientific research laboratory [16]. Its peculiar location with respect to the geodynamic context of the area, its frequent eruption events and the crustal movements connected to such events [1] led to an intense research activity aiming at defining a reliable physical model of the volcano. Furthermore, especially on the south and on the east sides, the Etna volcano area is intesely urbanized, with a relevant number of economical acivities, such as farms, factories and so on. So, the increasing interest of the Italian and international scientific community is justified   by the research aspect in itself and by the social and economical implications, these two aspects being connected in defining a risk evaluation model suitable for preventig from catastrophic events.

Hence, a global kinematical and dynamical analysis of the area under investigation would be of extreme interest. However, up to now, for many different reasons, the various scientific investigations which have been carried out on the volcano didn't result in a reliable physical model, since they have never been organically rearranged. This work aims at presenting, in a concise form, the principal research activities usually carried out on the Etna volcano, in order to collect an organic set of information which can be used for further analyses and as a basis to construct the aforementioned model. Particular care is devoted to the GPS technique which proved to be successful in describing the global features of the dynamical evolution of the Etnean area.

2. DIFFERENT INVESTIGATION METHODS APPLIED IN THE MOUNT ETNA AREA: AN OVERVIEW

Observations carried out to monitorize long term land deformations in the neighbourhood of the volcano have been collected since 1975. They were perfomed using the *clinometric technique* and *geodetic trilaterations*. The first kind of information are of continuous type in time, while the second are periodically distibuted, being collected once per year, and describe the planar deformations pattern, since horizontal reduced distances are used in the processing.

The clinometric network configuration was changed many times so that a comparative analysis among different periods is quite difficult. At the moment, there are two distinct nets [6]. The first net (Fig. 1) was established in 1975 and evolved up to the actual configuration which consists of 9 bore-hole type tiltmeters placed around the volcano sides. Observations coming from this net are simultaneously collected at a master station of the International Institute of Volcanology (IIV) of Catania, in order to have a real time monitoring. Furthermore, data are also locally recorded to be able to recover them in case of malfunctioning of the transmission system. The second net (Fig. 2), established in 1987, has observation stations between 800 m and 2950 m above sea level; most of the stations are located between 1400 m and 1850 m above sea level. In each station, tilts are observed both along the volcanic cone generatrix (radial direction) and perpendicularly to this line (tangential direction). The tiltmeters which are used in the two networks have the same precision, which is of the order of 0.1μrad.

The geodetic trilateration networks, covering at the beginning only small portions of the area under investigation, have been subsequently enlarged to ensure a better coverage of relevant volcanic activity areas [5] [9]. There are three trilateration networks placed on the south-western (SW), the southern (S) and the north-eastern (NE) sides of the volcano (Fig. 3). The net benchmarks (14 to 16 for each net) are at an altitude varying from 600m and 2800m above sea level; in each net, 46 to 47 distances are observed once per year, using an AGA EDM instrument, having st. dev. of 5 mm + 1 ppm.

From the mid seventies, altimetric investigation were also carried out, in order to improve the global description of the long term land variation of the volcano area [11]. *Spirit leveling measurements* were firstly performed by the English researchers Guest and Murray; nowadays, the spirit leveling network is managed by the Vesuvian Observatory (Naples) which started systematic campaigns from 1980, performing them periodically. The leveling Etnean network consists of nearly 100 benchmarks, from 1400 m to 1900m above sea level, mainly distributed along the road which leads to the volcano top (Fig. 4).

Seismological investigations have been mainly devoted to analyse the origin and the features of the microseismicity of the Etna Mount and to find anomalous areas in seismic waves propagation. Many different reaserch teams have worked on this particular investigation topic during the last years. One of the most remarkable effort in this field has been perfomed by the IIV which, form 1989, set up a new seismic monitoring system [14]. The main objectives of this system are: real time data acquisition and analysis; crosscorrelation between seismic and eruption phenomena to possibly define a model which allows eruption forecasts by means of seismic signal analysis; investigations related to seismic waves sources in volcanic areas. The IIV seismic net consists of 11 recording stations (equipped with Mark L4-c seismometer) and its geometry has been designed taking into account the seismicity features of the area under investigation. The observed data are collected at the IIV center in Catania via UHF signal transmission or via cable connection. Here seismic information are analyzed and compared to monitorize, through some relevant parameters, the volcano activities. This kind of investigations led to significant conclusions, since they allow to detect anomalous seismic bodies laying along a north-east/ south-west axis, in agreement with the main regional tectonical structures.

Gravity observations were also perfomed on the Etna Mount and in the surroundings [7]; they have been done to decribe long term variations of the gravity field which relates to mass reallocation phenomena, connected to the magma dynamic and to its chemical-physical variations. The microgarvimetric network, set up in 1986, has been slightly modified in the last years, to better decribe peculiar gravity anomalies which are present in some areas. Its 34 stations are placed along the same road used for the leveling measurements, from 1000 m to 1900 m above sea level (Fig. 5). La Coste & Romberg gravimeters are used to acquire the gravity data, which are observed along closed and chained loops. The main outcome of this kind of investigation has been the identification of one relevant gravimetric source which has a stable geometry in time and has its center of mass located in the southern part of the gravimetric network.

Parallel to the aforementioned methods, many *volcanologic* and *stratigraphical* researches have been performed in the Etneana area. Among the others, one can mention *spectrometric* analysis [4] (spettrometry of the SO flux on active volcanoes aims at identifying sudden variations of SO which are related to canghes in the eruptive activity) *petrogenetic* and *petrochemical* researches [8] (the chemical structure of the eruptive masses is an important information which allows a proper evaluation of the feeding system of the volcano) and, last but not least, *geological modelling* of the volcanic zone [12] [10].

All these investigations led to some partial but very interesting conclusions, particularly for what concerns the Etna Mount dynamic during the period 1990-1992.

From the planimetric point of view, one can observe no relevant deformation of the volcano structure (apart from small areal dilatations in the SW and NE zones) in the prior eruption period, up to 1991. On the contrary, during the eruption, a remarkable contraction took place which has been revealed by the observations carried out on the geodetic networks (the main deformations were detected in the SE and in the NE networks while in the S network no significant movements were observed). Spirit leveling proved that an uplift occoured in the south and north-east volcano areas, while a relevant drawdown was observed in the south-west. Clinometric data lead to conclusions which are in agreements with geodetic data; the tiltmeter signals coming from the south-west side have a decreasing trend while those from the south show an increasing trend.

Such a simple example proves the potentiality of an integrated analysis in order to better define the volcano eruptive dynamic. In view of that, the GPS positioning method can be used successfully to link together the geodetic observations. Up to now, the local geodetic networks were designed mainly to detect local interesting features of some areas, but a connection between them was not considered, either because it is time expensive and because it is intrinsecally problematic. The GPS technique can overcame the previously mentioned difficulties, allowing an interlinking between the local geodetic nets, and can provide also a global description of the Etna Mount. In the next paragraph, we shall describe in more details the GPS network , the data processing and the main results related to the 1990-1992 period.

3. THE ETNEAN GPS NETWORK

From 1988, GPS measurement campaigns are carried out to control the global movements of the volcano area [13]. Starting from the experimental GPS network established in 1988, the IIV researchers have continuously improved its design and acquired new data and experience on GPS technique; this led to the implemented GPS net which is nowadays used to monitor the entire Etnean area. In this paragraph, we want to analyse in more details the data and the processing results over the period 1990-1992 to prove the reliability and the effectiveness of the GPS method in describing the land deformations occurred in the Etna Mount during the eruption event of 1991.

The three GPS network measured in 1990, 1991, 1992 are shown in Fig. 6, Fig. 7 and Fig. 8. As one can see the network configuration varied from year to year and only a subset of points has been used in the three campaigns; this has caused some problems in the interpretation phase which will be discussed later.

Data have been acquired using four receivers: two L1 monofrequency Trimble 4000SL and two L1-L2 double frequency Trimble 4000STD (only 3 receivers has been used in the 1991 campaign).

Data have been observed according to the following scheme:

- 10 sessions for an average period of 3 hours (4 stations for each session) in the 1990 campaign;

- 8 sessions for an averge period of 1,5 hours (3 stations for each session) in the 1991 campaign;

- 15 sessions for an average period of 2 hours (4 stations for each session) in the 1992 campaign.

The large amount of acquired data has been firstly validated and then methodological studies have been carried out [2].

The phase measurements have been processed by means of the TRIMVEC program [15].

The same program has been used in multibaseline mode (both broadcast and precise ephemeridis), for each session, following four steps:

- pseudorange for each station;

- imposition of the "reference station" coordinates;

- L1 frequency analysis (Triple differences, cycle slips, double differences);

- L2 frequency analysis (Triple differences, cycle slips, double differences);

- Iono-Free (I-F) analysis (Triple differences, double differences).

The aim of these tests was to study the repeteability of the observations in connection to different processing methods which have been applied to the data. Considering the statistics of Tab 1. and Tab2., the following conclusions can be oulined:

- baseline differences computed year by year with precise and broadcast ephemeridis are of the order of 7-8 mm and don't show any sistematic effect, being M_Δ (the mean value of the differences) nearly zero,
- a satisfactory inner repeteability of the three networks is achieved if precise ephemeridis are used;
- the inner repeteability improves, as expected, if precise ephemeridis are considered instead of broadcast ephemeridis;
- a worse repeteability of the 1992 network is revealed in comparison with the 1990 and 1991 networks;

- the comparison between L1 solution and L1-L2 iono-free solution show the influence of the ionospheric effect on the baselines, both with precise and broadcast ephemeridis; this effect is sistematically negative and it is more relevant on long baselines.

	year	σ (m)	M_Δ (m)
Precise ephemeridis	1990	0.007	
	1991	0.009	
	1992	0.019	
Broadcast ephemeridis	1990	0.013	
	1991	0.014	
	1992	0.025	
Differences between precise and broadcast	1990	0.007	0.008
	1991	0.005	0.005
	1992	0.007	0.006

Tab. 1: Adjustments of GPS networks with precise and broadcast ephemeridis

| Adjustment type | 1990 network | 1992 network | | | | |
	RMS	RDOP	% of obs.	RMS	RDOP	% of obs.
L1 precise	0.021		90	0.033		90
L1 broadcast	0.022		90	0.033		87
L1-L2 precise	0.027	0.172	73	0.035	0.315	76
L1-L2 broadcast	0.029	0.172	77	0.032	0.312	76

Tab. 2: Comparison between 1990 and 1992 adjustment parameters
(the RDOP values for the L1 solutions have not been computed)

Furthermore, it must be noted that the percentage of observations used in the double frequency computations is sistematically less than the corresponding percentage in single frequency mode. This effect is probably due to a noisy L2 signal which, in our computations, is obtained by quadrature, since no P code was available. Another problem which occurred in the computations is the following: although, in the L1-L2 analysis, in the average, the number of observations used in the processing with broadcast and with precise ephemeridis is nearly the same (see Tab. 2), one can observe on some baselines strong differences in the percentage of used observations when considering precise ephemeridis instead of the broadcast, even if, for these baselines, the statistical indexes were the same (RMS, RDOP). The same effect, on the contrary, is not present when using L1 phase measurements only ; this seems to be again connected to the L2 noisy signal.

From Tab. 2 one can also see that the RMS, which is nearly the same using the two different sets of ephemeridis, is equivalent or smaller in the L1 analysis than in the iono-free solution. The RDOP value is also identical in the double frequency computations while using the two sets of ephemeridis. One should note further that the RDOP is overestimated in those sessions having a large number of skipped observations.

On the basis of the aforementioned analysis, some critical sessions have been selected and a more detailed analysis has been carried out on them; critical sessions leading to unreliable results in terms of control parameters have been skipped.

The subsequent compensation have been done year by year and led to results which allowed to investigate the spatial-temporal variations occurred in the volcano area.

We must firts of all underline that these results are not completely reliable, since the three considered GPS networks changed in time.

In view of the particular geodynamic situation of the area, the ideal condition would have been that of a time stable network, with observations carried out with the same instruments, the same procedures and during the same period in the three years. Furthermore, the network scheme should contain points which are not involved in the areal deformations which are under investigation; in such a case, reliable estimates of the displacements can be obtained if compensations are carried out with respect to the baricenter of the fixed points.

As mentioned before, the real situation is far from the optimal one and so only provisional conclusions can be traced.

The main results have been obtained in the central part of the area under investigation, i.e. the one coinciding with the top of the volcano which contains a common subset of points (this is mainly due to the fact that the 1991 network was bounded to the volcano top).

The TRIMNET [15] program has been used to adjust the GPS networks in order to:

- evaluate the consistency and the precision of the three different nets;
- define possible external reference points, which are not directly involved in the analysed deformations;
- estimate the possible displacements.

Since the TRIMNET program doesn't allow to fix the baricenter of a subset of network points and doesn't give the covariance matrix of the estimates, we computed the estimates on the basis of hypotheses and procedures which are suggested by intuitive considerations on the problem.

We have performed the following computations:

i) a free net adjustment for each year, to evaluate their inner precision of each net and the observation consistency year by year;
ii) the adjustemts of the 1990 and 1992 networks with respect to a fixed point (the point OBS in Fig. 6) to define, by inspection of the estimated displacements, external reference points;
iii) the adjustments of the 1990 and 1992 networks with respect to the fixed external points in order to evaluate the deformations at the volcano top.

The comparison between the 1991 and the 1990 networks doesn't show any systematic displacement of the "internal" points (those on the volcano top) either considering the 1990 free net adjustment or the 1990 net adjustment with respect to the OBS fixed point (see Fig. 9 and Fig. 10).

On the contrary, the comparison between the 1992 and the 1990 networks, either in the free net or in the fixed point adjustment, shows remarkable displacements (see Fig. 11 and Fig. 12), much more relevant than those obtained between the 1991 and the 1990 nets. Furthermore, a common displacement of the three external points BEL, CRP, MGN and of the STP point is clearly shown in the plot of the differences 1992-1990 (Fig. 11 and Fig. 12); one can also observe that this common displacement cannot be traced in the remaining points (some of them show displacements which have nearly this direction but the amplitudes differ).

Then, the decision of adjusting the net considering the three external points (BEL, CRP, MGN) as fixed (to analize the structure of the displacements in the top volcano area) can be justified, at least from an intuitive point of view (a more rigorous analysis on the displacements [3], confirmed such a choice). What we assume is that the three mentioned points have a common displacement which is independent from that of the top part of the volcano (point STP has indeed the same behaviour but it has not been fixed since it is in the area under investigation).

The plot of the results (Fig. 13) shows that in the top volcano area a remarkable contraction took place between 1990 and 1992 which is likely to be related to the eruptive activity; a global displacement of the whole volcano can be also seen, even though not so relevant. Furthermore, a more detailed analysis should be done on point TDF which displays an anomalous displacement as compared to the other points of the top area (this anomalous behaviour can be justified considering that the TDF point is close to an highly eruptive area).

4. CONCLUSIONS

The data acquired on the Mount Etna volcanic system have been globally analyzed and revised in order to define possible way of integrating different data sources to have a better insight on the volcano movements. In particular, the GPS campaigns which have been held in 1990, 1991, 1992 have been described and preliminary adjustment results were illustrated. This analysis showed that the GPS observation method can provide valuable information of global type about the volcano, which can be integrated with other data and can give a framework in which different data sets can be included. The adjusted data led to a displacement pattern which agrees with geological and geophysical results on the same area, showing that the volcanic activity is related to a common movements of the controlled points.

These results proved the effectiveness of the GPS method, even if applied to a data set which is not so satisfactory, i.e. with network configurations changing year by year. Hence, more valuable results can be expected in the future, when a stable network configuration and a standard observation procedure will be adopted.

REFERENCES

1. Barberi F., Carapezza M.L., Valenza M., Villari L.:
The control of lava flow during the 1991-1992 eruption of Mt. Etna
Journal of Volcanology and Geothermal Research , 1993

2. Barrile V.
Studi per il controllo delle deformazioni crostali nell'area Etnea
Tesi di Dottorato di Ricerca in Scienze Geodetiche e Topografiche
VII ciclo (1991-1994)

3. Barrile V., Crespi M.
Compensazioni ed analisi di rilievi GPS per il controllo di deformazioni nell'area etnea.
To be published on SIFET Bulletin

4. Bruno N., Caltabiano T., Romano R.
Misure COSPEC del flusso di SO2 dal vulcano Etna.
Rapporto Attivita' Etna 1988, CNR-IIV, Catania

5. Bonaccorso A., Velardita R., Villari. L.,
Ground deformation modelling of geodynamic activity associated with the 1991-1993 Etna eruption
Acta Vulcanologica, vol. 4, 1994

6. Briole P., Fortuna L., Nunnari G., Puglisi G.
Misure in continuo del tilt sull'etna: primi tre anni di attivita'
Bollettino GNV, 1989-1,107-125

7. Budetta G., Grimaldi M., Luongo G.
Variazioni temporali di gravita' nell'area Etnea
Bollettino GNV, 1989-1,137-146

8. Calvari S., Coltelli M., Pompilio M., Bruno N., Cara' M., Messina L., Scribano V., Turchio F.
I prodotti emessi dai crateri sommitali.
Rapporto Attivita' Etna 1988, CNR-IIV, Catania

9. Falzone G., Puglisi B., Puglisi G., Velardita R., Villari L.
Componente orizzontale delle deformazioni lente del suolo nell'aera del vulcano Etna.
Bollettino GNV, 1988,IV.

10. Lo Giudice E., Rasa' R.,
Very shallow earthquakes and brittle deformation in active volcanicareas. The etnean region as an example
Tectonophysics, 202 (1992) 257-268
Elsevier Scienze publishers B.V. Amsterdam

11. Luongo G., del Gaudio C., Obrizzo F., Ricvco C.
Movimenti lenti del suolo all'Etna mediante livellazioni di precisione
Bollettino GNV, 1989-1,345-361

12. Neri M., Garduno V.H, Pasquare' G., Rasa' R.
Studio strutturale e modello cinematico della Valle dl Bove e del settore nord- orientale etneo
Acta Vulcanologica, vol. 1, 1991

13. Nunnari G., Puglisi G.:
The Global Positioning System as a useful Technique for Studyng Ground Deformation On Mt. Etna
Journal of Volcanology and Geothermal Research (in corso di stampa)

14. Privitera E., Montalto A., Neri G., Patane' D., Pellegrino A., Scarpa R., SpampinatoS., Torrisi O.
Il controllo strumentale dell'attivita' sismica dell'Etna: La rete permanente IIV
Bollettino GNV, 1989-1,491-508

15. Trimvec-plus GPS Survey Software
User's Manual and Technical Reference guide
Trimble Navigation 1990

16. Villari L.
L'Etna
Le Scienze, n°4, 1983

Fig.1   Clinometric network (1) + EDM networks

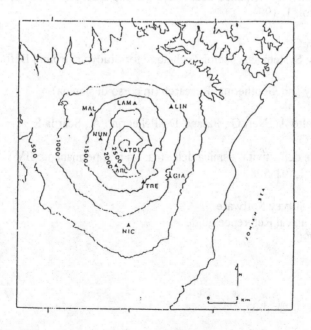

Fig.2   Clinometric network (2)

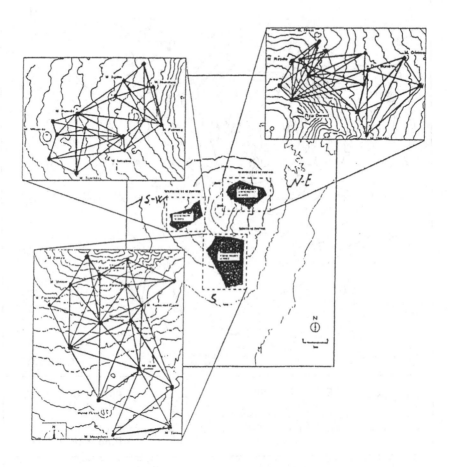

Fig.3   EDM Networks (N-E, S, S-W)

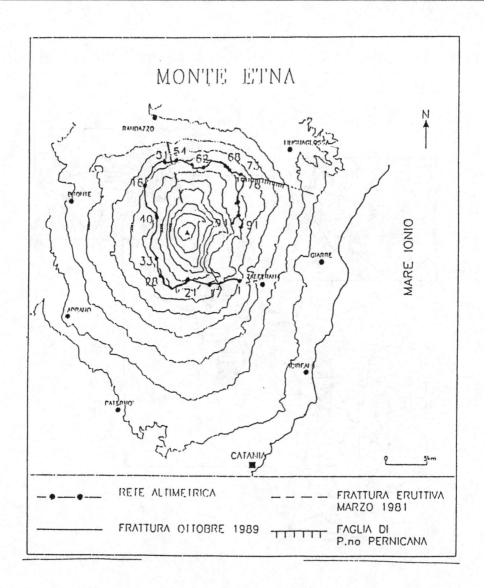

Fig.4   Altimetric network

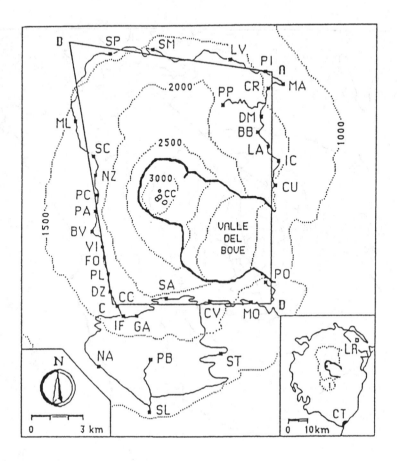

Fig.5   Microgravity network

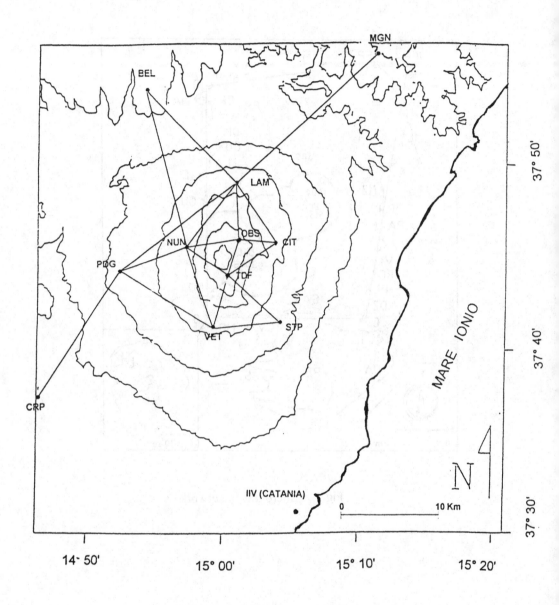

Fig.6   1990 GPS network

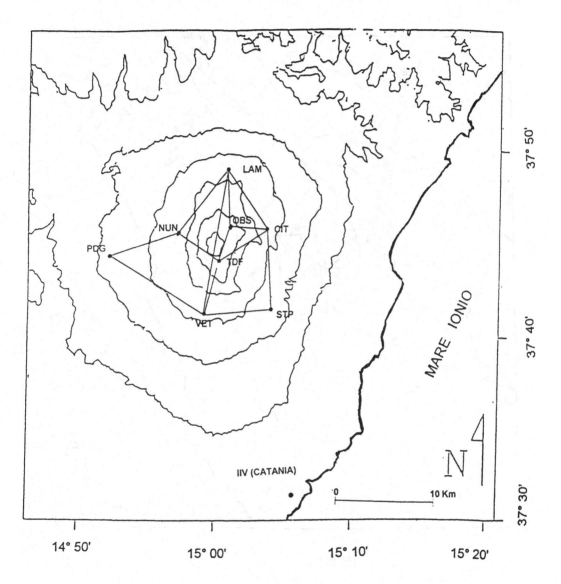

Fig.7   1991 GPS network

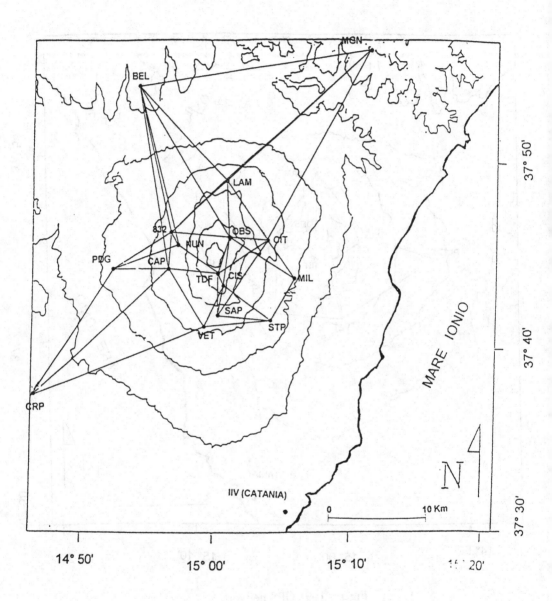

Fig. 8   1992 GPS network

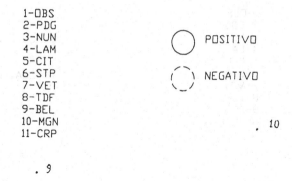

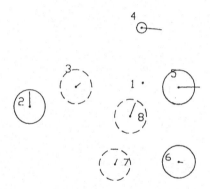

Fig.9 Δx 1991-1990 (free - net)

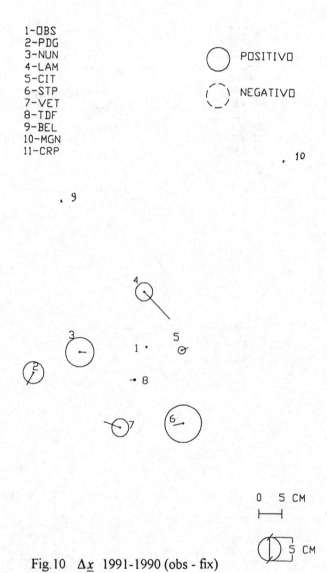

Fig.10 $\Delta \underline{x}$ 1991-1990 (obs - fix)

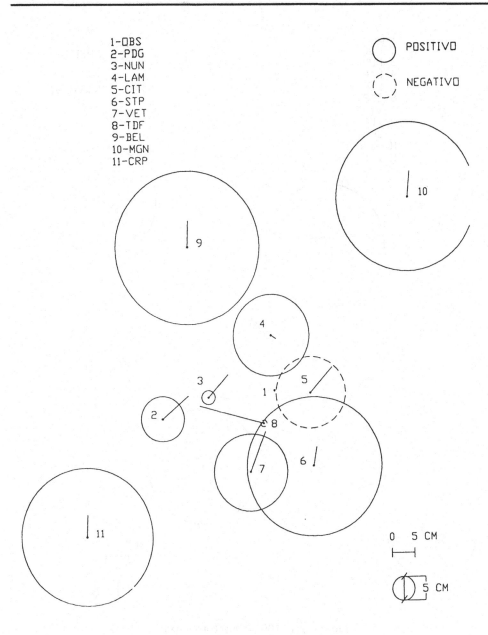

Fig.11 $\Delta \underline{x}$ 1992-1990 (free-net)

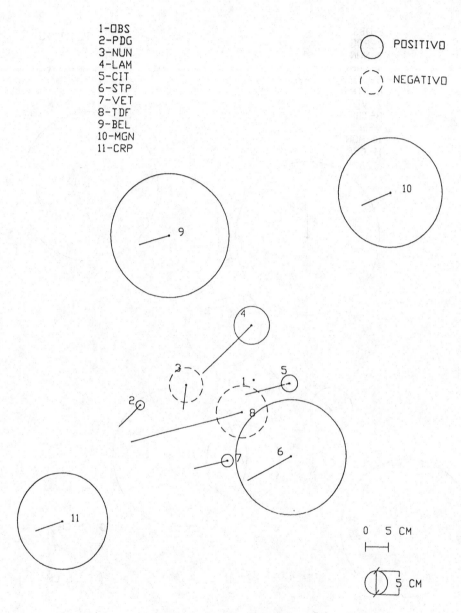

Fig.12 $\Delta \underline{x}$ 1992-1990 (obs - fix)

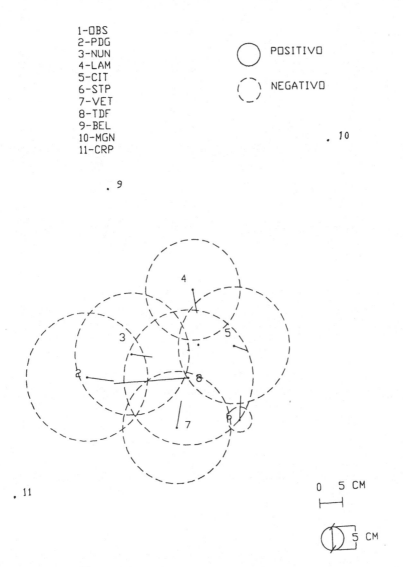

1-OBS
2-PDG
3-NUN
4-LAM
5-CIT
6-STP
7-VET
8-TDF
9-BEL
10-MGN
11-CRP

○ POSITIVO

◌ NEGATIVO

Fig.13 $\Delta \underline{x}$ 1990-1992 (3 fix point)

THE USE OF GPS DATABASE IN
CONTROL NETWORKS

D. Dominici

University of Bologna, Bologna, Italy

F. Radicioni

University of Perugia, Perugia, Italy

A. Stoppini and M. Unguendoli

University of Bologna, Bologna, Italy

ABSTRACT

Working with GPS, a large amount of data is often produced. In complex works, like the control networks, where many stations are involved, measurements are repeated more and more in time, and post-processing is performed with different softwares and/or options, in order to optimize the solutions; it is therefore very important to carefully archive all data, avoiding any possible mix-up. In practice, this is often not a simple job.
In present work, some general concepts are illustrated, which have to be followed archiving GPS data. Some typical problems are examined, referring to practical examples.

1. INTRODUCTION

The ultimate objective of a GPS survey for a control network is the accurate positioning of points on the surface of the earth. In order to obtain vertical as well as horizontal control points from the GPS survey, accurate data are needed to describe the area and the operation time in which the survey has took place.

Moreover, the extensive use of the GPS method in networks for monitoring ground movements by means of repeated measurements poses a series of problems regarding the correct, advantageous management of the vast amount of data received from satellites.

Without a suitable management, the said data gradually loses the potential inherent to the method, that is to say, the possibility of re-processing with new software, the use of more sophisticated atmospheric models, and the possibility of realistic comparisons based not only on the variance and covariance matrices, but also on the record of the number of satellites and the shape of the constellation in the various measurement periods, the length of the session and a series of details (cycle slip, ambiguity fixing, etc.) which provide a complete picture with regard to the reliability of one or other of the measurements.

For example,in 1990 the University of Bologna built a monitoring GPS network (Fig.1).

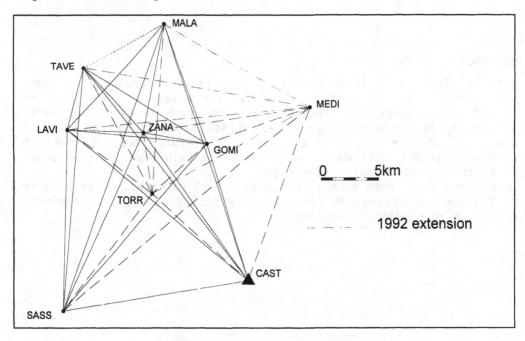

Fig.1 - Bologna 1990-92 network

In the first campaign a network of five control points in areas with great subsidence and two points near the reference benchmarks of the levelling network were surveyed. Two other points were introduced in 1992; the former located on the tower of the Engineering Faculty, in a stable area, and the latter located near the CNR Radio Astronomy Station,in Medicina, about 30 km away from Bologna. In the latter station VLBI, Doppler and Laser Ranging Systems are operating, so the WGS84 coordinates of the GPS point are known with a high accuracy: these values have therefore been utilised as origins of the coordinate system.

In the first time we used the software TRIMVEC, but in the last 5 years, as well known, there has been a great development of the GPS post-processing and network adjustment software. For instance, there is now a full utilization of computer graphics, more efficient algorithms for ambiguity determination, and so on; perhaps the new generation of software might have some other innovative components.

So, in the next survey of the control network we shall have new resolution strategies in the data processing, we shall be able to assume high accuracy orbits (better than 0.5 m) given by IGS service and use more accurate regional ionosphere models. For this reason it's impossible to make a simple comparison between the old results and the new ones, but it's necessary a complete re-elaboration of all GPS raw data.

In other words, a tool must be constructed, which will be the "heart" of the entire GPS data use and application chain; a module which allows the connections between the moments of data acquisition, data processing and the interpretation of results to be created and maintained.

This is mostly important when, in large-scale measurement campaigns, such stages are completed at different times within a single organisation, or even by separate organisations, each stage being entrusted to specialists in the relevant fields, often unacquainted with one another, therefore giving rise to difficulties in the simple, combined management of the data flow.

For this reason, amongst others, a great deal of research has been carried out in order to provide a simple solution to the problems described; simple, yet combined with the complete and correct nature of the data to be saved and used in a variety of ways.

We shall not describe the said research, but indicate several specific features of a GPS data base and provide some examples of how such a system can be of use in the repeated measurements sector.

2. CRITERIA FOR THE ORGANIZATION OF A GPS DATA BASE

The following is a description of the data useful for the correct management of repeated measurement campaigns for the study of movements and deformations, allowing the definition of the conceptual data model for a GPS data base.

The first outstanding feature is the need for unitary management in a format which allows the processing of data arriving from different instruments and with scientific programs.The RINEX format appears to satisfy quite well these requirements.

As well as the original observations (data files), in order to re-process data with new software, the storage of an entire series of parameters is required, allowing the assessment, for example, of the need for such re-processing, or the possibility of directly using those results obtained from previous surveys, as in traditional surveys. A set of data relative to the organisations which have commissioned and those which have effected the measurements must also be kept, as well as for the points of the network (monographs).

Summarised as far as possible, the data "entities" contained in a GPS data base may be classified as follows:

a) *Surveys*

Measurement campaigns effected over a given area, on a given number of points, in a given period, with certain aims.

b) *Organisations*

. Institutions which have commissioned and institutions which have effected one or more surveys.

c) *GPS Stations*

Points for the positioning of a GPS antenna. Description of the point and relative problems (three-dimensional point with autocentring, use of the tripod, etc.), state of wear, properties, monograph and 360˚ constellation visibility diagram.

d) *Set of observations*

A set of data which describes the observation effected on a single station, e.g.: observation start and end time, antenna height, sampling time, name of the operator, observables measured. May also contain meteorological files.

e) *Processing and adjustment*

Notions relative to algorithms for the processing and compensation of a survey. Must contain not only the measuring points and sessions involved, but also information on the program used, the atmospheric models and the reference system, as well as partial baseline results with relative error parameters.

f) *Adjusted positions*

File of adjusted co-ordinates with relative variance-covariance matrices and reference system.

g) *Datum transformations*

Geodetic datum and co-ordinate transformation: results and transformation parameters sets with relative variance-covariance matrices.

The model shown in Fig.2 provides an example of the relations between the data entities in a GPS data base and their complexity.

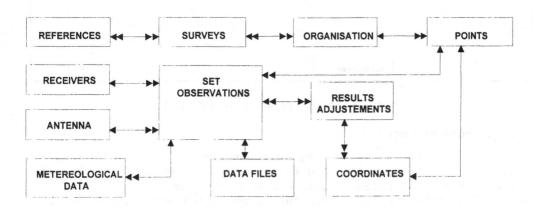

Fig.2 - Relational data model   (Bitelli, Vittuari, 1994)

It is, therefore, evident that, once created, a data base offers various possibilities for the correct use of GPS data, particularly for repeated networks.

3. SOME EXAMPLES

The "Tyrgeonet" project may be taken as an example, for which numerous measurement campaigns exist, effected using different instruments, in sessions of varying length, involving the re-occupation of all or part of the points with the addition of new points from year to year, and the inclusion of small local networks in the general network. It is easy to see how difficult, if not impossible, the management of such a vast quantity of data becomes without a simple, functional data base.

Moving on to tangible examples, the above mentioned GPS network of Bologna, realised in order to monitor subsidence, and a simple GPS network in Perugia (Fig.3) have been examined. Our aim is a global approach to improve the results of the surveys.

With all raw data we have attempted to test different methods of treatment. The first step has been processing data with different softwares, in the second test we have used different ephemerides (precise - broadcast), and finally changed the position of the fiducial (fixed) point of the network.

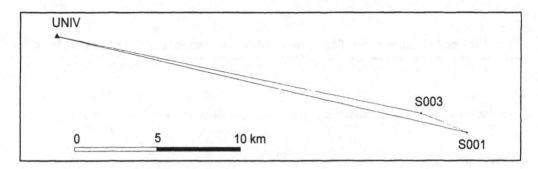

Fig.3 - Sample network in Perugia

3.1. Bologna network

In the first time, the GPS data have been treated with *Trimvec* software and the baselines so obtained adjusted with *Geolab* software. The adjusted coordinates so obtained have been compared with the re- sults of processing data (baselines computation and network adjust- ment) with *Topas* (or *Geotracer*) software. The differences found are given by the following table:

Point	ρ(m.)	λ(m)	h(m.)
01	-.008	-.008	-.010
02	-.008	-.008	-.002
03	-.011	-.004	.003
04	-.012	-.007	-.001
05	-.003	-.003	.003
06	-.007	-.006	.009

Tab.2 - Comparison of results with different softwares

The differences between the two solutions are not very large, still they are relevant, especially if we consider the particular aim of this network (subsidence monitoring).

3.1. Perugia network

This is a simple network (Fig.3) which connects two points of a landslide monitoring system (S001 and S003) with a control point (UNIV) located on the roof of the University building. The point UNIV is also connected to the Tyrgeonet network, so its coordinates in WGS84 reference frame are well known.

In one test, performed by means of the *Topas* (or *Geotracer*) software we analyzed three variable factor in data elaboration:

- broadcast versus precise ephemerides;

- use of different observables as L1, widelane or iono-free;
- use of different fiducial (fixed) points.

The first parameter that has been changed in data processing is the ephemerides type. Infact, in the first time in the data process-ing we have used the broadcast ephemerides, then we have compared the results (adjusted coordinates of the network points) obtained using the precise ones:

Points	φ (m.)	λ (m.)	h (m.)
UNI	0.000	0.000	0.000
S001	0.003	0.007	0.000
S003	0.003	0.006	0.000

Tab.2 - Broadcast versus precise eph

The second step has been the use of a different observable: L1 instead of Lc (iono-free). Referring to the adjusted coordinates, as in previous case, the differences are the following:

Points	φ (m.)	λ (m.)	h (m.)
UNI	0.000	0.000	0.000
S001	-0.036	0.030	0.004
S003	-0.032	0.025	0.005

Tab.3 - Lc versus L1

The third step has been the use of a different fiducial (fixed) point in the network adjustment. Assuming as fixed the point S001 in-stead of UNIV, the differences on the adjusted coordinates are:

Points	φ (m.)	λ (m.)	h (m.)
UNI	-0.002	-0.001	-0.002
S001	0.000	0.000	0.000
S003	-0.001	0.000	0.001

Tab.4 - Effect of changing of the fiducial point

4. CONCLUSIONS

In both sample networks, we have seen that varying some parame-ters in GPS data elaboration, the final results (e.g. the adjusted coordinates) have small but not negligible variations.

The importance of a correct archiviation of data is therefore evident, especially when a high precision is required, like in monitoring networks.

A reliable data base is required for the management of data relative to all measurement campaigns, because of the large volume of the GPS data and the continuous developing of new software and more efficient algorithms, that can give a better solution also for "old" data.

The data base must include not only GPS data, but also data of other kinds, like geometric or trigonometric levelling measurements, gravity measurements and increasingly refined models of the geoid, often constructed on the basis of the measurements themselves.

REFERENCES

1. Bitelli, G. and Vittuari, L.: Progetto di un sistema informativo per la gestione dei dati GPS, Boll. SIFET, 2 (1994), 31-42.

2. Date, C.J.: Relational Database - Selected Writings. Addison-Wesley, 1986.

GRID: A GEOGRAPHIC RASTER IMAGE DATABASE TO SUPPORT FIRE RISK EVALUATION IN MEDITERRANEAN ENVIRONMENT

P. Carrara, P. Madella, A. Miuccio and A. Rampini

ITIM - CNR, Milan, Italy

ABSTRACT

Forest fires, their frequency and intensity are one of the most important agencies of environmental degradation in the Mediterranean. Fire risk mapping is a powerful tool for the prevention of wildfires. The FI.R.E.M.EN. (FIre Risk Evaluation in Mediterranean ENvironment) project, sponsored by the E.E.C.- D.G. XII, purposes to define and develop the prototypal components of a decision support system for fire risk studies in the Mediterranean environment. It combines evaluation of remotely sensed data and geographic information, and by using dynamic modelling of the fire spread.

In the project framework, it has been decided to design and implement a database able to manage both still images and textual data; the main database purpose is to make as easy as possible the identification of files to be used or produced by the various steps of the risk map production process. In the database, called GRID (Geographic Raster Image Database), a friendly query environment is available; it has been designed in order to allow the user:

- to retrieve all images concerning an area by simply selecting this area on a map with the mouse;
- to retrieve all images "covering" a significant structure on a map (a river, a town, etc.) by selecting its name in a list.

The database architecture is relational and a traditional query environment allows expert users to deeply specify query constraints by expressing SQL queries.

GRID allows also easy communication with the other FI.R.E.M.EN. modules.

1. INTRODUCTION

The FI.R.E.M.EN project is supported by the Europen Community (D.G. XII); its main objective is to define, implement and validate the prototypal components of a knowledge based decision support system to produce fire risk maps in the following Mediterranean countries: Italy, Greece, and Spain. The system has to use multisource geographic data (such as remote sensing images, and various kinds of maps) selected from four test areas (two in Italy, one in Spain and one in Greece, respectively) to represent different features of the Mediterranean environment. The project leader is the Istituto per le Tecnologie Informatiche Multimediali (ITIM) of the CNR in Milan; the other project partners are the Istituto di Ricerca sul Rischio Sismico (IRRS) belonging also to the CNR in Milan, CAP GEMINI Italia (Milan), CINAR (Athens) and the INIA Institute of Madrid [1].

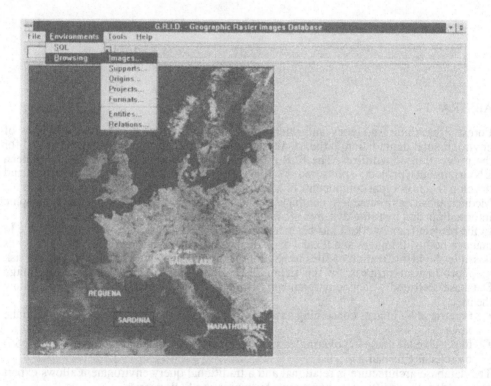

Fig. 1, Main menu of GRID

The multisource data used by the FI.R.E.M.EN. project in the decision support activity are of different formats, and acquired at different stages; though there are in literature examples of new approaches [2], [3], data are usually stored in a File System: in this way the user has to manage a lot of files which can be identified only by their names and hosted on different directories. This solution cannot be reasonably suggested for various reasons:
•   the identification of files is mainly dependent on the memory of the operator which created and archived them;
•   the file names have a low descriptive power as they are limited to 8 characters plus 3 characters for the extention;

- in many cases the file can be identified only by opening it, that is by activating the associated application;
- the majority of image formats do not provide a way to associate a georeferentiation and descriptive notes;
- the file system doesn't allow to esplicitely know the relations of files; for example it is not possible to link a classification map to the source files used to produce it.

Therefore, a relational database has been designed and developed to support classification and risk maps production within FI.R.E.M.EN.: it is able to archive and retrieve both images of various formats, and numeric or textual information which are mainly of geographic nature (georeferred geographic entities) as they are associated with the corresponding coordinates. This paper describes the database characteristics in terms of requirements with respect to the others FI.R.E.M.EN. modules and of design specification; a prototypal version (ready by May, 1995) has been produced and tested by the project partners. The last chapter describes the database future development: it will be redesigned as a hypermedia database allowing navigation facilities.

2. DATABASE REQUIREMENTS

From the data management point of view, the process of producing risk maps on the basis of different sources of data is a laborious task involving, for a single region, not only many source data collected at different time or carrying different information contents, but also intermediate data (usually images) produced during the various steps of the process [4]. From previous experiments, it can be estimated that to produce the fire risk map of a single geographic area, FI.R.E.M.EN. modules involve about 20 image files.

The traditional solution, that is file management by a File System, presents many drawbacks (see Introduction). It has been therefore decided to design and implement a database able to manage both still images and textual data; the main database purpose is to make as easy as possible the identification of files to be used or produced by the various steps of the risk map production process. This task is accomplished by three main guidelines:

- by enriching the files with associated descriptive data in order to clarify their content;
- by allowing to view images without opening any more application;
- by allowing a simple way to specify the user needs.

As a consequence of the last guideline, a friendly query environment is required: it has been designed in order to allow the user:

- to retrieve all images concerning an area by simply selecting this area on a map with the mouse;
- to retrieve all images "covering" a significant structure on a map (a river, a town, etc.) by selecting its name in a list.

A relational database architecture has been chosen and a traditional query environment is also made available which allows expert users to deeply specify query constraints by expressing powerful complete queries [5], [6].

One more aspect to be considered is the communication with the other FI.R.E.M.EN. modules: the files selected by the user have to be transferred (and sometimes converted) without leaving the application; on the other hand, produced maps or intermediate results have to be received and correctly archived in the database.

3. MANAGED AND ARCHIVED DATA

The decision support activity within FI.R.E.M.EN. requires the management of four main kinds of data: images, text files, entities with an intrinsic geographic nature as they can be easily associated to geographic coordinates (called *geographic entities*), and spatial relations among entities. The characteristics of the above data are described in this chapter.

Images.

Seven image types are managed:
*   satellite images;
*   aerial photos;
*   scanned cartography;
*   thematic maps (either digitised or produced by other FI.R.E.M.EN. modules;
*   classification maps produced by the classification module;
*   Terrain Digital Models (DTMs);
*   raster images produced by conversion of the corresponding vectorial sources.

As the other modules operate on pixels, all images are in raster format; furthermore they are georeferred (they are associated to the corresponding geographic coordinates in order to univocally identify their position).

The image files are external to the database, which directly manages the information on them that is:
*   textual descriptive data: they can be structured (such as the image file name, its format, its associated coordinates, its origin) or in narrative form (notes and information useful to the image identification and description);
*   new formats information: if the image format doesn't belong to the set of the database predefined formats, the new format description can be archived;
*   image relations: the database can archive the associations of classified images with images used to classify;
*   image-text file relations: the database can archive the associations of classified images with text files used to classify.

Text files.

They usually contain information which are used to support classification (therefore they are also called *supports*); the file content is often codified. The files are external to the database and managed by means of their associated information; besides the image-text file relations, each text file can be associated with a narrative free-format textual note in order to easily indentify its content.

Geographic entities.

The geographic entities managed by the database are: urban areas, water flows, lakes, natural parks, seas, mountains. Each entity is associated with one or more couples of coordinates (*vectors*), the number of which depends both on the entity nature and on the digitation accuracy. Information on the entities are contained in the database: the entity name, its type, a narrative free-format textual note.

Entity relations.

Although FI.R.E.M.EN. modules are "pixel oriented", for generality reasons it has been decided to allow the management of some large-scale relations among geographic entities (such as "rivers which pass through a town or viceversa", "towns which belong to a natural park", etc.): they cannot directly contribute by now to the classification, but they may be useful to provide a general overview of the characteristics of an area. There are no pre-defined relations: the user can decide which are the interesting associations for a particular area.

Georeferentiation constitutes the database unifying general framework, allowing uniform indexing both of images and of geographic entities; on account of it the database is able:
- to identify and retrieve all images and/or entities which fall within a given rectangle on a map;
- to identify and retrieve all images corresponding to a given entity.

Due to the fundamental role played by coordinate association (to images and entities) within the database, a special module guides the user in performing this task (see Figure 6).

4. DATABASE IMPLEMENTATION: THE GRID PROTOTYPE

The database implementation phase has been guided by the following preliminary considerations: a Personal Computer platform is best suited both to the cost constraints of the organisations for environment preservation and to the technological "maturity" of the potential users; the operating environment choice was oriented to Windows due to its standard charateristics in the PC world (its diffusion and Microsoft compatibility engagement are fair portability guarantees); an end-user development environment has been chosen to shorten code production time and to assure integration within Windows.

A database prototype has been implemented by May, 1995 and it is now running on a PC 486Dx66 with ET 4000 graphic board [7].

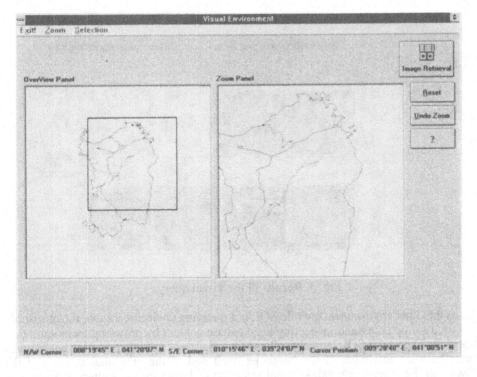

Fig. 2, Main window of the visual querying environment

The O.S. is MS Windows 3.1; the development language is MS Visual Basic 3.0 Professional and the database is implemented and managed by MS Access 1.1 which offers, in our opinion, best integration facilities with Visual Basic. Some external functionalities, which are particularly time and CPU consuming, have been developed by MS Visual C++ 1.0; they were linked as DLLs to the Visual Basic code.

The prototype is able to directly deal with same default file formats that is: .TXT for textual files (either structured or not); .BMP and .RSM for images; .MDB (Access); .VCT.

Next sections will describe the main characteristics, facilities and functionalities of the prototype, while the following chapter will illustrate the motivations which will carry to a further release.

4.1 The user interface architecture

In the prototype the user interface furnishes a main menu which allows direct access either to the test areas of the FI.R.E.M.EN project shown on an Europe map, or to a group of querying/browsing environments (see Figure 1). By clicking on an enhanced area of the map the user accesses the visual querying environment (see section 4.2).

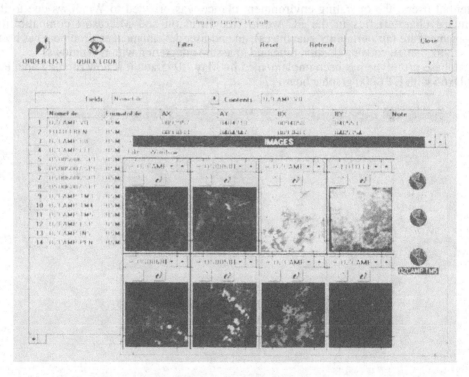

Fig. 3, Results of the visual query

As far as the other environments, while SQL is a querying environment too, the other ones can be used for updating and inspecting the database content by browsing techniques (see section 4.3). The use of SQL language allows formulation of complete queries but requires the user both a knowledge of the language itself and of the relational schema of the database: on the other hand, using visual selection or browsing techniques, the physical database structure is transparent to the user, so he may express his need in a simple, intuitive way [8].

4.2 Functionalities of the visual querying environment

Figure 2 reproduces the main window of the visual querying environment (supposing that Sardinia test area has been clicked by the user), which allows the selection of a rectangular area on a map, corresponding to the area containing the images which the user wishes to retrieve.

The area can be selected:

- by drawing its borders with the mouse;
- by manually inserting the coordinates of the opposite corners;
- by selecting an entity or an image which has to be contained in the area (the entity or image has to be selected in one of the browsing environment and then copied in this environment).

In Figure 2 the user selected the North-East area of Sardinia in the Zoom Panel, which is now highlighted in the Overview Panel: the coordinates of the opposite corners of the chosen area and the current pointer coordinates are also shown in the window.

After selection of an area, the Image Retrieval button opens a dialog box to specify the image cathegory to be retrieved; the result is then shown in the window "Image Query Results" (see Figure 3) which contains: a table with the data regarding the retrieved images; a content bar which shows the whole content of one single table cell (useful if the table has a lot of columns which are too narrow); a field list containing the list of column labels of the table, and an area containing the number of the selected row. It is possible to filter the content of one table column in order to further refine query result.

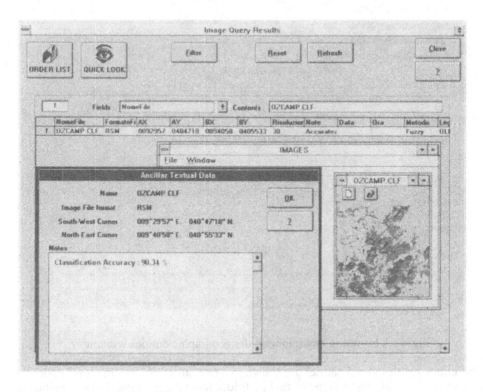

Fig. 4, Information associated to an image

The user can then select one or more images of the table, and the Quick Look button allows visualization of their thumb-nails; each image can then be magnified, or its associated textual information can be read (see Figure 4), or sent to other applications. In fact, another important characteristic of the user interface is to allow communication with the other FI.R.E.M.EN. modules: in this prototype a Dispatcher (an external server) manages the communication; after a selection, the user sends the list of the selected files to the Dispatcher which holds it in a Order List (a table containing names and formats of the selected files) (see previous Figure); if another FI.R.E.M.EN. module makes an explicit request, the Dispatcher copies the files of the Order List in the directories specified by users and, when necessary, converts the requested files.

4.3 An example of browsing environment: geographic entities

This section contains an example of browsing environment: it doesn't allow query formulation or result transfer to other FI.R.E.M.EN. modules, but it can be used to store information and to browse the the database content.

In particular, the geographic entities window allows to store a new geographic entity, to georefer it, to add notes, and to explore the entity content of the database (see Figure 5).

The Type list lists all entity types of the database; by selecting/deselecting one entity type, the corresponding data are shown/hidden in the Entities list area.

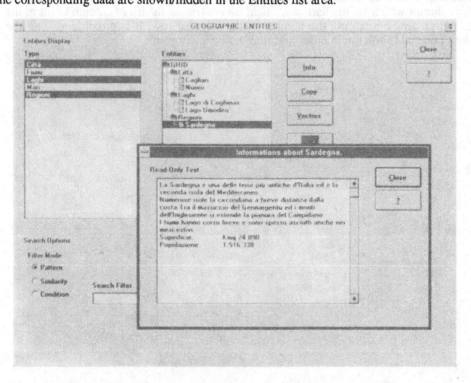

Fig. 5, A browsing environment: the geographic entities window

In the Entities list the stored entities are shown in a tree structure; the first level under the root contains the entity types (represented by folder icons); the second level contains the

entity instances stored in the database (represented by file icons); in the list it is possible to see:
- only entities the type of which is selected in the Type list;
- only entities satisfying the active filter defined in the Search Options area.

After selection of one entity in the list, the Info button shows the stored information on it as shown in Figure 5.

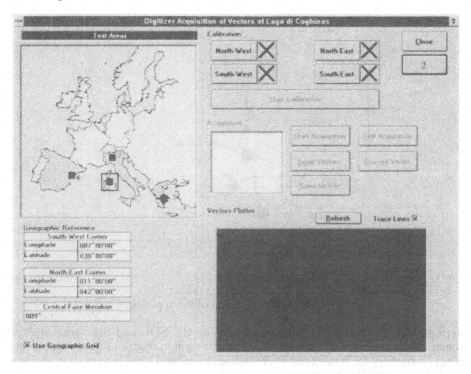

Fig. 6, Window for vectors management

Special buttons allow to store/cancel an entity of the database and to introduce also a new entity type. The Vectors button allows georeferentiation of a stored entity, i.e. its association with geographic coordinates in order to use it for retrieval of images; after selection of an entity, its clicking causes visualisation of the associated vectors (if any); it is also possible to introduce new vectors choosing to acquire them by digitizer (see Figure 6), manually, or importing them from a .vct file.

5. DATABASE IMPLEMENTATION: FUTURE DEVELOPMENTS

After its implementation, GRID has been tested in order to fix bugs and to check its usefulness and friendliness. During the testing phase it has been noticed that the presence of the same buttons in different windows, which was designed to enhance system flexibility, is a confusing factor for the user: in fact, in most cases he doesn't understand the real behaviour of these buttons (which corresponds to close the current window and open a new one in the main environment), but he may think that they are 'hypertextual' links letting access to information associated to the selected elements of the current window [9].

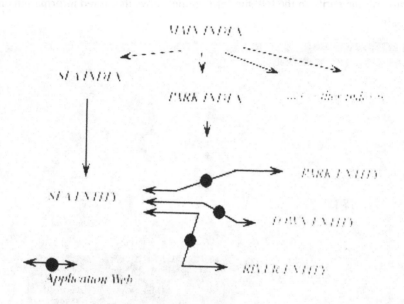

Fig. 7, Partial Author Schema of the application

Furthermore GRID allows two different query levels: SQL and the visual querying environment (see section 4.2); the first level reaches a high degree of query specification but is not so friendly for non-expert users, while the second level is very easy to use but it's not able to express flexible queries.

Therefore, a new intermediate access level has been introduced, which offers the user both a set of indexes for accessing the database content and navigation facilities to enlarge or better specify the user interests. To this aim, GRID has been designed again as a 'hypermedia database' adding the hypermedia concepts of connection and navigation to the schema based structuring of information and the powerful query mechanisms which are typical of the database management systems.

The new design phase has been supported by the HDM2 conceptual model [10], the basic concepts of which are:

- entities: that is the objects of the application (they can be structured);
- application webs: that is non-structural relations among entities (for example "a river flows_through a town", "a lake is_contained_in a natural park", etc.); they add domain information;
- index webs: that is ways to access application objects, the purpose of which is to give an overview of the application content and to allow the selection of the items of interest.

The result of the conceptual modelling of an application is its 'author schema'. A part of the GRID author schema is contained in Figure 7. HDM2 suggests a systematic approach to directly relate the author schema with the user interface: the application webs can be implemented with buttons; the index webs with lists of items or QBE editors [11].

With this new access level, an example of user session can be sketched in the following way. The user can have an immediate glance of the database content by looking at the list of indexes; then he can choose one of them, for example the index "Seas", declaring therefore an interest in knowing the seas of the database (in this way he goes from the index level to the entity level of the schema). After examination of the entities of the type "Sea", he can be interested, for example, in entities of other types which are associated to a particular sea (for example the natural parks or towns which are on the borders of the Tyrrhenian Sea): the application webs allow to satisfy this interest by guiding the user only to the instances of another entity which are linked (associated) to a selected sea. The application webs (hypermedia buttons) are the tools for navigating in the entity space, reaching therefore a sufficiently deep level of specification without using a query language.

GRID author schema is ready and it will soon be implemented as a propotypal release.

REFERENCES

1. A Knowledge-Based Decision Support System for Fire Risk Mapping, Environment CEC, Project N. EV5VCT9K0521, 12-month Technical Progress Report, May 1995.

2. Simpson, J.J. and Harkins, D.N.: The SSABLE System: Automated Archive, Catalog, Browse and Distribution of Satellite Data in Near-Real Time, IEEE Trans. on Geosc. and Rem. Sensing, 31 (2) (1993), 515-525.

3. Raafat, H.M., Xiao, Q. and Gauthier, D.A.: An Extended Relational Database for Remotely Sensed Image Data Management with GIS, IEEE Trans. on Geosc. and Rem. Sensing, 29 (4) (1991), 651-655.

4. Binaghi, E. and Rampini, A.: Fuzzy decision making in the classification of multisource remote sensing data, Optical Engineering, 32 (6) (1993), 1193-1203.

5. Elmasri, R. and Navathe, S.B.: Fundamentals of database systems, The Benjamin/Cummings Pub. Co. Inc., Redwood City (CA) 19891.

6. Korth, H.F., Silberschatz, A.: Database System Concepts, McGraw-Hill Ed., 1986.

7 Miuccio, R.A.: Progettazione ed implementazione di una base di dati per la gestione visuale di dati geografici, Tesi di Laurea, Università Statale di Milano, Corso di Laurea in Scienze dell'Informazione, A.a. 1993/1994

8. Batini, C., Catarci, T., Costabile, M.F. and Levialdi, S.: Visual Query Systems: A Taxonomy, Visual Database Systems II, North-Holland, 1992.

9. Conklin J.: Hypertext: An Introduction and Survey, IEEE Computer 20 (9) (1987), 17-41.

10. Garzotto, F., Mainetti, L. and Paolini P.: HDM2: Extending the E-R Approach to Hypermedia Application Design, in: Proc. 12th Intern. Conf. on the Entity-Relationship Approach, Springer-Verlag 1994 (to appear).

11. Garzotto, F., Mainetti, L. and Paolini, P.: Using and Developing Hypermedia Information Points: Lessons Learned., in: Proc. ENTER '94 - Int. Conf. on Information and Communication Technology in Tourism, Innsbruck (Austria) 1994, 102-109.

MULTIMEDIA IN MONUMENTS MANIPULATION

A.D. Styliadis

TEI of Thessaloniki, Thessaloniki, Greece

ABSTRACT

Churches and other structures in monuments family are quite complex objects technically, functionally and visually. This kind of structures has always be described by the Architects through a variety of traditional means, like drawings, writing and scale models.

The computer based multimedia technology has begun to automate architectural designing giving promizes in monuments CAAD manipulation and better chances for monuments presentation using authoring tools.

The monuments CAD models could be combined with raster images (photos), video clips, recorded sounds, etc, to show in a realistic way the proposed ideas of the Architect, and then Archaeologists could use the new presentation technology and particularly authoring tools for monuments promotion.

In our pilot application and in order to create the prototype, a 3D church model based on existing old drawings (in scale 1:50) was used. The church model was enhanced by a number of scanned photographs which have been manipulated with Aldus Photostyler image-editing software. Finally a navigation system, based on MacroMind's Director and IBM's Storyboard Live!, was used to link everything together.

Obviously the prototype could be expanded into a complete CD-ROM library of relative monuments for eventual distribution to promotion agencies, educational establishments, etc.

Keywords: Multimedia, Monuments, CAD, Architectural Design, Authoring Tools.

1. INTRODUCTION

Multimedia technology could be thought as a combination of text, computer graphics, video, scanned photographs, rendered and animated computer models and sound. The introduction of multimedia in monuments manipulation has been tested in a church pilot-system. The Byzantine church of St.Panteleimon-Thessaloniki Greece, was chosen as the pilot-monument to apply multimedia in its manipulation. The manipulation system based on a MicroStation PC CAD-platform having the usual H/W requirements for multimedia PCs. The vector 3D file of the church is separated in a number of levels/layers devoted to distinct parts of the church. Each part can be considered as a separate segment with its own description, history and documentation. The key-characteristic of the monuments manipulation system is the combination of visual displays (eg. arbitrary cross-sections, multi views), text information-documentation from historical resources, digital video, raster images and an audio description (sound) for the objects found in the selected sections/views.

The database connection of church's segments to corresponding text information was implemented using SQL statements. Particularly a SQL table with a number of sixty (60) records-entries (i.e. as many as the distinct segments/layers of the monument) has been constructed. Each SQL-record holds a text description of the segment in an ASCII format, relative information and three pointers to corresponding video clips (AVI files), raster images (TIF, BMP files) and audio-sounds (wave, MIDI files). Alternatively an additional pointer to text information (WRI, DOC files) could be used in the SQL-record.

The church pilot-system integrates full motion video at 30 frames per second in AVI format. Users can point to church's segments on-screen to pop up videos of that segments. Video play is controlled from a dialog box to allow for instant access of data. Digital video combined with 3D monuments models gives system users a visual edge in illustrating a proposed development. A number of photographs of St. Panteleimon church have been taken using a conventional camera. Then these photos have been scanned using HP ScanJet II+ scanner in resolution 300 dpi.

Digital samplings of church ceremonies are incorporated
into the pilot system. To create the audio data, church
ceremonies have been recorded and measured loudness at
various geographic points around the church. The
recordings were then digitized and stored, to be played by
the pilot system at various volumes. In addition, the
noises were calibrated to match actual observed conditions
[Figure 1].
For the recording procedure and for 1 min and 40" recording
duration for WAVE form (22KHz/16-bit), a disk space of 10MB
is required. For the same requirements there is a need of
100KB disk space for MIDI form recording.

2. Multimedia in Architectural Design (CAAD)

2.1 Using Multimedia

   For centuries, Architects and Engineers have conveyed
disign ideas through a variety of drawings, writings and
scale models. Nowadays they have begun to benefits of the
latest in computer-based multimedia technology [1].
Two main categories of multimedia applications regarding
architectural design exist so far. The first is related to
creation of multimedia presentation, to show prospective
clients-house buyers of a building society 3D models of the
houses for sold. The Architects as well could use
multimedia technology to show prospective clients not only
how a house will look, but also how the design process will
work. The design process is related to a line-by-line house
model development with doors opening and closing to
simulate realism.
The second application is related to architectural
education where famous buildings are used as examples of
teaching architectural principles [2]. For instance, one
building could be a good example of natural ventilation,
whilst another one could be a good example of daylighting.

2.2 The Byzantine Church of St. Panteleimon - A prototype case study.

   To create the prototype 3D computer-based model of the
church, a digitizing process of old drawings in scale 1:50
has been implemented. The 3D model created in MicroStation
PC 4.0 PC-CAD platform [3]. The structural net is explained
in multi-views, diagrams and cross-sections; the physical
site is portrayed in an animation sequence of the changing
sun using scanned photographs (backgrounding). The history

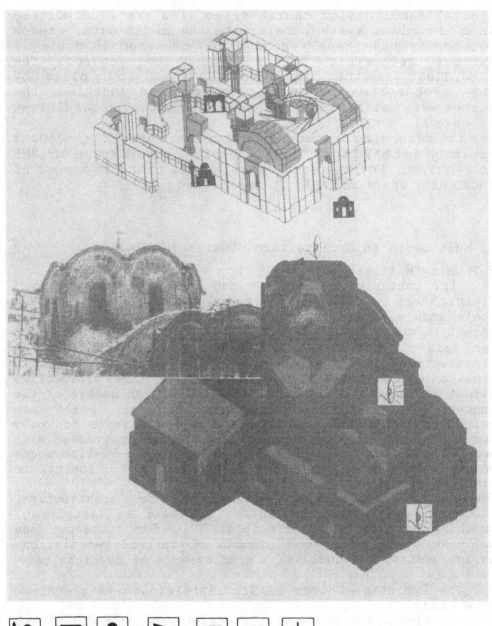

Figure 1. Sounds connected to particular segments

and critical analysis of the church is given in text;
materials and building techniques are shown in a sequence
of contributed photographs. The acoustics of the church is
demonstrated through recorded sounds and the church
ceremony is simulated through music. The connections
between such characteristics have been programmed and
implemented in MDL, the MicroStation Development Language.
As a result of the above a user could look at every aspect
of the church in a multimedia way. For instance, details or
text information for particular pillared columns, domes or
archades could be displayed.
The pilot-church documentation has been required about 250
hours of work plus the obstacles for obtaining permissions
to use copyrighted drawings and photographs. A prototype
library of case studies that presents famous monuments
through scanned photographs, technical diagrams, text,
video, animated and rendered models and recorded sounds
could be implemented. This library might be stored in a CD-
ROM for eventual distribution to museums, tourist offices,
schools,etc.

3.0 Multimedia Programming and Tools for specific purposes

3.1 Multimedia Programming for MS-Windows
    In monuments manipulation when the CAD system which is
used as the platform, doesn't support multimedia links to
vectored data there is a need to implement such as links
using a programming language. For our pilot-system this
need has been fulfilled using C++/MDL programming; an
event-driven programming language for the MicroStation CAD
platform.
Actually, the pilot system uses C++ when the user is
interested in monuments manipulation on MS-Windows 3.1x
platform [4]. So far this programming language has been
used to implement the links between church segments and
recorded sounds saved as wave files and used as program
resources in an object-oriented approach. This kind of
programming based links could be found useful when we are
interested in church segments located somewhere inside the
church [Figure 2].

Figure 2. Church segments located inside the church

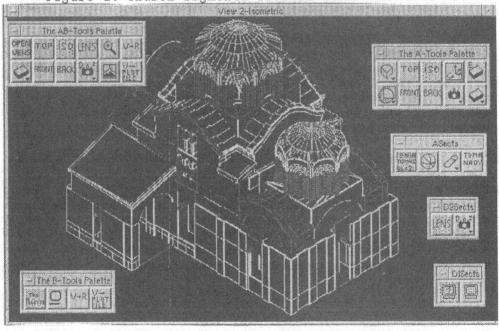

Figure 3. A Wireframe Church Model

3.1.1 Playing WAVE files in MS-Windows programming

Wave files have numerous controls allowing the determination of audio quality. For instance, wave file sounds could be played in mono or stereo, and at frequency ranges from a wispe beep to a compact-disk quality. Wave files can be stored as individual files on disk, or as elements in the resource file of a Windows Programming Object Oriended Application. We will be able to have wave files play when we click on a monuments part/segment.

In Borland's C++ 3.1 for Windows, there are four classes of functions to deal with playing wave files : MessageBeep, sndPlaySound, MCI calls and waveOut calls. In sndPlaySound class for instance, we can store wave files as program resources and load them using the LoadResource() call. In this way, a Windows application can have a number of canned sounds linked to monuments segments; load them to memory using LoadResource(); then using LockResource() which will return a pointer to that wave file in RAM and finally play this wave file using sndPlaySound() when an event (eg. a mouse click) appears.

Eg. (Borland's C++ for Windows)

```
.....................
HANDLE sound;
LPSTR psound;
HANDLE handle;
if ( (handle=FindResource(hInst,"waveFileName", RT_RCDATA)) != NULL )
    if ( (sound=LoadResource(hInst,handle)) != NULL )
        if ( (psound=LockResource(sound)) != NULL )
            sndPlaySound(psound,SND_ASYNC | SND_MEMORY);

.....................
sndPlaySound (NULL,0);
if (psound != NULL) UnlockResource(psound);
if (sound != NULL) FreeResource(sound);

.....................
```

3.2 Multimedia Programming for PC-CAD Environment

The following technical notes and fragment coding have been used to link multimedia (text, images, video and sound) to St. Panteleimon 3D model [5][6].

A. Inputing Data

- In a Xbase environment (eg. dBASE III+, dBASE IV) create the main table with the name **MSCATALOG**; this is actualy the control table. In its table and in the field <u>Tablename</u> we put the name of the SQL table which will be created, for instance **naos**.

- In a SQL window, i.e. a dialog box in MicroStation environment, we have to create the SQL table with the name **naos**.
    .CREATE TABLE naos (mslink integer, textInfo char(50),
        iconInfo char(12), clipInfo char(12), soundInfo char(12))
    .submit
- Alternatively to SQL table **naos**, the Xbase table with the name **naos.dbf** could be created using a Xbase editor.
- Then we have to insert the data to that SQL table
    i=1,60
    .INSERT INTO naos (mslink, textInfo, iconInfo, clipInfo, soundInfo)
        VALUES (i, 'The main dome of ...', Dome1.gif, Dome1.avi, Dome1.wav)
    .submit

B. Implementing the Links
In order to implement the links between the various disk files holding text, images, video clips and sound, and the church segments, we have :a) to programming with MDL to link the current **mslink** value to a particular church item, and in turn for all the mslink values, or b) to use the MicroStation's Database Palette to implement these connections graphically.

C. Displaying the Multimedia information related to specific church segment.
After the implementation of links, it is possible to get the multimedia information of a particular church segment using a pointing device (eg. mouse)[7]. The following MDL code is needed:
        sprintf(sqlStatement1,"SELECT textInfo FROM naos WHERE mslink=%lu", link.mslink);
        sprintf(sqlStatement2,"SELECT iconInfo FROM naos WHERE mslink=%lu", link.mslink);
        sprintf(sqlStatement3,"SELECT clipInfo FROM naos WHERE mslink=%lu", link.mslink);
        sprintf(sqlStatement4,"SELECT soundInfo FROM naos WHERE mslink=%lu", link.mslink);

        ....

        status1=mdlDB_sqlQuery(text,sqlStatement1); status2=mdlDB_sqlQuery(icon,sqlStatement2);
        status3=mdlDB_sqlQuery(clip,sqlStatement3); status4=mdlDB_sqlQuery(sound,sqlStatement4);

        ....

        sprintf(text_file, "mdl l dbTEXT; display %s", text);
        sprintf(icon_file, "dialog display %s", icon);
        sprintf(clip_file,"mdl silentload movie;movie display %s", clip);
        sprintf(text_file, "mdl l dbSOUND; play %s", sound);

        mdlInput_sendKeyin(text_file,0,next++,NULL); mdlInput_sendKeyin(icon_file,0,next++,NULL);
        mdlInput_sendKeyin(clip_file,0,next++,NULL); mdlInput_sendKeyin(sound_file,0,next++,NULL);

D. Updating Data
- In SQL Window-Dialog Box
.UPDATE naos SET image=imageFileName.GIF WHERE mslink=17
.submit
- In Ustn level (in order the database to be updated)
ustn>|reload

4. Promoting Monuments with Multimedia Presentation and Authoring Tools

To explain the spatial organization and technical aspects as well as proposed additions of the Byzantine church of St. Panteleimon a multimedia "real-world" presentation has been created using MacroMind's (San Francisco) Director program and IBM's Storyboard Live! authoring tool.

This presentation could be used not only for how the church will look, but also how the reconstruction design process will work. The presentation begins with three scanned photographs of the site taken from different points of view; then shows a basement 2D plan developing line by line and having the bases of pillared columns, the position of doors and entrances, the width of walls, etc [Figure 2]. Putting these primitive church elements the drawing gradually grows in detail, and then is projected into a simple 3D wireframe [Figure 3], which is then rotated, rendered and the final church model is displayed with the site photographs in the background [Figure 4].

The presentation continues with phto-realistic renderings and walkthroughs/flythroughs on selected church's paths. In reconstruction procedures technical ideas could be virtually implemented through animations that show the monument gradually development piece by piece and illustrate how heating, cooling and other utilities work [Figure 5].

All the figures used in this paper have been captured using the dynamic screen capture utility described in next chapter and they are used in Director and Storyboard Live! authoring tools.

Figure 4. The Final Rendered Church Model

Figure 5. The Church's Natural Ventilation System is shown
in a cut-away section.

5. A Dynamic Screen Capture Utility for raster Images
supporting multi File Formats.

There is a number of screen capture utilities in PC-Market
working in a Terminated and Stay Resident (TSR) base.
Nevertheless, as far as I know, there is not a utility to
capture the whole screen-image in resolution 1024x760 and
true color/24-bit; whilst the MicroStation 5.0 CAD s/w is
loaded and a rendering procedure is performed. Actually, a
memory of 2.3MB is required for this captured screen
(1024x768x24/8 = 2.3Mbytes).

To avoid this difficulty and in order to print-out the
Figures 2 and 5, a screen-capture utility has been
implemented in MDL/C++ using huge pointers, i.e. _Far/32-
bit pointers capable to access extended RAM. This utility
following a TSR logic is loaded in extended RAM just behind
the memory location of MicroStation PC 5.0 for DOS [8].
This utility does not work with the MicroStation version
for MS-Windows due to different memory locations used by
this version. Future enhancements might include this
extension.
While the MicroStation is loaded - by the Phar Lap 386|DOS-
Extender s/w embedded into the system - in extended RAM and
particularly when a rendering procedure is working a
selected key-combination activates this TSR utility; holds
the screen and asks the user for a raster format to be used
for the saving. Up to now there is a ability to select one
of TIF, GIF, PCX and BMP raster file formats. The captured
screen is saved firstly in extended RAM, due to its size,
and thereafter in hard disk.
The key-combination, known as hot-key, is selected in such
a way to avoid conflict with MicroStation function keys.
Finally, it is important to note that, in order to use
extended RAM the TSR turns DOS/processor into the 32-bit
protected mode operating environment.

Conclusions - Future Enhancements
       The recent marriage of Geographical Information Systems
(GIS), Computer Aided Architectural Design (CAAD), and
Multimedia Technologies is taking users and decision making
personnel from a simple world of static maps and manual
presentations to an on-screen world, that is visual and
animated.
A multimedia GIS environment for monuments can allow users
to experience the sight and sound of their data, creating a
"virtual monuments base" of information.
The ability to integrade Text, Sound, Video, Images and
Animation in 3D monuments models results in a "real-world"
presentation of the monuments. The multimedia presentation
could be found useful to Architects and Archaeologists
whilst an authoring tool might be used for demonstration
purposes.
Future enhancements mainly related on introducing Virtual
Reality in monuments' presentation [9]. This experience of
"being in space" could be given from real-time animation
and high-quality, rendered walkthroughs/flythroughs.

Hardware and Software Suggestions

A standard IBM-compatible PC with Intel Pentium microprocessor and multimedia support is suggested. A color inkjet printer is needed for color printouts. The Intergraph's MGE PC for windows 95 together with the Vista Map s/w would be the ideal platform. Alternatively for customized multimedia s/w the Bentley's MicroStation PC CAD s/w accompanied with the MDL is suggested.

Acknowledgements

I would like to thank colleagues at AUT/FRSE, TEI of Thessaloniki and Univ. of Melbourne for useful discussions. Particularly I am grateful to Prof. John Paraschakis, Prof. Jim Kleftouris and Prof. Cliff Ogleby.

REFERENCES :

1. Novitski, B.J.: Constructive Communication, Computer Graphics World, June 1993, 35-38.
2. Millet M., Cohan P., and Hildebrand G.: ArchiMedia - A prototype library of building case studies. Univ. of Washington, Dept. of Architecture, 1991.
3. Styliadis, A.D.: Monuments Manipulation with an easy-to-use GUI, 1995 (in press).
4. Rimmer, S.: Multimedia Programming for Windows, McGraw-Hill/Windcrest, 1994.
5. Jones, B.W., Graves R., and Sterling H.J.: MicroStation Database Book, OnWord Press, 1992.
6. Mach N. Dinh-Vu: Programming with MDL, Pen & Brush, 1991
7. Steinbock, B.: 101 MDL Commands for MicroStation, OnWord Press, 1991.
8. Taylor, B.: C/C++ Programmer's Guide to using PC BIOS, The Lance A. Leventhal Microtrend Series, Carlsbad, CA, 1993.
9. Ogleby, C.:Digital Technology in the Documentation and Conversation of Cultural Monuments, Proc. of 17th ICA Int'l Conference, Barcelona-Spain, 1995, 1150-1155.

THE BEHAVIOUR OF THE LEANING TOWER OF PISA IN THE PERIOD 1993-1995

A. Colombo and A. Giussani
Polytechnic of Milan, Milan, Italy

G. Vassena
University of Brescia, Brescia, Italy

ABSTRACT

This paper deals with the analysis of the results obtained from three years of geodetic controls of the Leaning Tower of Pisa; during the last three years, 1993-1995, an International Multidisciplinary Commission, appointed by the Italian Government, has planned important works in order to stabilise the Tower. The studies presented in this paper try to understand the Tower behaviour and the deformations of its foundations due to a lead counterweight placed on the North external side of the Tower base, in order to decrease the Tower leaning towards South.
Finally three methods to evaluate the Tower tilt in the North-South direction are described and compared. The deformation axes of the graphics here presented are not scaled and only few deformation data shown. This is because the results of the control are still under the study of the International Commission and for this reason, at the moment, cannot be published.

1. INTRODUCTION

The I.I.A.R. Department of the Politecnico di Milano is involved, since May 1993, in the supervision works of the geodetic control of the Leaning Tower of Pisa. Some papers and reports have already be written describing the levelling network and all the others geodetic control measurements [1] [8] [9] [18]. The aim of this paper is to describe some methods used to study the Tower deformations in the period 1993-1995 and to answer to some important questions about its behaviour in the last three years. The Tower movements and deformations are investigated using the following measurements:
- periodic high precision levelling measurements to check vertical movements of the Tower basement and of the surrounding square;
- daily high precision levelling measurements used to check the daily vertical movements of the benchmarks placed inside the Tower;
- periodic geodetic observations of 10 control points placed on the Tower body to check the spatial deformations and its deviation from the vertical.

Picture 1.1 shows the positions of the benchmarks analized.

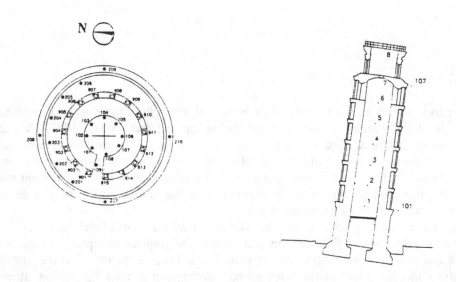

Picture 1.1 The positions of the benchmarks.

Analysing all these elevation and leaning data, measured before, during and after the displacement of a lead counterweight in the north external part of the Tower base, it is possible to try to give an answer to the following two important questions:
- is the Tower basement tilt rigid or is it characterised by some deformations too ?
- is the value of the Tower tilt, evaluated with the geodetic measures, according with the values obtained from the inside Tower levelling measures ?

All the analysis described in this paper follow a static point of view; some results about the kinematic behaviour are given in [2].

2. VERTICAL MOVEMENTS IN THE SURROUNDING AREA

The square surrounding the Tower is subjected to a movement composed by a cyclic yearly deformation (a vertical sink about 1-2 mm per annum) [7] [18], due to the seasonal changes of the environmental conditions, especially of the groundwater level.

Some studies carried on in the past [10] [12] [13] have shown that is also present a subsidence process involving all the Pisa area. This large sinking movement is not of course readable from the levelling measurements observed in the square and here studied. Infect the elevation movements of subsidence in the Pisa plane were studied using as constrained some benchmarks placed in a very stable area close to the mountains [16], and the values of subsidence obtained were very varied, from a value of few millimetres to 16-20 mm per annum.

The University of Brescia and the Politecnico di Milano are now planning a GPS campaign to control the elevation movements of the Pisa plane in order to estimate the subsidence values and to know more in detail the vertical deformations.

3 TOWER VERTICAL MOVEMENTS

The Tower is sinking with respect to the Piazza dei Miracoli medium level. The Polvani Commission [7] in 1971 studied this trend of deformation and evaluated that the Tower has been subjected to a sink of 3,28 millimetres per annum in 795 years, from 1173 to 1965; this value has been obtained dividing the mean value of the deepness of the 15 benchmarks placed on the external Tower basement, numbers from 901 to 915 in picture 1.1, (the medium base deepness has been evaluated with respect to the level of the square ground) for the total years of the Tower life (795 years). A more up to date and reliable value of the sink speed for the years between 1986 and 1991, has been evaluated in 0,04 mm per annum by I.G.M.I. (Italian Geographic Military Institute) dividing the mean sink of the 15 benchmarks for the number of years of this period.

Between May 1993 and November 1995 the high precision levelling network along the inside and outside Tower base. has been measured 14 times. All the measures are referred to a metallic benchmark with its bottom fixed to a 60 meters deep sand layer (the top of the benchmark is at ground level). Picture 3.1 compares the mean vertical behaviour of the Tower and of the square. The picture shows:

- the presence of a cyclic elevation movement in the square;
- the Tower sink with respect to the surrounding square.

The picture shows the difference between the Tower mean vertical behaviour (as the average of the movements along the 15 external levelling benchmarks) and the medium vertical deformation of the square close to the Tower, (as the average of the deformation value of 4 benchmarks placed 15 meters far from the Tower respectively in North, West, South and East direction).

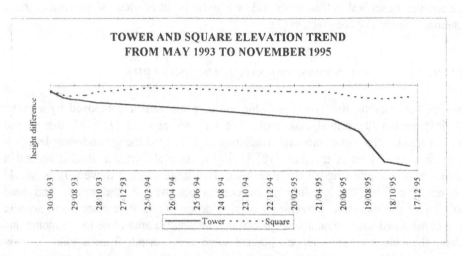

Picture 3.1 Tower and surrounding square vertical behaviours.

Also picture 3.2, that shows the lines of iso-sink from 03/05/1993 to 21/02/1994, points out clearly the Tower sink with respect to the square. It is clear that the effects of the counterweight displacement are different from the subsidence observed in the past and that they make changes in the behaviour trend observed in the past.

The counterweight displacement on the North side (see picture 3.3), carried on from June 1993 in 5 steps for a total weight of 635 t, is the cause of a new kind of deformation process. Infect the eccentric position of the counterweight involves different vertical movements between the external base benchmarks.

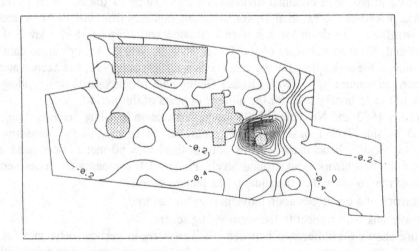

Picture 3.2 Vertical deformations from 03/05/1993 to 21/02/1995 in the square

The benchmarks on the North side sink more then the ones on the South and this behaviour is caused by a tilt towards North of the Tower basement. This means that for the first time after 7 centuries the Tower has rotated towards North, in a safer position, instead of the usual 7" per annum dangerous tilt towards South.

The altimetrical section in North-South direction is the most interesting deformation-view. Since the Tower leaning is mainly in this plane towards South, the counterweight has been placed on the North side Tower basement and so the tilt due to it has the main components in this plane. Otherwise in the West-East direction the leaning components are much smaller.

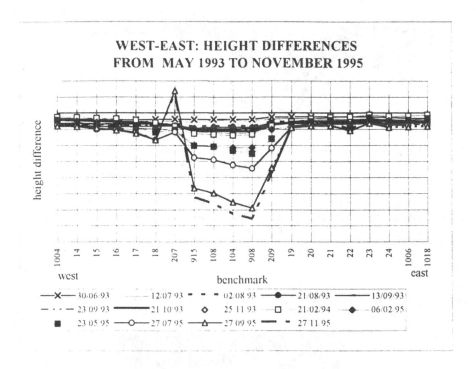

Picture 3.3 The counterweight position on the North edge of the Tower

Pictures 3.4 and 3.5 show the trend of vertical movements of the benchmarks placed respectively along the North-South and West-East directions. As said before the deformation axes of picture 3.4 and 3.5 are not intentionally scaled.

Benchmarks 904 - 908 - 911 and 915, installed by Professor Polvani in 1965, are placed on the external basement respectively on North, West, South and East position; benchmarks 102 - 104 - 106 - 108, respectively on North, West, South and East position consist of invar rods, 30 centimetres long, hanged to the internal walls of the Tower in

1993 (see picture 1.1). Furthermore pictures 3.4 and 3.5 show how the load displacement has also much increased the sinking process.

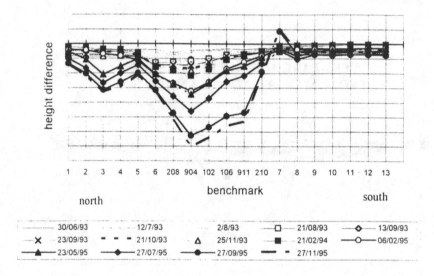

Picture 3.4 Vertical movements trend along North-South direction

Picture 3.5 Vertical movements trend along West-East direction

4. VERTICAL MOVEMENTS OF THE INSIDE LEVELLING BENCHMARKS

The inside Tower levelling network (8 stadia rods), placed at Tower groundfloor level, is measured about every day with an automatic Zeiss level.

Picture 4.1 shows how the height difference between benchmarks placed on the North side (102) and on the South side (106) changes from 25/06/1993 to 23/05/1995. This graphic shows a decrease of the height difference due to the decrease of the Tower leaning.

In order to study if there are dangerous deformations in one day time, during one week after one step of load positioning, several measurements (6 or 7) of the internal network have been carried on at different times of the day. No deformations have been observed; infect the results show movements having always the differences in elevation smaller or similar to their m.s.e. value.

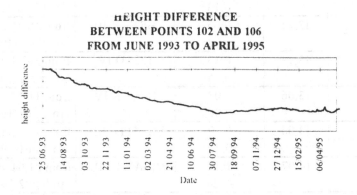

Picture 4.1 Height difference between benchmarks on North and South side

5. GEODETIC CONTROL OF THE TOWER

In order to know the spatial deformations in X, Y of 10 control points placed along the outside Tower "body", from the bottom to the top, angular and distance measures are regularly taken from 3 monuments placed in the square [8] [9]. By these measures it is possible:

- to study the tilt of the Tower elevation part, as rigid body, and then compare it to the tilt of the foundation plane;
- to analyse if there are deformations along the Tower body.

BENCHMARK	H_{104}--H_{108} (mm)	SPEED ($mm10^{-2}$ /day)	SPAN TIME	PERIOD
104	13,19	0,33	BREAK 0	from 14/06/1993
108		0,30		to 13/07/1993
104	13,22	-0,12	LOAD I	from 14/07/1993
108		0,53		to 30/07/1993
104	13,35	0,08	BREAK I	from 31/07/1993
108		0,50		to 23/08/1993
104	13,24	0,25	LOAD II	from 24/08/1993
108		0,50		to 08/09/1993
104	13,17	-0,10	BREAK II	from 09/09/1993
108		-0,14		to 29/09/1993
104	13,16	0,13	LOAD III	from 30/09/1993
108		0,13		to 15/10/1993
104	13,21	0,05	BREAK III	from 16/10/1993
108		0,02		to 26/11/1993
104	13,14	0,33	LOAD IV	from 29/11/1993
108		0,25		to 10/12/1993
104	13,06	0,03	BREAK IV	from 11/12/1993
108		0,00		to 10/01/1994
104	13,06	0,70	LOAD IV BIS	from 11/01/1994
108		0,30		to 20/01/1994
104	13,07	-0,18	BREAK IV BIS	from 21/01/1994
108		-0,24		to 03/02/1995

Table 2

7. THE LEANING TOWER BEHAVIOUR

Taking as true the hypothesis of the Tower basement as rigid body, it is possible to study the tilt movements from the vertical speed informations and to analyse the movements coming from the benchmarks placed in the directions North-South and East-West. In such a way the analysis of the tilt speed is done separately for the two components North-South and East-West. The data are taken from the vertical speeds of the couples of opposite benchmark (N-S) and (E-W) and from their distances. By using of the vertical speeds of benchmarks 106 and 102 and of their distance, the tilt speed of the Tower basement in the North-South direction is obtained as follow:

$$\vartheta = \arctan(\frac{a_{106} - a_{102}}{distance}) \ (7.1)$$

In table 3 the Tower basement tilt speeds towards North and East are shown.

The table shows clearly that the values obtained by the measures of the inside network are in a very good agreement with the ones obtained by the measures of the outside network; this means that there are not deformation inside the Tower walls, at least at the groundfloor level. This results is very significant if we consider the structure of the Tower walls. Infect the Tower is an allow masonry cylinder composed by an external and an internal wall both in San Giuliano marble. These walls are filled up with a rubble cemented with San Giuliano mortar. To notice almost the same value of tilt using the inside or the outside benchmarks means that there are not important shifting between the two walls.

The previous table shows that the difference between the tilt evaluated with the geodetic measurements body and the one obtained from the levelling ones are not negligible. This means that there is a sort of shift between the lower part and the higher part of the Tower.

7. INTERPOLATING PLANE BY LEAST SQUARE ADJUSTMENT OF THE HEIGHT VARIATIONS OF THE TOWER BASE BENCHMARKS.

Since the results using the inside and the outside levelling network are in a good agreement (Par 6) the next study is carried on only for the measurements of the outside network that is characterised by more benchmarks (15) then the inside one (8).
The height variations and the co-ordinates of the benchmarks placed on the external Tower base, have been used to obtain the equation of the interpolating plane of the height variations for each measuring time.

The parameters A, B and C of the interpolating plane

$$z=Ax+By+C$$

are estimated by least square adjustments for each time of measurement.

Tables 7.1 shows the results obtained for the parameters:

OUTSIDE BENCHMARKS												
Date	03.05.93	30.06.93	12.07.93	02.08.93	21.08.93	13.09.93	23.09.93	21.10.93	25.11.93	21.02.94	06.02.95	23.05.95
A	0.00000	0.00000	-0.00001	-0.00001	-0.00001	-0.00001	0.00000	0.00000	0.00000	0.00001	0.00002	0.00006
B	0.00000	0.00001	0.00000	0.00005	0.00006	0.00009	0.00009	0.00012	0.00012	0.00018	0.00025	0.00024
C	0.00000	-0.00581	0.01089	-0.04208	-0.04359	-0.07689	-0.07936	-0.10792	-0.11833	-0.16952	-0.24344	-0.27479

Table 7.1

In order to understand if there is a deformation trend in the period 1993-1995 we have looked for a variable *VARH* that shows in each time the "level" of agreement between the real altimetric deformation of the Tower base and the interpolating plane.

For each benchmark of the 15 considered, is evaluated the difference between its vertical deformation and the value, in the same planimetric position, of the deformation plane (see Picture 7.2). The variable *VARH* is the sum of the modulus of the difference between these two values. The growth of *VARH* means that the basement is subjected to a deformation that is moving the base elevation shape away from the original (May 1995) plane.

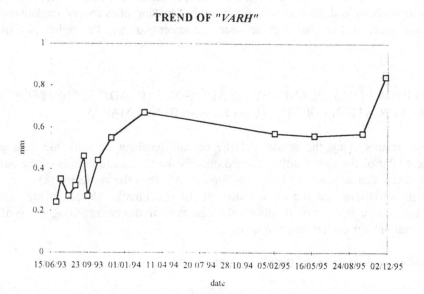

Picture 7.1 Deformation trend of the foundation plane.

Picture 7.1 shows how change the agreement variable *VARH*: it is present an increase of shift from 1993 to February 1994 and from September 1995 to November 1995; the tilt of the Tower base is not perfectly like a rigid tilt but there are some deformations inside. Picture 7.2 shows, benchmark by benchmark, the differences between the height differences calculated (with the interpolating plane) and the high differences measured.

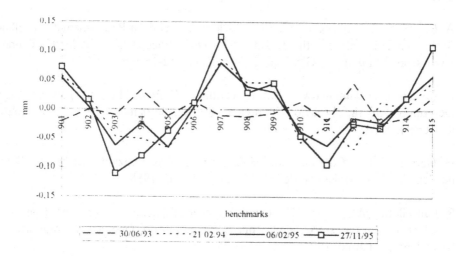

Picture 7.2 Difference between the heights calculated and the heights measured

8. CLOSING REMARKS

The analysis before described prove the presence of small deformations of the Tower basement. Therefore the results show the beginning of a detachment process between the Tower body and its basement (see Table 6.1). Unfortunately the load displacement, made in different steps, and the works on the soil around the Tower basement modify continuously the Tower trends of deformation. Therefore at the moment it is impossible to propose a mathematical model able to give a satisfactory description and prediction of the whole Tower behaviour. Now it is only possible to try modelling short periods of time [2], when the soil conditions and the counterweight values can be considered as stable, or highlight some particular deformations.

REFERENCES

[1] A. Colombo: "Analisi delle deformazioni della Torre di Pisa dalla data della sua fondazione a oggi", Tesi di laurea in Ingegneria Civile per la Difesa del Suolo e la Pianificazione del Territorio, Politecnico di Milano, facoltà di Ingegneria, anno accademico 1993/1994.

[2] A. Colombo, G. Vassena, C. Monti: "The leaning Tower of Pisa: kinematic levelling of the foundation plane in the period 1993-1994", Proceeding of the ISPRS Comm. I Workshop "Multimedia GIS Data", Cism, Udine, 12-15 June 1995.

[3] D. Costanzo, M. Jamiolkowski, R. Lancellotta, M.C. Pepe: "Leaning Tower of Pisa: Description of the behaviour", Invited Lecture-Settlement '94, Texas A&M University, 1994

[4] G. Geri, B. Palla: "Considerazioni sulle più recenti osservazioni ottiche alla Torre pendente di Pisa", Bollettino della SIFET, n. 2, Roma ,1988.

[5] R. Lancellotta, M.C. Pepe: "The Leaning Tower of Pisa. a report on movements of the Tower and the surrounding square", Laboratorio Geotecnico, rapporto di Ricerca n°2.3, Torino, 1991

[6] Istituto Geografico Militare: "Oggetto: misurazioni geodetiche e topografiche di alta precisione per il controllo dei movimenti della Torre di Pisa e del suolo circostante, effettuate dall' IGMI nella ultima decade di Maggio 1991 e nella prima decade di Luglio 1991", Documento n°3 del 24 Luglio 1991, Firenze, I.G.M., 1991.

[7] Ministero dei Lavori Pubblici: "Ricerche e studi sulla Torre pendente di Pisa e i fenomeni connessi alle condizioni d'ambiente", Vol. 1-2-3, Firenze, Istituto Geografico Militare, 1971.

[8] C. Monti, G. Vassena: "Le operazioni geodetiche di misura delle deformazioni altimetriche e delle variazioni di pendenza della Torre di Pisa", in Rivista del Dipartimento del Territorio, Roma, n.1 1995.

[9] C. Monti, G. Vassena: "The leaning tower of Pisa: the geodetic approach to the control of the deformations", I[st] Turkish International Symposium on Deformations, 5-9 September 1994, Istanbul, 1994.

[10] B. Palla : "I movimenti verticali del suolo nella città di Pisa e nel territorio circostante",Istituto di Geodesia, Topografia e Fotogrammetria dell'Università di Pisa, 'Dinamica dei littorali' del Progetto Finalizzato'-Conservazione del suolo', Pisa 9-10 novembre 1978

[11] B. Palla: "Le variazioni di livello del suolo nella strettoia di Ripafratta tra Montuolo, (Lucca) e Rigoli (Pisa)", Bollettino della SIFET, n°3, 1974, pgg. 1-16.

[12] B. Palla , T. Cetti, M. Poggianti, E. Mengali, A. Bartolini : "I movimenti verticali del suolo nella pianura pisana dopo il 1920 dedotti dal confronto di livellazioni", Provincia Comune di Pisa, Pisa - aprile 1976.

[13] B. Palla: "Livellazione per il controllo della subsidenza nella pianura pisana a sud dell'Arno e ad est di Pisa", Comune di Pisa, Pisa - ottobre 1978

[14] B. Palla, M. Poli: "Linee di livellazione di precisione per il controllo della subsidenza nelle tenute di S. Rossore e Migliarino (Pisa)", Comune di Pisa, Pisa - ottobre 1978

[15] B. Palla: "Altimetria relativa: la scelta del riferimento. Applicazione ad una indagine sui movimenti verticali del suolo in terreno alluvionale ed in roccia a nord-est di Pisa", da Atti del 2° Convegno Nazionale di Geofisica della Terra Solida, vol II, C.N.R., Roma 12-14 dicembre 1983

[16] B. Palla: "Altimetria relativa: la scelta del riferimento. Applicazione ad una indagine sui movimenti verticali del suolo in terreno alluvionale ed in roccia a nord est di Pisa, Atti del 2° Convegno G.N.G.T.S., pgg. 671-682, Roma, 1983.

[17] B. Palla, R. Nardi, P.C. Pertusati, M. Tongiorgi: "Movimenti differenziali lungo la dorsale media toscana all'altezza dei Monti Pisani", da Atti del 4° Convegno Nazionale di Geofisica della Terra Solida, vol II, C.N.R., Roma 29-31 ottobre 1985

[18] Politecnico di Milano - Dipartimento di Ingegneria Idraulica, Ambientale e del Rilevamento, Comitato per gli intervento di consolidamento e restauro della Torre di Pisa. "Supervisione ed interpretazione dei rilievi geotecnici - topografici nella Piazza dei Miracoli e sulla Torre di Pisa", Milano, 1994.

EXAMPLES OF ARCHITECTURAL DATABASES

L. Colombo
University of Salerno, Salerno, Italy

G. Fangi
University of Ancona, Ancona, Italy

G. Tucci
Polytechnic of Milan, Milan, Italy

ABSTRACT

The typological knowledge of an architecture, its documentation and filing for monitoring, safeguard, maintenance and management is strictly connected to a pre-defined survey of the constructive systems, with their components and characteristics, like shape, morphology, dimensions, materials and so on.

Surveying techniques are therefore a basic tool for each plan on a monument, able to perform understanding of different significative aspects.

Besides, every architectural heritage is part of environmental and urban sites; so the range of operations has to be widened, taking in account the reciprocal actions between building and its surroundings.

Paper deals with these themes and presents some examples related to the process of vector and raster information in hybrid databases.

TASKS OF SURVEY

The architectural survey is to be planned in relation with the main features of the object, being investigated.

It needs 'therefore to define program by studying all the aspects regarding the understanding of an architecture, keeping in mind the following points of view:

1. for the paste and for the present,
   • the history;
   • the morphological and functional aspects (volume distribution and functions of the construction): one has to assess the layout of the building;
   • the structural point of view, etc.;
2. for the future,
   • recovery, restoration, preservation;
   • the relationships with the surrounding environment.

In a certain sense, it should be known what all the operators are looking for; a detail of the aim on which an architectural object might be inquired, is in figure 1.

Architecture investigation

♦ **Historical**

♦ **Morphological and functional**

♦ **Structural and constructive**

♦ **For recovery and preservation**

♦ **Urban environment analyses**

fig. 1

LINK WITH ENVIRONMENT

The investigation includes therefore the urban texture assessment, looking for common character going from the single architectural element until the neighbouring ones.

From geometrical relationships it can be derived a series of thematic maps; for example:

• technological networks;
• material prospects;
• representation of structural defects, building disease or pathology;
• the plan of colours, for the urban restoration.

NEW SURVEY TECHNIQUES

Once the needs for knowledge have been established, it can be started the geometrical survey; nowadays, new surveying facilities can make the phase easier. In fact, for this operation, in addition to the classical instruments, such as:

- geodetic techniques;
- classical stereoscopic photogrammetry;

new tools are now available, namely:

- photogrammetric monoscopic systems, letting the survey be more versatile and economic;
- digital photomosaics and orthoimages.

With the help of all this equipment, technicians can cope with their requested and planned tasks.

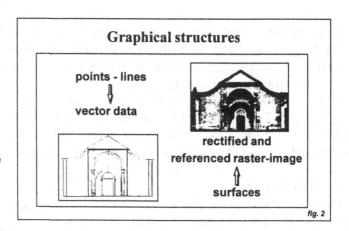

Graphical structures

points - lines

⇩

vector data

rectified and referenced raster-image

⇧

surfaces

fig. 2

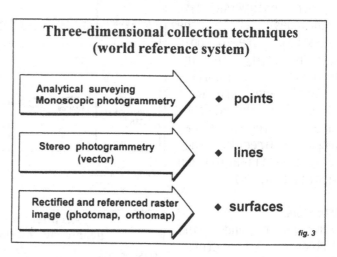

Three-dimensional collection techniques (world reference system)

Analytical surveying
Monoscopic photogrammetry   ◆ **points**

Stereo photogrammetry
(vector)   ◆ **lines**

Rectified and referenced raster
image (photomap, orthomap)   ◆ **surfaces**

fig. 3

RECORD OF ARCHITECTURE STATE
The aim of the survey is generally:

- to testify, to record and to ascertain the actual state of the architectural object, in a specific moment of its life;
- in order to let it be monitored in time;
- to assess and verify its decay cycle;
- to control the building functions, that can vary in different periods;
- to plan its restoration and maintenance.

For all these purposes it is necessary an efficient filing system: so, data collection has to be oriented, in advance, to the subsequent general Information System.

TYPE OF GEOMETRIC ITEMS
During the survey, items were collected, according to the graphical primitives:

- points, lines, surfaces (fig. 2).

It is assumed that a polyline, composed by the first two items, points and lines, can represent an object with its shape and geometry: this is stated by convention because of thematic description is demanded to a list of relational attributes.

Points and lines are acquired and recorded in a vector form, whereas surfaces are better represented by rectified and referenced raster images, such as scale digital-mosaics and orthoimages. Figure 3 deals with gathering facilities according to the requested primitives.

COLLECTION TECHNIQUES

Geodetic survey and monoscopic photogrammetry are pointwise data techniques, while the classical stereoscopic methods supply line information; finally, raster images are well suited to record surfaces and therefore can support directly thematic classification and mapping (fig. 4).

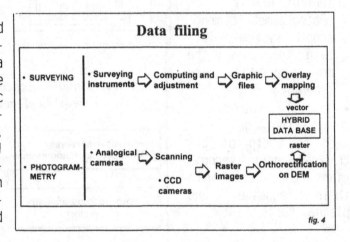

DATA FILING

From point and line collected, as vectors, it is possible to build up graphical files related to the metric description of each object, with its elements and a DEM; the typical organisation of the vector data is by different overlaid layers.

For image collection it can be used analogue photographs, developed and scanned, or taken by CCD cameras, to get at once raster information.

The digital images are then orthogonalised and resampled with the help of a previous DEM, acquired in form of TIN or GRID; both graphical vector data and raster ones are then linked together to create hybrid archives (fig. 5).

This is in the opinion of the authors the easiest and fastest way to build a reliable and efficient database for architecture management.

DATA UNDERSTANDING AND CLASSIFICATION

In an architectural object one can select "multiple geometry" for the graphical structures, that is, sequences of points, lines, surfaces with their typology.

Multiple geometry may be defined as:

- *Shape Geometry* (vector data), a set of items and features of the object obtained by geodetic ways, like surveying and

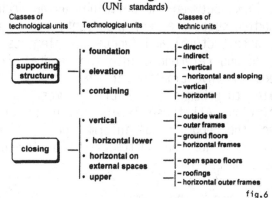

Architectural analysis: geometric elements
(UNI standards)

fig.6

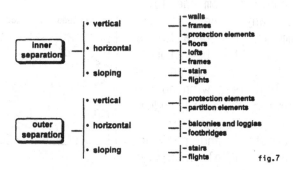

Architectural analysis: geometric elements

fig.7

photogrammetry, and classified according to Italian UNI standards; it can be regarded as a Constructive System Geometry. The figures 6, 7 show how constructive and closing elements may be recorded: the supporting structures with foundations, elevations, the frame elements; whereas the closing elements comprehend walls and roofs (horizontal, vertical and inclined), etc.;

- *Material Geometry* (vector data), regarding chemical and physical aspects;
- *Pathology Geometry* (vector data), related to the state of masonry decay;
- *Raster Geometry*, comprehends all character necessary to realise syntheses, according to each of the previous identified aspects.

For material and pathology geometry, the operator of the sector, himself, can collect directly every information, digitising shape geometry; a classification of pathology, suggested for Italy by CNR (National Research Council), is still in development.

Urban multi-level database

The links between the building, as an architectural object, and the surrounding space are very important, due to the relationships among different architectural and functional elements; so, the database must be connected and referenced into the general datum. There exist different degrees of investigation in the storage of data, from the geographical scale, to urban scale, until the most large scale, useful for detail description. All these levels should be organised in a kind of pyramid, where the geographical information is at the top of pyramid and the detail scale at the bottom, with different degrees of accuracy. In any case, all these archives must be congruent, keeping the same co-ordinates for the same point. What is changing from one layer to any other is the amount of details or points to be stored. Taking in account topology is a winning solution to avoid redundancy of geometric information and perform, fast spatial inquiries into thematic aspects.

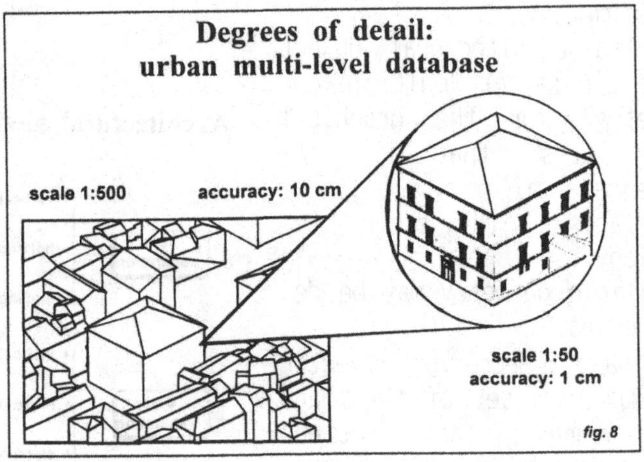

The referenced system utilised can be:

- for a specific object;
- for more than one object.

In every case, it must be possible to go throughout from any local reference system to the general datum. What is good for this Information System is the capability to pass from

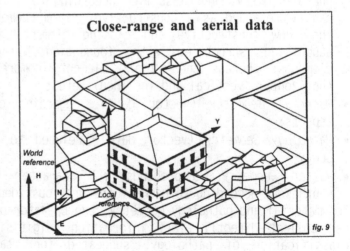

fig. 8

fig. 9

one system to another, from the general system to any particular reference and vice versa. (Figs. 8, 9).
Cartographic and architectural objects are described through their
semantic character: a set of alpha-numeric attributes stored in
relational cards (Figs. 10, 11).

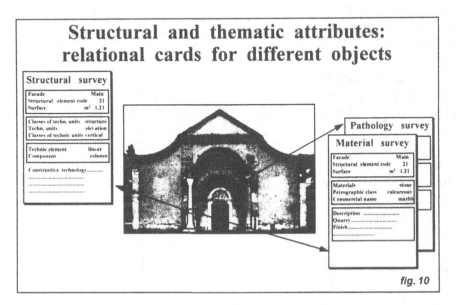

EXAMPLES
Italian application regarding these topics is pointed out in the
following pages: the survey and database collection for St.Sofia's church
in Benevento and Castle of Casape.
St.Sofia's church is an ancient construction of Longobard origin, well-
known for its historical events since eighth century to nowadays.
First, the ground section of the building has been represented by means
of a set of surveying measurements; later, investigating on internal and
external spaces, the masonry thickness was identified, together with the
arrangement of column orders bearing the vault. The intrados of the
covering structure was reproduced through longitudinal and cross
sections, according to the symmetry of the building; this aim has been
performed by a pulse distance meter, collecting thousands of points.
The shape and the attitude of the walls were described, on the contrary,
in a photogrammetric way, taking stereoscopic and monoscopic photos,
suitable for joint restitution: for instance, in a Siscam Stereometric
System.
The image quality regards high-definition colour films; besides, a Rollei
camera 6006 has been still preferred to digital one for its large
reliability.

Every wall of the church is visualised at 1:50 scale by means of photo-mosaics, realised with merging of elementary rectifications(Siscam Archis) and printed on raster devices (Epson Stylus colour).

A DEM for main façade was provided, as the necessary support for orthophoto production, both in a manual way and by automatic image correlation. The vector plotting has been located in a general reference and connected, finally, in a three-dimensional model(wire-framed) of the church; this base has allowed operators to carry out analyses on the relationship among elements, on cross-sections, materials and pathology (figs. 12, 13, 14, 15, 16).

Castle of Casape is part of a medieval castrum sited close to Rome; the main elements of his-

St. Sofia's church in Benevento: cross section on façade

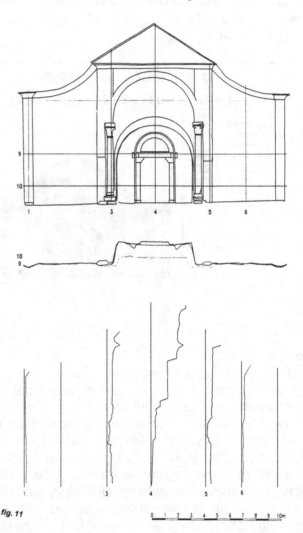

fig. 11

0 1 2 3 4 5 6 7 8 9 10m

torical interest are just the castle, at the beginning of an ancient nucleus, and the central road of the village.

From maintenance point of view, the buildings show many static problems, due to interventions occurred during the centuries and partly to their different utilisation.

Besides, the general lack of restoration and the earthquake effects have caused a dangerous decay with collapses. The photogrammetric survey of the village, raising as an island over the top of a hill, was performed

Three-dimensional wire-framed view of geometrical plotting for main elements

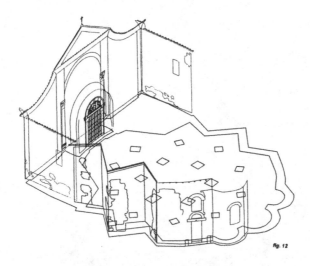

fig. 12

Digital photomap of the façade

fig. 13

taking images with nearly horizontal attitude, by means of a 35 mm wide-angle camera. So, it can be regarded as a half-way experience between close-range architectural works and aerial photogrammetric application. Analytical restitution and scale print-outs were carried out on different layers, operating with monoscopic graphical-systems (figs. 17, 18).

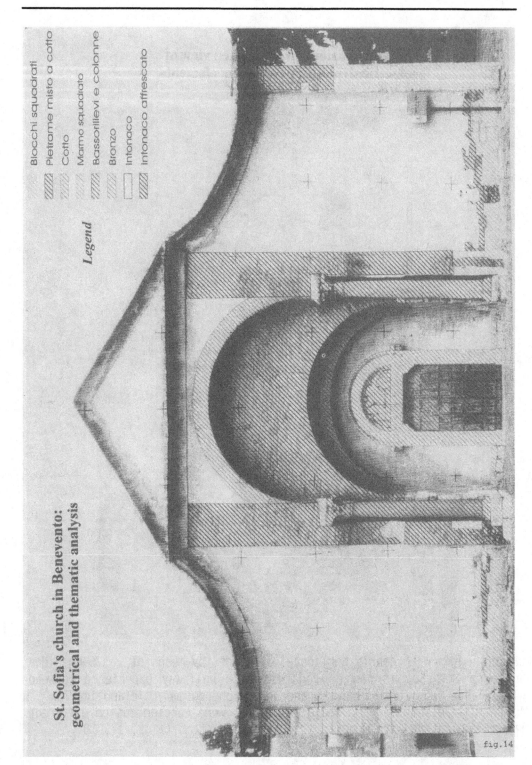

St. Sofia's church in Benevento:
geometrical and thematic analysis

Legend

Blocchi squadrati
Pietrame misto a cotto
Cotto
Marmo squadrato
Bassorilievi e colonne
Bronzo
Intonaco
Intonaco affrescato

fig.14

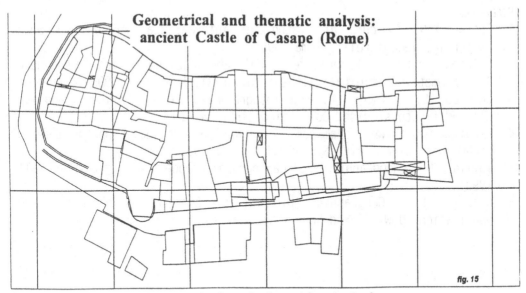

**Geometrical and thematic analysis:
ancient Castle of Casape (Rome)**

fig. 15

Geometrical and thematic analysis

fig. 16

Final evaluations of the accuracy for geometrical information stored in databases, were carried out through surveying measurements of distance and point determinations over the elements. The discrepancies between compared values are generally homogeneous and under the standard t $[t(mm) = 0.4 * n_i]$ pre-defined at different scales $1:n_i$.

NOTE: A topological organization seems to be very heavy when the application field is significantly three-dimensional; however, topology could be neglected in architectural applications because scales are always very large and therefore each graphical object is described directly through its co-ordinates.

REFERENCES

1. Bottoni, M. (a cura di): Studi di informatizzazione per il bene culturale, Vecchiarelli, Roma 1990.
2. Colombo, L.: Esperienze nell'applicazione di procedure fotogrammetriche digitali ai beni culturali, Pixel, 6 (1995).
3. Colombo, L. - Tucci, G.: La fotogrammetria digitale per l'analisi dell'edificato storico, Bollettino SIFET, 4 (1993).
4. Fangi, G.: Two examples of non-conventional photogrammetry techniques, CIPA - XIV Symposium, Delphi 1991.
5. Tomlin, D.: Geographic and cartographic modelling, Prentice Hall 1990.
6. Worral, L.: Geographic Information System: development and application, J.Wiley and Sons 1990.

THE LEANING TOWER OF PISA: KINEMATIC LEVELLING OF THE FOUNDATION PLANE IN THE PERIOD 1993-1994

A. Colombo and C. Monti

Polytechnic of Milan, Milan, Italy

G. Vassena

University of Brescia, Brescia, Italy

ABSTRACT

The aim of this paper is to analyse kinematically the rigid tilt of the Pisa Tower foundation plane in the period 1993-1994, when a counterweight was displaced on the North side of the basement. The kinematics is studied using a kinematic levelling approach. In the first four paragraphes the kinematic levelling method and the counterweight application history are briefly described; moreover the results obtained are briefly analysed with some closing remarks.

The eight benchmarks under study are hanged to the Tower inside walls and can give a good idea of the Tower basement deformation trend. The levelling measurements have been made daily since 14/06/1993 and this fact allows to use these informations as a valuable way to check the reliability of the electronic instruments that observe continuously the Tower behaviour.

1. INTRODUCTION

The Leaning Tower of Pisa has been subjected, since its erection in 1173, to a progressive tilt towards South, reaching nowadays the alarming value of 5° 28'; in the last five years a great concern has grown on its structural integrity due to the yearly leaning increase that, at the present, is about 5-6 second of arc'.

In view of the high risk of a structural collapse, increasing with the leaning value, an International Multidisciplinary Commission has planned to stabilise the Tower and to avoid the increasing danger. The first preliminary work was planned in the way to move the Tower in safety conditions; it consisted on placing in a provisory and removable way 6000 KNewton of lead on a concrete ring built on the North side basement of the Tower. The aim of the lead displacement was to tilt the Tower towards North, in more stable conditions, permitting to work in safely conditions during the long term stabilisation works.

2. THE LEAD COUNTERWEIGHT APPLICATION

The lead counterweight was placed on the North side of the Tower basement in the period 1993-1994 in five phases. This temporary and reversible work has been made gradually keeping the Tower under constant monitoring. These five loading steps were planned and realised with a period of span stabilisation time planned after each of them.

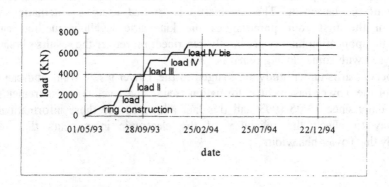

fig. 1. The counterweight displacement history

Figure 1 shows the loading history between 6/05/1993 and 21/01/1994. The counterweight was placed on a prestressed concrete platform built all around the Tower basement, in order to support the lead. After 21/01/1994 the total weight of the lead was about 6000 KN, corresponding to the 4.8% of the Tower total weight.

3. THE TOWER MOVEMENTS MONITORING

The counterweight displacement is the main origin of the deformations of the foundation plane and the main cause of the entire Tower rotation towards North in the period here studied. The Tower movements are checked using different kinds of methods; some of them control only the elevation part of the monument, some others observe only the basement deformations.

Historical studies [7] have shown that the upper part of the Tower (over the basement) is subjected to a well observable deformation due to the environmental conditions like wind, temperature changes, groundwater oscillations, sun position and so on. On the other hand the basement of the Tower is much less influenced by these effects. For this reason, in order to study only the movements due to the counterweight loading, becomes very interesting to analyse the Tower basement behaviour.

In 1911 fifteen levelling benchmarks were placed on the external base of the Tower; nowadays they are a part of a wider high precision spirit levelling net placed all around the Tower and on the surrounding Square. In 1993 nine new benchmarks, shown in figure 2, were placed in the inside part of the Tower and fixed to the stone walls. This small levelling net, composed by 30 centimetres long invar rods, can be easily measured also several times a day. For this reason it was decided to measure this network at least ones a day, at 9.00 a.m.. During the loading periods these measurements were fitted reaching the number of 7-8 per day. Due to their position, these benchmarks are practically not effected by environmental changes and in this way they are suitable to describe the evolution of the Tower basement rigid tilt.

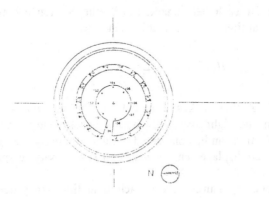

fig. 2. The position, inside the Tower, of the levelling net benchmarks.

4. KINEMATIC LEVELLING METHOD

In order to better understand the behaviour of the Tower basement it was decided to analyse also the kinematic trend of the levelling benchmarks. This study can easily show the different deformation speeds and movements of the different benchmarks after each lead displacement.

The kinematic levelling is a method used to obtain kinematic informations about vertical movements, from the measurements of height difference done during several levelling campaigns of the same network. By a kinematic model that links, in a motion equation for each benchmark, high and motion, it is possible to make a reconstruction of the vertical movements for the entire net. When the number of time of measurement doesn't permit to make a time series analysis, it is possible to built a functional model by a polynomial, function of opportune degree of time t, to describe the vertical shift of the benchmarks. After the adjustment, the parameters of the model can be analysed as samples from a realisation of a stationary stochastic process. If the vertical shifts of the network are regular they can be modelled by a polynomial function of time like this:

$$H_i = H_0 + c_1 \cdot \Delta t + \frac{1}{2!}c_2 \cdot \Delta t^2 + \frac{1}{3!}c_3 \cdot \Delta t^3 + \dots \tag{4.1}$$

where the symbols meaning is:

H_i = height at time t_i H_0 = height at time t_0 Δt_i = spantime
c_1 = speed at time t_0 c_2 = acceleration at time t_0

The parameters c_i with $i > 2$, have not physical meaning but can be interesting in a statistic analysis.

By this model, done for each benchmark, the height difference between a couple of benchmarks (A and B) at time i is modelled in such a way:

$$\Delta H_{A,Bi} = H_{Bi} - H_{Ai} = H_{B;0} - H_{A;0} + \frac{1}{j!}\sum_{j=1}^{m_B}c_{jB} \cdot \Delta t_i^j - \frac{1}{j!}\sum_{j=1}^{m_A}c_{jA} \cdot \Delta t_i^j \tag{4.2}$$

The kinematic levelling, that makes a model for the movements of the net by a polynomial, works in the right way only during spans time without counterweight changes when vertical shifts can be considered regular. If, during the span time, there are some changes in the lead displacement or weight, it is necessary to split the total gap in several spans.

All the height difference measurements in each span time are processed in only one adjustment; in this way all the parameters are estimated only ones for each span. The estimation of H_0, H_i and c_i for each benchmark and in each span time is done by a least squares adjustment using this *FUNCTIONAL MODEL*:

$$l_i + v_i = A_{h_i} \cdot h_i + A_{1_i} \cdot c_1 + A_{2_i} \cdot c_2 + \dots + A_{J_i} \cdot c_J$$

$$\tag{4.3}$$

and this *STOCHASTIC MODEL:*

$$\Sigma_{ll,j} = \sigma_0^2 \cdot Q_{ll,j} \qquad\qquad (4.4)$$

where the symbol meaning is:

l_i = observation vector v_i = residue vector $A_{...,i}$ = design matrix h_i = height vector
c_i = parameter vector $\Sigma_{ll,i}$ =covariance matrix $Q_{ll,i}$ = cofactor matrix σ_0^2 = variance factor

The reference system is chosen taking at the start time zero, at least one constrained benchmark (at high arbitrary chosen); this is like taking its height constant and the others motion parameters equal to zero during the span time.

5. KINEMATIC LEVELLING OF INSIDE NETWORK MEASUREMENTS

The period analysed, from 14/06/1993 to 03/02/1995, is composed by 588 measurement dates of the levelling network placed inside the Tower. The measurements, taken every day with a Zeiss Ni002 level and composed by 8 level measurements connecting 8 benchmarks, have been already adjusted using a classical least squares approach obtaining a 0,01 mm medium r.m.s. The Kinematic adjustment is carried on using the program *Level* (Politecnico di Milano) [4], [5], [6] taking the benchmark 101 constrained. Program *Level* asks as input two files, the first with the height observations and the second with the benchmark of the network and the steering parameters. The first file contains, for each observed height difference: the two benchmark names, the observed height difference, the value of σ_0^2 , the length of the levelling segment, the time of measurement and the group to which the observation belongs (when variance factors have to be estimated for group of measurements).
The second file contains: the benchmark name, a parameter to define if the benchmark is a constrained one or not, the value of preliminary height and the degree of the polynomial. This program gives in output point by point and for each span time: the values of the height, of the motion parameters and of their r.m.s..
Using the program it is necessary at first to choose the right degree of the polynomial equation, as described in par. 4.
The second step consists on dividing the whole time studied in several spans of time; in each of these the deformations must be considerable as regular. These choices can be done processing the measurements in different combinations and looking for the less r.m.s. of the parameters. For these reasons several tests were carried on, before the data processing of the measurements coming from the Leaning Tower inside levelling network. In this way, it was defined that the best choice was the use of 1st degree polynomials. Therefore it was decided to divide the total measurements time in 11 spans time, so to make the model of the Tower behaviour in each loading step and in each

break between two loadings. The final 1st degree polynomial model used to describe the motion of each benchmark j in the span time i appears as follows:

$$H_{j,i} = H_{j,0} + a_{j,i} \cdot (t_i - t_0) \qquad (4.5)$$

where the parameter $a_{j,i}$, constant during the all span time i, is the speed of the benchmark j.

The second step of work is to process the entire data base with the combination work above described to estimate all the parameters.

6. RESULTS OF KINEMATIC LEVELLING

The Kinematic study of the Tower behaviour during the counterweight loadings is extremely important for the link between the counterweight displacement and the Tower motions.

The results of the kinematic analysis are shown in table 1, span time by span time, for the benchmarks 102 and 106 that are considered the most interesting ones for their position. 102 is placed in the North side and 106 in the South, that means along the direction on the Tower leaning.

Looking at table 1 it is necessary to remember that the speed is taken positive if the benchmark sink and that the movements and the speed are referred to point 101, taken as constrained benchmark, having for this reason speed and movements always equal to zero. The difference of level between 102 and 106 it is obtained as a result of the kinematic levelling analysis. The results show that after the first loads the Tower basement has immediately a quick reaction. After the first load the benchmarks are characterised by a non regular speed; from LOAD II the benchmarks speed is always more regular, that means a more coherent reaction of the ground to the lead displacement.

BENCHMARK	$H_{102}-H_{106}$ (mm)	SPEED $(mm\,10^{-2}/day)$	SPAN TIME	PERIOD
102	106,44	0,07	BREAK 0	from 14/06/1993
106		0,50		to 13/07/1993
102	106,32	0,18	LOAD I	from 14/07/1993
106		1,06		to 30/07/1993
102	106,01	0,04	BREAK I	from 31/07/1993
106		0,13		to 23/08/1993
102	106,05	-0,13	LOAD II	from 24/08/1993
106		1,50		to 08/09/1993
102	105,73	-0,14	BREAK II	from 09/09/1993
106		0,10		to 29/09/1993
102	105,71	0,00	LOAD III	from 30/09/1993
106		0,44		to 15/10/1993
102	105,59	-0,05	BREAK III	from 16/10/1993
106		0,02		to 26/11/1993
102	105,55	-0,50	LOAD IV	from 29/11/1993
106		0,33		to 10/12/1993
102	105,40	0,00	BREAK IV	from 11/12/1993
106		0,00		to 10/01/1994
102	105,36	0,10	LOAD IV BIS	from 11/01/1994
106		1,00		to 20/01/1994
102	104,75	0,09	BREAK IV BIS	from 21/01/1994
106		-0,43		to 03/02/1995

Table 1

A similar analysis can be done in the direction East-West using benchmarks 104 and 108 (104 on East side and 108 on West side). Table 2 shows the results:

BENCHMARK	H_{104}--H_{108} (mm)	SPEED (mm10^{-2}/day)	SPAN TIME	PERIOD
104	13,19	0,33	BREAK 0	from 14/06/1993
108		0,30		to 13/07/1993
104	13,22	-0,12	LOAD I	from 14/07/1993
108		0,53		to 30/07/1993
104	13,35	0,08	BREAK I	from 31/07/1993
108		0,50		to 23/08/1993
104	13,24	0,25	LOAD II	from 24/08/1993
108		0,50		to 08/09/1993
104	13,17	-0,10	BREAK II	from 09/09/1993
108		-0,14		to 29/09/1993
104	13,16	0,13	LOAD III	from 30/09/1993
108		0,13		to 15/10/1993
104	13,21	0,05	BREAK III	from 16/10/1993
108		0,02		to 26/11/1993
104	13,14	0,33	LOAD IV	from 29/11/1993
108		0,25		to 10/12/1993
104	13,06	0,03	BREAK IV	from 11/12/1993
108		0,00		to 10/01/1994
104	13,06	0,70	LOAD IV BIS	from 11/01/1994
108		0,30		to 20/01/1994
104	13,07	-0,18	BREAK IV BIS	from 21/01/1994
108		-0,24		to 03/02/1995

Table 2

7. THE LEANING TOWER BEHAVIOUR

Taking as true the hypothesis of the Tower basement as rigid body, it is possible to study the tilt movements from the vertical speed informations and to analyse the movements coming from the benchmarks placed in the directions North-South and East-West. In such a way the analysis of the tilt speed is done separately for the two components North-South and East-West. The data are taken from the vertical speeds of the couples of opposite benchmark (N-S) and (E-W) and from their distances. By using of the vertical speeds of benchmarks 106 and 102 and of their distance, the tilt speed of the Tower basement in the North-South direction is obtained as follow:

$$\vartheta = \arctan(\frac{a_{106} - a_{102}}{distance}) \ (7.1)$$

In table 3 the Tower basement tilt speeds towards North and East are shown.

SPAN TIME	SPEED OF TOWER RIGID TILT TOWARD NORTH ("/day)	SPEED OF TOWER RIGID TILT TOWARD EAST("/day)	LOADING COUNTERWEIGHT (KN)
BREAK 0	0,12	0,01	1020
LOAD I	0,25	0,08	2374
BREAK I	0,04	-0,18	2374
LOAD II	0,45	-0,07	3855
BREAK II	0,07	0,01	3855
LOAD III	0,12	0,02	5366
BREAK III	0,02	-0,05	5366
LOAD IV	0,23	-0,16	5997
BREAK IV	0,00	0,01	5997
LOAD IV BIS	0,25	0,11	6896
BREAK IV BIS	-0,14	0,02	6896

Table 3

The figures 3 and 4 show how the two components of speed   (respectively in N-S and E-W direction) of the Tower change increasing the total applied counterweight.

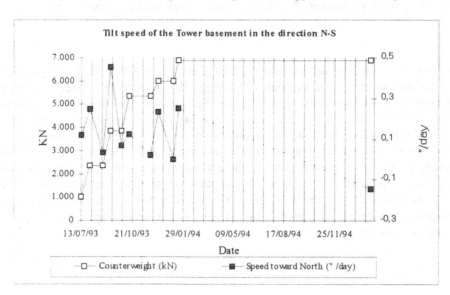

fig. 3. The speed of tilt towards North and the counterweight displacement history.

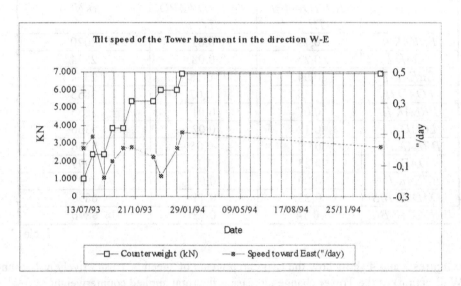

fig.4. The speed of tilt towards East and the counterweight displacement history.

The Tower behaviour in the two different directions N-S and W-E is quite different. In the N-S direction the speeds are much higher and this is a natural consequence of the load positioning on the North side. While in the direction N-S there is an important movement towards North, that decreases the original Tower leaning component (that is about 5° 28' towards South), in the direction E-W the Tower swings between East and West. In E-W direction the motion is like an oscillation and it doesn't change the little originally E-W Tower leaning. For all these reasons (important tilt movements and original mean direction of the Tower leaning) the N-S direction seems at the moment much more interesting than the W-E one. In the N-S direction each loading step involves important increases of tilt speed, whereas during the breaks span time important decreases of speed are present.

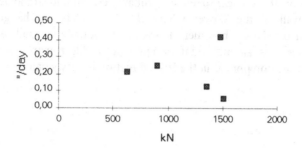

fig. 5. The relation between counterweight and tilt speed increases.

It is not easy to define a model that links vertical speed and counterweight displacement.

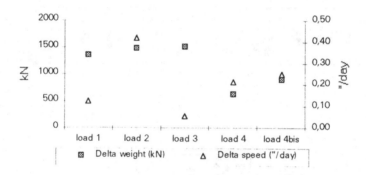

fig. 6. The link between weight and speed increases, loading step by loading step.

Figure 6 shows how changes, step by step, the link between weight and speed increases. At the moment it isn't possible to define any kind of relationship between these two parameters, but in order to find one vertical speeds of all benchmarks are compared span time by spine time, looking for a congruence between all the values.
During loading steps the speeds are quite high but they change regularly along the Tower basement and this means that there are not important deformations in the basement in the period studied but only a tilt towards North of the entire basement (see figure 7). Only the first loading step involves some deformations inside the basement, in fact benchmark

103 has vertical speed very different from benchmarks 104 and 102 next to it. The irregular values for the vertical speed are probably due to a local movement of the stones of the inside wall of the Tower where the invar rods are hanged and to a first redistribution of the Tower basement loads on the foundation soil. Anyway during the loading steps there is an important component of tilt rowards North of the entire basement, while the component in the W-E direction is very small.

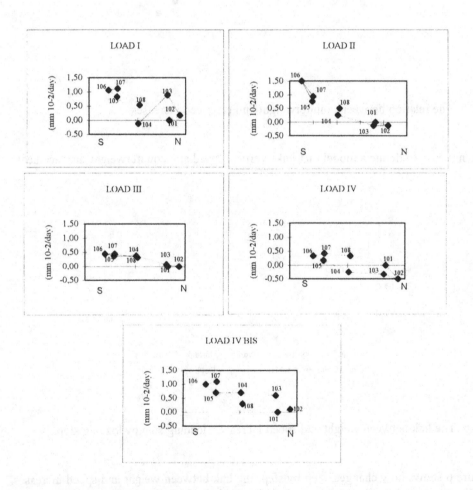

fig. 7. Vertical speeds during loading steps

On the contrary, during break steps, vertical speeds are very small (see figure 8) and this means that the component of movement due to the increases of the counterweight applied go quickly to zero; all these little vertical speeds are settlement movements of the basement due to the previous counterweight increase. It is very important to notice that during breaks all the vertical speeds are quite regular and this means that there are not important deformations inside the basement during break times (except break 0) when benchmark 108 has a vertical speed very different from benchmarks 107 and 101 next to it.

During the breaks span time the W-E component of the Tower tilt is similar to the value in N-S direction.

This analysis shows that the hypothesis of rigid body for the Tower basement in the period 14/06/1993 -03/02/1995 is quite right at least disregarding few local movements.

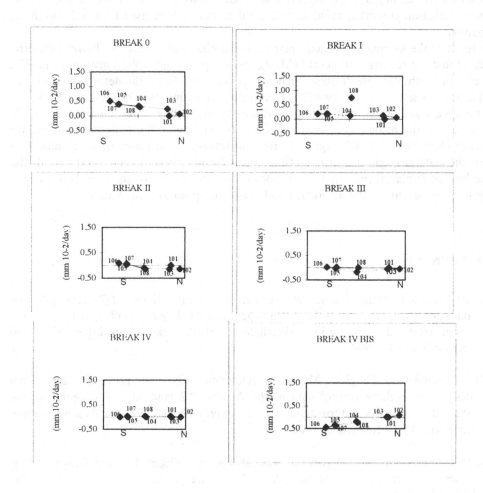

fig. 8. Vertical speeds during break steps.

7. CLOSING REMARKS

By a kinematic levelling it is possible to make a very simple model of the Tower behaviour.

This model is easy to make, to use and describes in a right way the Tower motions under the counterweight loading. Since a soil geotechnical model is very difficult to do, this model even if doesn't explain the geotechnical problems, is very important and useful to understand what happens to the Tower under the counterweight displacements and to plan next works. In particular important informations are obtained about the soil reaction to the stabilisation works; additionally these data can be later used for geothecnical soil analysis.

A result of the kinematic levelling method, is the observation that the Tower basement, after a first not regular reaction (LOAD 0), show a quite regular deformation speed. This means that until now aren't present dangerous discontinuites in the deformation process. This analysis show that the hypotesis to consider the counterweight displacement just as a temporary and not resolutive remedy is almoust right. In fact the total counterweight produced at the beginning a decrease of the Tower leaning but it doesn't changes the Tower characteristic trend (regular increase of leaning toward South); the results show that the counterweight causes rapid tilt towards North only during the short period after the loading displacement. Later the tilt speed decreases until zero and then Tower begins again to move with its characteristic trend of low tilt speed towards South.

REFERENCES

[1] A. Colombo: "Analisi delle deformazioni della Torre di Pisa dalla data della sua fondazione a oggi", Tesi di laurea in Ingegneria Civile per la Difesa del Suolo e la Pianificazione del Territorio, Politecnico di Milano, facoltà di Ingegneria, anno accademico 1993/1994.

[2] L. Colombo, G. Fangi, L. Mussio, F. Radicioni: "Kinematic processing and spatial analysis of levelling control data of the Ancona '82 landslide", Proceeding of the symposium on height determination and recent crustal movements in Western Europe, Hannover, 15-17 September 1986.

[3] D. Costanzo, M. Jamiolkowski, R. Lancellotta, M.C. Pepe: "Leaning Tower of Pisa: Description of the behaviour", Invited Lecture-Settlement '94, Texas A&M University, 1994.

[4] A. De Haan, G. Forlani: "The system of Program LEVEL for Kinematic Levelling Adjustments", Bollettino di Geodesia e Scienze affini, year XLVIII, n° 4, Ottobre-Novembre- Dicembre, 1989.

[5] A. De Haan, G. Forlani: " Kinematic Levelling Adjustments with Polynomials and Polynomial Splines", Bollettino di Geodesia e Scienze Affini, year XLVIII, n° 3, Luglio- Agosto- Settembre, 1989.

[6] G. Forlani, L. Mussio. "The Kinematic Adjustment of levelling networks", Bologna, ed. Unguendoli, 1989.

[7] C. Monti, G. Vassena: "The leaning tower of Pisa: the geodetic approach to the control of the deformations", Ist Turkish International Symposium on Deformations, 5-9 September 1994, Istanbul, 1994.

[8] Politecnico di Milano - Dipartimento di Ingegneria Idraulica, Ambientale e del Rilevamento, Comitato per gli intervento di consolidamento e restauro della Torre di Pisa. "Supervisione ed interpretazione dei rilievi geodetici - topografici nella Piazza dei Miracoli e sulla Torre di Pisa", Milano, 1994.

PROCEEDINGS OF THE ISPRS WG III/4 TUTORIAL ON "SPATIAL DATA ANALYSIS THEORY AND ALGORITHMS"

Whitin the workshop activities, a full tutorial was organized by the ISPRS working group III/4.

The tutorial covered some fundamental arguments in GIS theory, as well as some advanced topics in Dynamic GIS and data presentation.

The following papers report the contribution given by the invited lecturers, coming from four european schools in GIS.

F. Crosilla

DATA STRUCTURES AND DATA MODELS IN GIS

N. Bartelme

Technical University Graz, Graz, Austria

1 Introduction

1.1 Outline

In Geographic Information (GI) technology, it is essential to find suitable mappings between the – one and only – 'real world' and – several possible – 'virtual worlds' that constitute greatly simplified copies of the original, omitting aspects that are irrelevant for a certain range of applications [1]. It is also of prime interest to design data structures that are powerful and rich enough to support such applications, yet lean and flexible enough to adapt to today's restrictions, as they are imposed by the hardware with respect to data volume and performance [2]; other restrictions – widely unrecognized, but far more severe – are due to the lack of experience as to which goals can be and which shall be achieved by the use of geographic information systems (GIS).

The first issue leads us into discussions about modeling and about the different approaches that can be taken when creating a model of reality. Once a model has been agreed upon, the second issue comes up, addressing logical and technical matters of structuring data [3], setting up a data schema and utilizing such general tools as database technology and standardization processes.

The subsequent discussion of modeling and structuring approaches, although focussing primarily on GIS data (we might use the terms 'geographic data', or 'geo-data'), must naturally fit into the more general context of data that are being made available to the ever-growing appetite of our information-oriented society; data of quite different kinds, one of them representing objects that are being referenced to space, to the earth's surface, to some coordinate system or other spatial reference frame agreed upon. Although in their earlier days, GIS for obvious reasons started out with representations of the real world that closely resembled traditional maps with additional semantic appendices, the systems of today reach far into other domains.

While conventional systems administer a static view of a real-world situation (cadastre, landuse, road networks, ...), so-called activity systems (route planning and guidance, urban mobility and commuter managing) extend the scope into dynamic worlds where a great number of long-lived and short-lived aspects influence each other, both on a global perspective and in their individual behavior. Multimedia systems integrate images, video and sound, conveying information to the user that, in its combination, is far more expressive than in any of its parts. Hypermedia adds another dimension by allowing the user to navigate through paths in the virtual world and even to set up her/his own paths. Naturally, such extensions call for new types of data, forcing us to re-think about conventional modeling and structuring strategies.

1.2   From Real World to Data

By abstraction, the real world is mapped into a model. By adding organizational detail, the model is further formalized into a schema [1] (fig. 1). *Data* are results of such abstractions and formalizations, as in the following examples:

> Coordinates as abstractions of position,
> a polygon as abstraction of a parcel,
> the name (of the parcel owner) as abstraction of a person.

A simple geometric model of a river is a polygon, i.e. a sequence of coordinate pairs or triples being connected by straight line segments. (There are many alternatives to this simple approach.) The geometry is accompanied by a semantic classification (RIVER) and additional semantic attributes, leading to a so-called *geographic feature*:

Feature classification:	RIVER
feature identification:	Name
feature characteristics:	Additional names of local importance
	Depth min. and max.
Geometry:	Area, given by 'outer' polygon;
	Islands, given by 'inner' polygons;
	heights are not considered;
Constraints:	No loops allowed;
	Depth min. < Depth max.
Relations:	RIVER flows into LAKE, SEA or RIVER

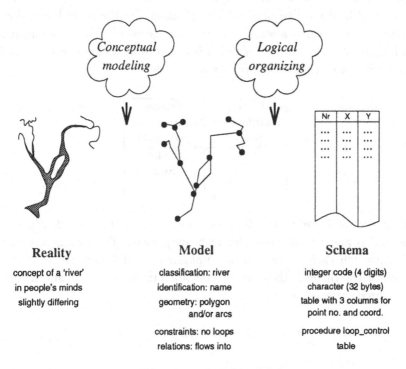

Figure 1: Reality, Model and Schema

Such a model is comparable to a first conceptual draft. For making it work in a computerized environment, it must be made more precise. This leads to *data schemas*. They are prerequisites for the implementation and usage of software, database and other tools. A *data schema* is a comprehensive description of the structure and the contents of data and of rules to be observed when dealing with such data. It must be expressed in a formalized way, textually and – if possible – also graphically. A *schema* for our river model could be based upon a *feature table*:

Surrogate	Feature identification (external)	Feature classification (code)
1	Danube	1005
2	Rhine	1005
3	Po	1005
4	North Sea	2117
5	Mediterranean	2117
...	...	...

A *surrogate* is a unique number or string that can be used for building relations or pointer structures. It is also called *internal identification*. On the other hand, the *external identification* can be used as a search criterion for the user in his/her application world ('natural identification'). The *classification* is achieved by a unique code describing the feature class; it is a surrogate on a higher *(meta-)* level:

Feature classification (code)	Feature classification (external)
1005	RIVER
2117	SEA
...	...

For creating a virtual world, structural components must first be identified and then assembled. There are several choices for such components. Three main approaches may be listed. Although each of them puts emphasis on a different aspect, they are not mutually exclusive. Instead, they can be combined to some degree, and many systems do so. The three approaches are

- entity–attribute–relationship approach (EAR approach)

- layer approach

- object-oriented approach

In the EAR approach [3], *entities* are uniquely identifiable; each entity has characteristics that make it distinguishable from others. Entities are classified into *entity classes (entity types)*. *Relations* both between individual entities *(instances)* as well as between entity classes are defined. *Attributes* are assigned to entities as well as relations. As an example, think of entity classes like CITY and AIRPORT. The CITY entity class may be associated with an attribute POPULATION. The AIRPORT entity class may have an attribute CAPACITY. AIRPORT and CITY may be in a relation SERVEDBY, which, in turn, may be attributed by DISTANCE or TIMETRAVELED. There might be another relation DIRECTFLIGHT between entities of the same class AIRPORT being attributed by FLIGHTSPERWEEK.

VALUE DOMAINS can be specified for each attribute. As an example, DISTANCE may be a real number greater than zero; FLIGHTSPERWEEK may be a positive integer number. The *degree of a relation* tells us how many partners are involved in a relation. Most relations are *binary*, i.e. between two partners. The *cardinality* tells us if the relation is unique on both sides *(1:1-relation)*, multiple on one side *(1:M-relation)* or multiple on both sides *(M:N-relation)*. Examples for the three different cases are PROVINCE – CAPITAL (being 1:1), PROVINCE – CITY (1:M), PROVINCE – METROPOLITANAREA

(M:N). Value domains, relational degrees and cardinalitities are special cases of *integrity constraints*.

The EAR Approach may be replaced – or complemented or refined, depending on the point of view – by the *layer* approach. By sifting the extract of the world that we are modeling according to different interests, we get a pile of layers in a way similar to putting several transparencies on top of each other. This approach has its origins in the map-making techniques. It is only feasible if the situation is known in all locations of mapped space. Either a phenomenon is absent or it is present. There are no 'white spots'. (Strictly, this is only valid for raster data, not for vector data.) Also, all layers must use the same *metric*, the same *scale*, and their data must have the same *accuracy*. This is rarely the case in GIS. On the other hand, the hierarchical separation into independent layers makes it is easier to distribute work loads, competence and access rights. Still, there may be conflicts, since associations in the real world are not always hierarchical.

The third alternative is the *object-oriented* approach where entities may be managed on different complexity levels. Additionally, such an approach makes it possible to define a *behavior* repertoire and to *encapsulate* objects and their accompanying aspects.

2   Data Models

2.1   Worlds

2.1.1   World of Pixels: Raster Models

A raster model is based upon the assumption that space can be split up into areas with homogeneous semantic content (*tesselation*). Mostly, this is a square or rectangular decomposition, a *mesh* or *grid* of raster *cells* (fig. 2). The term *pixel* comes from image processing and may be used in a more general meaning also in GIS. A decomposition into irregularly shaped triangles is called *TIN (triangulated irregular network)*. GIS technology uses it mainly for *digital terrain models (DTM)*.

Figure 2: Raster model

The (conceptual) raster may be mapped into a (logical) *matrix*. Semantic evaluations result in numerical values for matrix elements. Raster models are highly compatible with the layered aproach. The overlay of several thematic layers in raster mode is called *composite map*. As long as the decomposition is the same in all layers, it is easy to compare and overlay them. Basic algorithms of image processing – e.g. logical comparisons and combinations – can be used. The *grey value* must be seen as a secondary information encoding a semantic fact.

2.1.2   World of Lines: Vector Models

Vector models build upon points and lines. Areas (regions) are modeled by closed loops of lines – there may be inner loops to exclude 'islands'. Simple geometric elements *point, line, region* are furnished with semantic *attributes*, thereby creating *features*. This must be done explicitly (while in raster models, grey values are inherent in the pixel pattern). Therefore, a vector model is also called *geo-relational*. Another term is *feature-based model*. If each geometric element carries exactly one semantic meaning, the term *single-valued map* is used [4]. In practical applications, this simplifying assumption rarely holds.

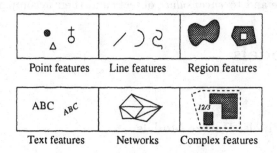

Figure 3: Vector model

In a vector-based environment, we have *point features, line features, region features, complex features* and *text features* (fig. 3). Complex features result from combining simpler features. A special form of complex features are *line* and *region (area, surface) networks*. Text features do not appear in every GIS. Texts on a map – and also in a GIS – are strictly speaking only visualizations of attribute values (except for some special cases). Still, some GIS look at texts as if they were features on the same level as points, lines, etc.

Point features build upon coordinates; line and region features use also coordinates; in a topological concept they used *edges, boundaries, faces* and the like. The vector concept may be generalized to three dimensions. For points, a third coordinate is necessary; for lines and regions, interpolation methods must be defined. GIS are often called *2,5dimensional*, because height plays a minor role (in most, but not in all GIS). For a truly 3-dimensional GIS, other modeling techniques must be used that borrow from CAD methods.

2.1.3   Other Worlds

Vector models are built upon discrete elements (points, lines, regions) that do not completely cover the domain of interest. Raster models do better when such a coverage is asked for – at least as long as we do not approach an accuracy that goes near the pixel resolution. In the strict sense, a *continuous coverage* with semantic information is given only when there are functional recipes for creating it. For lines we could use circles, trigonometric functions and the like, for surfaces we could take a sphere, an ellipsoid or anything that can be expressed by a mathematical formula. For practical purposes, we have to resort to piece-wise definitions, as is is done for interpolating a *digital terrain model (DTM)*.

The tools used for creating terrain and height models can be generalized to models of temperature, precipitation, per-capita-income, etc. We can efficiently use such tools whenever there are nearly-smooth variations of semantic aspects across a surface. In GIS of nowadays, such models usually are not integrated; we rather talk about two different systems (GIS on one side, DTM on the other) that exchange data.

Cognitive approaches in GIS make use of insights gained in psychology, philosophy, linguistics, anthropology and informatics to find ways to handle *symbolic* references such as 'in front of', 'behind', 'between', 'quite near to' etc. Such references are too *fuzzy* to be considered in nowadays GIS. But this fuzzyness is their strength when they are used in everyday life. We give up accuracy requirements which are unnecessary – and we come to conclusions much faster. A typical characteristic of cognitive models is the absence of a unique scale, a common metric, a space-filling coverage. (Note that all these aspects are deemed essential in nowadays GIS!)

2.2   Geometry and Topology

In vector models, *geometric information* is carried by points. All higher-level structures like lines, polygons (regions), . . . as well as derived properties (length, area value, distance, azimuth, . . .) are built upon points (fig. 4). *Coordinates* are used to model the geometry of points.

Geodesy can contribute well-defined strategies to map points on the earth's surface to reference planes that give the foundation to any GIS geometry. No such strategy can claim to be the ultimately correct one. Geometry can always be improved, say, by even more accurate measurements, leading to continuous transformations between

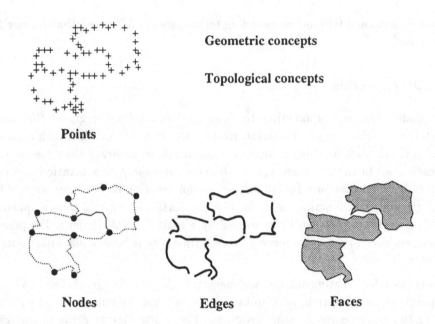

Figure 4: Geometry and topology

geometries of points and of 'everything else' that is built upon points. We are therefore interested in finding geometric relationships that remain invariant under such transformations. This is the essence of *topology*.

Topological properties cannot be measured directly such as metric properties (lenghts, angles, ...). In conventional maps, they are 'subconsciously' respected; they are noticed only when violated. That is why it is important to model them explicitly in a GIS. *Graph theory* and its tools can be used to describe, classify, check, and manipulate topological structures. Their main building blocks in 2D space are *nodes*, *edges* and *faces*.

2.2.1 Nodes and Edges

Nodes are connected by *edges*. The property of nodes being connected by an edge is called *adjacency*. Edges meeting in a node are called *incident*. Topology is not affected by the geometric shape of edges such as straightness, arc or interpolated curve. Edges that differ only by their *shape parameters* are called *topologically equivalent* since they can be continuously mapped to each other.

A well-known example for a purely topological representation is a public transport network (fig. 5). It only carries information how to get from point A to point B and where to change to another bus line, tram or subway. Those stops where at least two lines meet take on the role of *nodes*; the line segments between nodes are *edges*. Metric properties are not important. The position of nodes, their distance from each other, the shape of edges on such a representation does not reflect real-world properties. If metrics are changed (because a bus line is transferred to a neighboring street), the topology remains the same – unless there are new intersections with other lines resulting from such a transfer. Street curves are not relevant for topology, but interruptions or loops are of topological relevance.

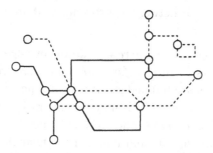

Figure 5: Topology of a public transport network

Edges may be *directed*. Directions do not reflect the internal order of storage for the end nodes of such an edge; they rather reflect application needs. We need directions in a water supply or sewage network, but not in a telephone network. In public transport, directions may be necessary if bus lines take different routes for the way back. Of course, this is also a matter of *scale*. If an entire city and its suburbs must be represented, the requirements are different from the ones in a large-scale display of a $200m$-zone around a bus stop, where the left and the right traffic lane (and maybe a separate public transport line) should be represented by different edges.

If the scope of attention is broadened from a local area containing one edge and its adjacent nodes or from a few incident edges to a large *network*, global aspects like *planarity*, *connectivity* and/or *tree structure* come in. In a planar network, edges can be arranged in a fashion that avoids overlaps. In a connected network, for two arbitrary nodes, a path of consecutive edges can be found. In a tree, there is only one such path for each pair of nodes. If the network is not a tree, there are non-trivial *circuits* (round-trips). There may even be the need to ask for a network containing nothing but circuits (no dead-ends). Such topological aspects are usually application-dependent.

The geometry of edges may be supported by intermediate points (so-called *vertices*) which do not contribute to topology since they are not on the same level as nodes; they only improve metrics. In a way, such intermediate points may be seen as a general version of a shape parameter.

Although metric properties are not of primary concern in such a graph *(edge-node-structure)*, it is natural to impose some metric restrictions: Usually for one location in space at most one node is postulated. We may also postulate that edges may not have identical (or partly overlapping) 'locations': Two different bus lines using the same street segments should be modeled by a single edge. Two neighboring parcels are supported by the same boundary edge. Edges may not intersect each other; at the point of intersection, a node must be introduced, and the edges must be cut in two. Such explicit constraints on topology help us to avoid redundancies, to make updates safer and faster, and to make queries simpler. Simplifications may go too far; each application may draw the line between constraints that make sense and constraints that do not.

Besides introducing the concept of *where* a node or an edge can be found in space, metrics call also for answers as to *how far apart* nodes are from other nodes (this is simple), how far apart nodes are from edges (this is in general not as simple), and how far apart edges are from other edges (this may become rather complicated). Euclidean metrics are not sufficient, and some other metrics must be defined that satisfy the constraints of being positive definite and fulfilling the triangle inequality.

A topological structure is desirable in many GIS applications. However, especially during *data capture*, there are phases when GIS data do not (yet) satisfy topological criteria. This means that each line segment is isolated from its neighbors. At the end point of such a segment, we know nothing about continuing segments. Line segments may intersect, there may be dead ends; points may approximately coincide. Such data are called *spaghetti* (fig. 6). Typically, they represent a pre-stage of topologically consistent data. This consistency is achieved after some measures of point location averaging and intersection have been taken, and *undershoots* (lines being too short) as well as *overshoots* (lines being too long) have been eliminated.

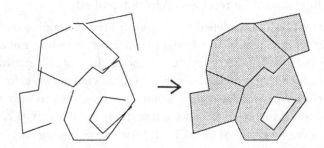

Figure 6: Transforming spaghetti into topologically consistent structures

2.2.2   Faces and Islands

If a graph contains *circuits* (non-trivial round-trips starting in one node and returning to the same node) then it becomes meaningful to consider *faces* that are surrounded by minimal circuits. This, of course, depends on the application. We arrive at a triplet of entity types *node*, *edge*, and *face* (fig. 7).

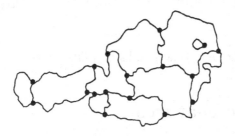

Figure 7: Graph for data of type area

Islands ('holes') inside regions are topological phenomena. There is no continuous mapping transforming a region without holes into a region with holes. Such holes are exceptions; their semantics are different from those of the parts surrounding them. In fig. 7, the city and province of Vienna is an 'island' to the province of Lower Austria. There are many examples for topological islands. Consequently, a face in general has one *outer boundary (ring)* and one or several *inner boundaries (rings)*. Such an island inside a face A can be made up by one or several faces B_1, B_2, etc. There are edges that belong to inner rings of face A and, at the same time, to outer rings of faces B_i.

Broadening the focus of attention to the whole network, we arrive at a *tesselation*. A well-known approach to modeling a tesselation is DIME (an abbreviation for *Dual Independent Map Encoding*). It contains nodes (*0-cells*), edges (*1-cells*) and faces (*2-cells*). All cells are identified by unique numbers. Adjacent nodes – being connected by an edge – and adjacent faces – being separated by an edge – can be put into an order as in the following quadruple:

node 1, *node* 2; *face A*, *face B*

Expressed in words, this means that nodes 1 and 2 are connected by an edge; if we start at 1 and proceed along this edge towards 2, then face A is to the right and face B is to the left. There is a *dual* counterpart to this: If we cross the border from face A into face B, node 1 is to the left and node 2 is to the right.

For each edge (1-cell) we can find a dual 1-cell expressing the crossing of the (primal) 1-cell. 0-cells and 2-cells exchange their roles: In the dual graph, regions change into nodes and vice versa. (In fig. 8, the edges of the dual graph are dashed.)

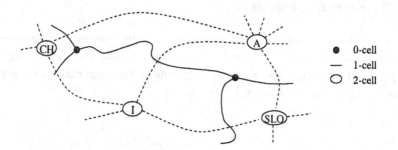

Figure 8: Dual graphs in a DIME structure

3   Geographic Features

3.1   Feature Types

3.1.1   Simple and Complex Features

In most GIS environments, the real world is mapped into *point features*, *line features* and *region features* that – in addition to geometric properties – carry *semantic* marks (fig. 9). Geometry is seen as a primary characteristic determining the type of the feature that emerges; semantics are attached at a later stage. The 3rd dimension (height) is not as important as the other two dimensions; that is why GIS are called *2,5dimensional*.

In some approaches, however, coordinates are treated just like any other semantic attributes. As an example, take a feature of type BUILDING. In such approaches, no conceptual distinction is made between the property of a building being spatially related to a certain part of the earth's surface and other properties like height, usage, or construction date. A uniform treatment of all attributes – spatial or purely semantic – clearly brings advantages. It can also handle features whose geometry is not (yet) known or which may be controversial. (Note that a building may be spatially represented by a point or pixel in a map 1:25000, but by a polygon in larger scales!) However, regardless of advantages, such an approach is rarely followed in GIS of nowadays.

In any case, *metric properties* as the position of points (x/y/z), the shape of lines (straight, arc, interpolated, ...), length, azimuth, direction, angle, area, curvature, slope, ... are expressed by measures (numbers) and related to each individual feature. *Topological properties* (being connected with, being contained in, being adjacent to, overlapping, ...) usually involve a pair of features.

Semantic properties express the affiliation of the feature to a certain class of features ('...is a ...'). They map the set of features into classes. Examples:

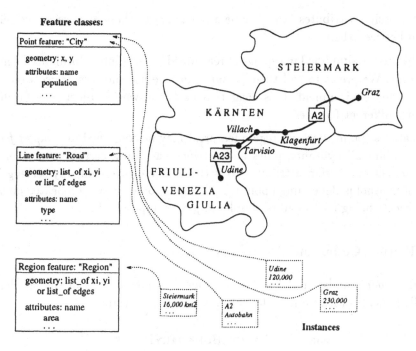

Figure 9: Point features, line features, region features

▷ a certain point feature is a HYDRANT, i.e. it belongs to the class of hydrants;

▷ a certain line feature is a ROADELEMENT;

▷ a certain region feature is a PARCEL.

Additional semantic properties *(attributes)* describe the feature in more detail. They map the set of features into value domains:

▷ the ADDRESS and PHONENUMBER of the utility company serving the hydrant;

▷ the NAME of the road;

▷ the PARCELNUMBER for the parcel.

Some remarks on terminology are important to note. Firstly, features are sometimes called *objects*. However, this implies a connection to object-oriented techniques which might not be correct. Secondly, the term *entity* is used as a synonym. Since this usually denotes any composition of elementary parts that must be handled, it should be used in a more general meaning, being applicable also to topological nodes, edges, faces, ..., whereas features bear some semantics.

We also have to distinguish between real-world objects (resp. features, entities) and model objects (resp. features, entities). Real-world objects are manageable 'portions' of reality helping us to divide the world into small parts that we can handle in a mental model. We assign attributes (and maybe a behavior) to them and we notice relations that hold between them.

Model objects are simplified mappings of real-world objects into a formalized computer environment. We prefer to call them *features* and we use the term *objects* only in an object-oriented environment. It is clear that each real-world object may be mapped into several different features.

Simple features (point, line and region features) may be combined to *complex features*. We can combine all hydrants, pipes and network parts in a certain area to one such complex object and assign attributes to it (the name of the company serving it, the number of housholds depending upon it, ...). This process may be repeated on several stages (objects being composed of simpler objects composed of ...).

3.1.2 Feature Codes and Attributes

The *feature code* stands for the partition of features into *classes* of similar elements, where all elements of a common class are mapped to one code:

$$
\begin{array}{rcl}
\text{aaa} & = & \text{PARCELBOUNDARYPOINT} \\
\text{bbb} & = & \text{PARCELBOUNDARYLINE} \\
\text{ccc} & = & \text{PARCELPOLYGON} \\
\text{ddd} & = & \text{PARCEL as a complex feature}
\end{array}
$$

Using such codes, GIS functions become more efficient and less error-prone, because we can attach to each feature code certain constraints, rules, and procedures (the *'behavior'*) holding for all representatives *(instances)* of the corresponding class.

Attributes are semantic characteristics, just as feature codes. In contrast to them, attributes usually do not control constraints, rules and procedures, so that they may be omitted without destroying the identity of the feature. Attributes may also be classified, and classes of attributes can be attached to classes of features. Examples:

- The attribute class OWNER may be attached to the feature classes PARCEL and HOUSE, but not to the class STREET.

- The attribute class ADDRESS may be attached to the feature class HOUSE, but not to the classes STREET and PARCEL.

- The attribute class STEEPNESS may be attached to the feature classes PARCEL and STREET, but not to the class HOUSE.

Just as there are features being individual instances of a feature class, attribute values are *instances* of attribute classes. Domains for attribute values are special kinds of integrity constraints.

Remark: In a somewhat sloppy diction, 'attribute' sometimes stands for a whole attribute class and sometimes only for a value.

3.1.3 Other feature types

Raster cell: Here we imply that an area of interest may be divided into elementary cells (many times regular, rectangular or even square). They are called *pixels*. We can assign a *grey value* to each of them. All these terms originate from image processing, but we extend their meaning. As an example, 'pixels' might be cells of a regular DTM (digital terrain model), and the 'grey values' might be the heights. Alternatively, we can map the set of cells into some semantic attribute domain: each cell may be assigned a value holding the vegetation type, population density, per-capita-income etc.

Text: There is no consensus about the usage of text features. While some GIS assign to them the same status as to all other feature types, others assert that texts are just visualizations of attribute values, since they are only meaningful as long as these attribute values exist. Examples are:

▷ a PARCELNUMBER being visualized on a cadastre plan,
▷ a NAME of a town or of a landmark on a road map,
▷ a STREETNAME or a HOUSENUMBER on a city map.

Note, however, that for some visualized names like 'Alps', 'Rocky Mountains', 'Lowlands', it is hard (if not impossible) to create features that may carry appropriate attributes. As an example, we cannot draw the boundary line (ring) for the 'Alps' region feature.

The list of feature types is still incomplete. It can be extended by:

▷ *PointGroup* and *Grid*
▷ *TextGroup*
▷ *PixelGroup* and *PixelChain, Voxel*
▷ special cases of lines (arc, spline, clothoid, etc.)
▷ *Graph, Tree, Network*
▷ *Tesselation*
▷ *Surface* in 3D space
▷ *AttributeObject*

Furthermore, the border line between 'genuine' GIS feature types and other data types used in information technology may be arbitrarily drawn, accommodating for associations with records of external files and databases, photogrammetric and remote sensing images, audio and video recordings etc.

3.2 Relations

Metric-topological relations between features are of great relevance in GIS. They have been discussed in detail in section 2. Ohter relations arise as a consequence of semantic characteristics of features. *Aggregating relations* are 'consists of'– or 'part of'–relations and mostly hierachical. As an example, a HOUSE consists of its FOUNDATION (region feature), its WALLS (line features), the ROOF (several region features). *Associative relations* create connections mostly between features of the same type and class that share some characteristic of special interest. (Example: all houses that have gas heating.) Another kind of an associative relation is the connection of a feature to its visualization. (Example: a feature from the CHURCH class is associated with a feature from the CHURCHSYMBOL class.)

While such relations hold between individual features, there are others that connect entire feature classes. We mention *generalizing* ('special case of'– or 'general case of'–) *relations*. As an example, HOUSE and FACTORY are special cases of BUILDING.

4 Organizational Aspects

The data models developed in the preceding sections – although of great help for structuring ideas of *what* shall be and what can be achieved with GIS – are not yet precise enough for putting them to work. The *'what'* question must be supplemented by the *'how'* question. For this purpose, information technology in general and database technology in particular provides the means. The *data models* designed in the preceding sections are now elaborated, introducing organizational detail and technical aspects, and arriving at *data schemas* that can be implemented in a *database* [5].

Most databases – not only in GIS, but also in other application areas – are based on the relational concept. A *relational database (RDB)* consists of tables whose entries may be connected by common values. As an example, for each feature class, a table is set up. An individual feature is represented by a row *(tuple)* in this table. The columns *(domains)* stand for the attributes of the feature. The order of entries is irrelevant; this is of great benefit to long-term data management tasks. The class of parcels results in a PARCEL table:

PARCEL					
ID	NUMBER	VALUE	ONSALE	AREA	...
1	171/2	ATS 340.000.-	yes	1981 m^2	...
2	285	ATS 425.000.-	no	1024 m^2	...
...	...	...	...	...	...

The *geometry* is taken care of by introducing node tables and preferably also edge tables. In the node table, each node occupies a row:

Node						
Id	Name	X	Y	z	Quality	...
node-12	P113	12.12	21.21	–	...	...
node-34	T01A	34.34	43.43	–	...	...
...	...	...	...	...	...	...

In a very simple implementation, the geometry of each parcel can now be assigned by creating a table that associates each PARCELID with the corresponding NODEID's. Since relational database management systems do not implicitly keep track of sequence (as it would be the case if ordinary sequential files were used for storage), we have to take care of sequence explicitly, by assigning sequence numbers of nodes as they appear when we travel around each parcel in a clockwise fashion.

NodesPerParcel		
NodeId	ParcelId	Sequence
node-12	7	4
...		
node-34	7	2
...		
node-34	15	5
...		

Note that the ordering of rows in this table is arbitrary and that a node being part of m boundaries appears m times. Note also that the 'original' two tables PARCEL and NODE are – as far as their internal structure is concerned – completely independent from the *join* table NODESPERPARCEL.

For most applications, this approach would be too simple. First of all, it neglects topology. There is no explicit consideration of the boundary line between to adjacent polygons; islands cannot be detected; different shapes of lines cannot be taken into account. By introducing an EDGE table, most of these shortcomings can be eliminated. But of course, in such a case the NODESPERPARCEL table would have to be replaced by at least two tables, a NODESPEREDGE and an EDGESPERPARCEL table. It is evident that for large data volumes, the time required to find and to sort all 'pearls on a string' may grow unacceptably.

Also, since many GIS functions depend upon geometric queries and algorithms, they always have to start with the elementary building blocks, the nodes, resp. their geometry. There is a lack of *abstraction* capabilities; it is not possible to deal with parcels or features that are even more complex on a level where basic geometric questions can be taken care of once and for all *(principle of encapsulation)*.

Our simple example is therefore general enough to point out the greatest advantages and disadvantages of the relational approach:

- *pro:* robustness, local behavior, ease of applying standardized techniques (see e.g. *SQL - Structured Query Language* [6]);

- *contra:* lack of performance, stress of details vs. neglect of global, higher-level aspects, like e.g. encapsulation.

The performance argument leads towards the discussion of *non-standard databases* that allow data types being more general. This means that a feature attribute may have a data type VECTOR or POLYGON or an even more complex structure. This also represents one step towards an *object-oriented* approach (see [7], [8]) that – besides providing general data types – enhances the concepts of *encapsulation, generalization* and *inheritance* of attributes.

References

[1] Bartelme, N.: Geoinformatik. Modelle, Strukturen, Funktionen. Springer Verlag, Berlin Heidelberg 1995.

[2] Bartelme, N.: Data Analysis in GIS, in: I Sistemi Informativi territoriali – fondamenti e applicazioni (Ed. F. Crosilla, L. Mussio), International Centre for Mechanical Sciences, Udine 1992, 87–106.

[3] Laurini, R. and D. Thompson: Fundamentals of Spatial Information Systems. Academic Press, London 1992.

[4] Molenaar, M.: Data Models and Data Structures in GIS, in: I Sistemi Informativi territoriali – fondamenti e applicazioni (Ed. F. Crosilla, L. Mussio), International Centre for Mechanical Sciences, Udine 1992, 41–74.

[5] Zehnder, C.A.: Informationssysteme und Datenbanken. Teubner, Stuttgart 1989.

[6] Melton, J. and A.R. Simon: Understanding the new SQL: a complete guide. Morgan Kaufmann, San Francisco 1994.

[7] Milne, P., S. Milton and J.L. Smith: Geographic object–oriented databases – a case study, Int. J. Geographical Information Systems, 7 (1993), 39–56.

[8] Worboys, M.F.: Object–oriented approaches to geo–referenced information, Int. J. Geographical Information Systems, 8 (1994), 385–399.

CLASSIFICATION OF SPATIAL DATA ANALYSIS TOOLS

R. van Lammeren and A. van der Meer
Agricultural University of Wageningen, Wageningen, The Netherlands

This paper describes the spatial data analysis theory as it is teached at the agricultural university of Wageningen.

Digital geographic information is stored in databases or information systems to make them available and accessible for users. Accessibility implies that the user can manipulate the data in a way that information can be obtained. There is no univocal terminology in the 'GIS community' about the manipulation of data. In Wageningen the term **geographical data processing** is reserved for all operations where data can and will be manipulated. Data processing is divided into queries, transformations and operations. Each class exists of a collection of various operators and functions.

The classes have the following meaning:

Queries - are used when you're only interested in the available data. Data don't change when they are queried. If you consider the available data as a set according to the set theory, a query will give a subset, empty set or the whole set.

Transformations - are used when you're only interested in the available data. The meaning (semantics) of the data don't change, the structure or the form of expression (syntax) will. If you consider the available data as a set according to the set theory, a transformation will give the same elements of the set described in another form.

Operations - are used when you want to deduce data from the available ones. If you consider the available data as a set according to the set theory, an operation will deliver a new set of data.

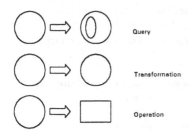

Query

Transformation

Operation

Figure 1: Queries, transformations and operations

1. Introduction

Many disciplines focus on the study of processes that are located in time and space (a spatial situation). Properties and processes are described, modelled and analysed, but also changed through design and construction of new properties and simulation of new processes. Those new properties and simulated processes are evaluated on their effects in the end.

The previous sentences show a methodological main line of application-oriented research. In this three main phases are distinguished: analyse (or definition), design (or construction) and evaluation.

In each of these three phases GIS can be used. The actual use of GIS depends on disciplinary knowledge assignment, the availability of geographical data and the translation of this knowledge into discipline dependent GIS applications. The success of this translation is considerably dependent of the knowledge and understanding of the classes of operations.

The disciplinary knowledge assignment can focus on the phase of analysis. In this phase the knowledge and understanding of the spatial situation is the point of main attention. This includes the description or definition of:
- a (statistical) typology of spatial objects;
- spatial distribution: a clustering of spatial objects according to the relation between spatial position and their accompanying properties. For example in demographic studies the difference between areas with a high or low population density. Or in landuse studies the difference between spatial concentrations of specific functions like agriculture, cattle breeding or industry;
- spatial patterns: a spatial distribution with a recognisable structure, which can be described by a number of parameters. For example population densities which decline from the centre with high densities to the surroundings with lower densities. Or road patterns;
- spatial correlation: a relation between several properties. For example a relation between altitude and rainfall, between soil type and vegetation, or between landuse and soil pollution;
- spatial dynamics (spatial process): the development, change in time of a property of an object. This development can also have a spatial dimension. Spatial patterns can therefore also have dynamics. Spatial dynamics of an object or a pattern always influences the direct surroundings of the object or pattern. This happens for example with the dispersion of vegetation types, dispersion of diseases, water streams and the growth of cities.

The disciplinary knowledge assignment can also focus on the phase of design. In this phase a possible, wanted or thinkable spatial situation is constructed via adjustment of:
- the (application) properties of spatial objects, distributions, patterns, correlations or dynamics;
- the (mutual) location of spatial objects or patterns;
- the form of spatial objects or patterns;
- the size of spatial objects or patterns;
- the effects of spatial processes.

Finally the disciplinary knowledge assignment can focus on the phase of evaluation. In this phase changes are traced and valued, so consequences of adjustments can be determined. These consequences apply to:
- the spatial objects;
- the spatial distribution;
- the spatial patterns;
- the spatial correlations;
- the spatial dynamics.

These possibilities of applying GIS signify the enormous potential of GIS in the diverse disciplines. Particularly GIS data processing plays a role in these phases of discipline dependent assignments. This paper describes the most important aspects of the three classes of data processing: queries, transformations and operations.

2. Queries
Queries focus on a set of data, where the whole set, a subset or an empty set of data can be the result of the query. The original set of data is not used to get new data.
A query searches data that meet certain conditions. Queries are carried out with the goal of getting information from data in a table or database, without changing these data.

A query has the following formal structure:
  find the data of type T
  that fulfils condition ( )
  and carry out operation ( )

This formal structure makes it clear that:
- A specification of the data type leads to the finding of data by indicating which columns (attributes) are of interest. This is called **projection**.

- Formulating a condition leads to a **selection** of rows, representing entities or objects.
- An operation can be carried out with the selected data.

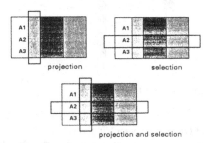

Figure 2: projection and selection

Figure 2 gives an example of projection and selection in an abstract drawing of a table structure.

When a query only is used to find data from a database, we call this a **plain query**. In this case the selected data are only shown or written to a file. As we see in figure 2, there are three types of plain queries: a projection, a selection or a combination of projection and selection.

In other cases we talk of **queries in a broad sense**. This implies that the query is followed by further processing of the selected data. When this further processing change the data set, we no longer talk about queries, but about operations (see paragraph 4).

2.1 Conditions
Conditions that are formulated in queries refer to the values of one or more attributes. When:

> a is the value of attribute A for a certain entity

and:

> w is a value where a is compared with,

then the basic conditions are:

> $a = w$, $a \neq w$

or in terms of logics:

> a eq w, a ne w

If the values of A have a ranking order, than the following basic conditions can be added:

> $a < w$, $a \leq w$ (logic: a lt w, a le w)
> $a > w$, $a \geq w$ (logic: a gt w, a ge w)

This type of condition is called **simple conditions**. An example of a simple condition is: Find the raster cells of database Ecology with a organic matter value of 5 mg and show the positions on the screen. In the condition is the attribute a organic matter and the wanted value 5.

When two conditions are combined in one query, the condition is called assembled. An example of an **assembled condition** is: Find the raster cell of database Ecology with position (i eq p) and (j eq q) and assign the value of the attribute organic matter value to w.

To form an assembled condition two simple conditions are joined by means of a logical operator. Suppose C1 and C2 are simple conditions; the following logical operators can be used to form assembled conditions:

> CS1 = C1 and C2 (both conditions have to be true)
> CS2 = C1 or C2 (one or both of the conditions have to be true)
> CS3 = C1 xor C2 (one of the conditions has to be true)

Furthermore the is a logical operator that doesn't join two conditions, but only uses one:

> CS4 = not (C1) (anything but C1 is true)

2.2 Querying thematic data
When thematic data are queried, no difference is made between raster or vector data. The underlying assumption is that the geometric data are not used with thematic queries. In this case geometry is mainly used for displaying query results.

A thematic query aims at:

- one or more attributes of all objects (type [1]): a projection;
- all attributes of one or more objects (type [2]): a selection;
- the specific values of one or more objects (type [3]): a combination of projection and selection.

Examples of these query types are:

Type 1: Which organic matter levels are present in the study area?

Type 2: How many objects have a organic matter level > 10 gram?

Type 3: What is the landuse of the objects that have a organic matter level > 10 gram?

More than one dataset

There are some queries that apply more than one dataset. This is done by means of relational algebra.

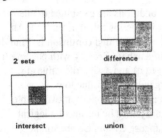

Figure 3: elementary relations between databases

If the definitions of the tables' attributes are identical, three elementary operations can be distinguished:
- union: the new set contains the joined original sets.
- intersect: only the values that occur in both sets will remain in the new set. This is mostly done with the operator AND.
- difference: only the values of one original set that don't occur in the other set will remain in the new set. A difference is mostly made with the operator AND NOT.

When the data definition differ, but there is a relation between two identical column structures, for example based on the object identification, columns will be added (natural join). This is in fact also a union, but the result differs due to the different prior condition.

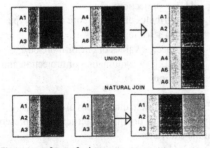

Figure 4: two forms of union

Besides the three elementary combinations between databases, there is a fourth elementary operation: cartesian product. This operation results in all combinations of the dataset values. In this case no similarity is necessary.

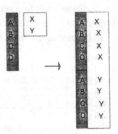

Figure 5: Cartesian product

2.3 Querying geometric data

On the basis of the geometric properties of data structures the following query domains are possible:
1. The whole area as described by the geometric data.
2. Specific objects or a group of raster cells by a selection or a projection and selection (of geometric properties);
3. The relation between objects or cells. This concerns the querying of topological properties.

Discussion of the last query domain follows in paragraph 2.4. In the first and second domain one can focus on the whole study area or on the following metric properties:
- position (1.1 or 2.1)
- length or perimeter of objects (1.2 or 2.2)
- size (area or contents) of objects (1.3 or 2.3)
- form type of the objects (1.4 or 2.4)

Examples:
Type 1.1: which thematic properties occur.
Type 2.1: which thematic properties occur on a specific location.
Type 1.2: what is the greatest length.
Type 2.2: which area has a perimeter lesser than 4 kilometres.
Type 1.3: what is the mean object size.
Type 2.3: which areas are bigger than 3 ha.
Type 1.4: what types of forms occur.
Type 2.4: which areas are rectangular and have a length-width proportion of 1:4.

Information about perimeter, area and contents is available in a database when the software automatically calculates this. These aspects can not be determined of points. A line only has length, a polygon has perimeter and area and a volume has area and contents. The calculation of these aspects is done based on co-ordinates of points, the georeference and the relations between co-ordinates. When the software doesn't calculates it automatically, it is necessary to have additional applications.

There is a possibility to use more than one database in geometric queries, just like in thematic queries. The combination of databases mostly changes the geometry and is done with overlay-operations. Prior condition for overlay operations is that they have a common area, otherwise there would be nothing to overlay. Overlay operations are discussed in paragraph 4.4.

Object oriented queries
A fifth type of geometric queries (type 2.5) is object oriented queries. When object oriented approach is used it might be of interest to query the nature (e.g. the object type). Examples of this type of queries are: What kind of object is object number 7? Where is object number 7 located?

2.4 Topological queries
A spatial object is a separate object, but is has locational relations with other objects. Words we use in daily life fore this are: left, next to, above, under, between, in, crossing, etc. Those relations are described by means of topology and can be divided in two:
- relations between geometric data types: these relations are described by the relation between graphs in vector structure and by the relation between cells in a raster structure.
- relations between objects: these relations can be derived from mutual geometric elements and the relation between those elements and objects.

The last relation is conceptual and in a conceptual model the following object relations are distinguished:
1. area-area relations (aa-r);
2. line-line relations (ll-r);
3. point-point relations (pp-r);
4. area-line relations (al-r);
5. area-point relations (ap-r);
6. line-point relations (lp-r).

Queries can be based on each of these topological relations. Such a query can be stated in two different ways. For example the relation between area and point can be found with the question "Which points are positioned in an area" but also with the opposite question "Which areas contains points".
Topological relations can barely be described with rasters. Mostly tricks are used like using more raster cells to describe an object. In this way one raster cell can represent a point, adjacent cells a line or an area.

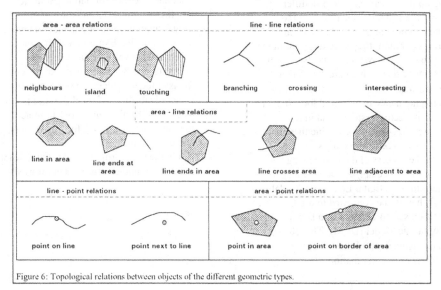

Figure 6: Topological relations between objects of the different geometric types.

Examples of topological queries are:
Which areas are adjacent to the city?
Which areas does the railroad pass by?
Which areas does the railroad cross?

3. Transformations

Transformations or conversions are operations
on geographical information that change the
form of expression of the data, with the
maintenance of the semantics as much as
possible.
The same remarks around the world are
recorded in different ways. Compare this with
the translation of a Dutch text in English. There
are no new objects or new attributes created, as
with overlay or buffer operations. No new
knowledge or information is added. Different
from queries, where according to certain
criterion's subsets from the whole dataset are
derived, will transformations and conversions
submit all the data to their processing.

Transformations and conversions are necessary
because:
1. Data are delivered in different formats by
   the different software houses and GIS-users.
   When we want to use these other data, we'll
   have to convert them in another format.
   This process can be simplified by using
   international standard exchanging formats.
2. Geographic data are captured in different
   co-ordinate systems. When we want to
   combine those data, for instance in an
   overlay operation, they have to be put in the
   same co-ordinate system. This also counts
   for thematic data, which are coded in a
   different way.
3. Geographic data can be stored for different
   reasons in vector, raster or quadtree
   structure. To be able to combine these data
   they'll have to be in the same structure.
4. Within the process of data processing the
   form of the data changes. Descriptions of
   the terrain in a Digital Landscape Model
   (DLM) have to be converted to digital
   models on screen or paper in a Digital
   Cartographic Model (DCM). This changes
   both the geometry and the attributes,
   separately or in combination.

Common structure of transformations
There are three parts important to carry out
transformations:
• an input file with the geometric and/or
  thematic data;
• a set of transformation rules and/or
  translation tables;
• an output file with the transformed
  geometric and/or thematic data.

The transformation rules and/or translation
tables are included in the calculation software.
They are not totally specified; this relays on the
data of the input file and of the wanted data
structure of the output file. Some transformation
parameters have to define those variables.

The transformation parameters of geometric
data are often determined by means of a number
of measure points. These are points in the
terrain, with known co-ordinates in the input
file and in the output file. The number of
measure points depends on the complexity of the
transformation (1e grade or high-order-
transformation) and the wanted degree of
control.

The process of transformation of geometric data
is in outlines:
1. Determine the co-ordinates of a number of
   measure points of the input-file and of the
   output-file. Use more points than necessary.
2. Calculate the transformation parameters.
3. Use these parameters and the
   transformation rules on the measure points
   of the input-file. Examine whether the
   difference between the calculated coordinates and the co-ordinates of the output-
   file isn't too big (there are some statistical
   standards for this). If the difference is too
   big, search for the causes, delete some
   measure points and look for some others,
   and go back to 2.
4. Apply the transformation rules to the whole
   input-file (another control is not wasted).

3.1 thematic transformations

Transformation of thematic data is possible following different ways. The method that is most close to the geometric method is rescaling. Suppose you have two adjacent remote-sensing-images, recorded at different times. The radiation value in the border area can be quite different, but you want it at the same level. What you'll have to do is search common pixels and determine the radiation value. With this you can calculate parameters for a linear or higher-order-transformation of the radiation values, in order to carry out this reclassification.

Another transformation is reclassification based on statistical analysis. The input file is investigated at how many times (or at what area) a certain quantitative attribute value occurs. This gives a frequency histogram for the attribute values. You cut this histogram into equal classes with equal frequency, called quantiles. Because the borders of the quantiles are known, a translation table can be made. For each original attribute value in this table is given what the new value should be. With this table the whole input file can be transformed. Thematic transformations only affect the attributes of a database, not the geometry. An example of a thematic transformation is the reclassification of the values of one information class:

> Areas with ground water level 1 and 2 are judged suitable. Areas with ground water levels between 2 and 6 are barely suitable and the other ground water levels are not suitable.

With reclassifications the original attribute values can be replaced with the new value classes. In order not to lose information, it is wiser to create a new attribute to store the new values. This implies that a new attribute has to be defined. Reclassification is the determination of new value classes. Those new classes can be in the same measure scale or a lower one.

3.2 geometric transformations

Transformation of geometric transformations is also possible in different ways. In this paper we'll discuss: congruence transformation, affine transformation, rubber-sheeting, projective transformation, map projection, vector-raster conversion and raster-vector conversion. With congruence transformations object descriptions are changed from one orthogonal

co-ordinate system to another, in such a way that corners are similar. A triangle will not change in shape. At most a rotation and a displacement (translation) will take place.

Affine transformations are, among other things, applied on 'deformed' data. They are important when digitising paper maps or photographs. The fact is that mechanised produced paper has a direction in which the paper feeds. This means that the paper has a different elasticity in both ways. After drawing or printing the scale of the map can become different in different directions. When you start digitising the size-holding is lost. This can be compensated by affine transformation.

Both congruence and affine transformations are first degree or linear transformations. In the process of putting the terrain on map they are often used. Sometimes data are so much deformed that more complicated methods have to be used, for example with remote sensing images from irregular moving planes. Sometimes a deformation is deliberately put in, for example in town plans where the city should be displayed bigger than the rest. These applications call for higher-order transformations, also called rubber sheeting. Disadvantage of higher-order transformations is that the borders have a freakish pattern. That is also the reason why it is no actual practice.

Projective transformations are used in photogrammetry and when three dimensional DLM's has to be displayed on two dimensional screens in shape of a perspective drawing.

Map projections are geometric transformations from a description of geographic co-ordinates λ and φ to rectangular co-ordinates X and Y. Every map projection transforms the smallest parts in principle affine. There are some groups of map projections though, which have special properties. Among those are conformity and equivalence.

Conformal map projections picture the smallest parts (square, circle) identical: a square on earth will be a square in the projection plane. The Mercator projection is an example of this type. Equivalent map projections picture the smallest parts affine is such a way that every part is presented in right proportions. Every square kilometre of the earth gets after projection the same surface in the projection plane. This leads to controversial results like the Peters-projection (just proportions).

3.3 raster to vector and vector to raster

In many spatial analysis studies it is necessary to combine raster data with vector data. There are some possibilities to do this and the most obvious is coverting the data from one data structure to the other. In order to transform data from one structure to another you'll have to know exactly what the difference between the two structures is.

In raster structures thematic data is stored as a function of position. This means that each position (i,j) has a thematic value stored in a certain attribute scale. In a vector structure the earth is described by objects: area-, line- or point-objects. Area-objects are spatial connected areas, geometrically described by surrounding polygons. The nature of this object is described by a label, which refers to the object and not to a position (like with rasters). Line-objects have a small width compared to the length and are described by a chain. The width can be added as attribute to the line, but isn't geometrically stored. The nature of the object is described by a label. Point-objects are geometrically described by position and thematically by a label. Spatial properties are described by attributes.

When rasterdata are transformed into vector data for each raster cell has to be determined to what kind of object this cell will belong. This means that based on decision rules the area-, line- and point-objects has to be identified. For each raster cell has to be decided not only what kind of object it belongs to, but also to what specific object. After that for each object the spaciousness can be determined and translation to a vector structure can take place.

Transformation from vector to raster is mostly easier, because now only the position of a raster cell has to be determined. Now the object can be deduced from the object label and placed on the position.

4. Operations

Operations are used when you want to deduce other data from the existing ones. If you consider the available data as a set according to the set theory, than an operation will deliver a new set of data. This applies to the geometric and the thematic data.

In Wageningen three groups of operations are distinguished:

- local;
- focus;
- global.

In the following discussion the word element is used. In this paragraph the word element has the meaning of a geographic entity (raster cell, region, feature or object).

A local operation aims at single elements. This means that the operation uses the thematic or geometric properties of each single element. The result of such an operation implies a change in geometric or thematic properties.

Interactive editing of databases is a form of geometric local operations.

A focus operation focuses on the thematic and geometric properties of one or more elements and their direct surroundings. This means that this type of operations uses attribute values of a source element and of nearby elements. Focus operations are divided in two general classes:

- the geometry is deduced from a certain element (zoning);
- the thematic properties are deduced from a certain element (filtering).

A global operation focuses on all elements that are present in the database. That type of operations considers the original thematic and geometric properties of the elements. The result of this operation gives a new set of data.

4.1 Local operations

Local operations aim at single elements. In this paragraph we'll focus on the thematic local operations.

The structure of thematic local operations is identical to that of queries:

*Find the data of type T
that fulfils condition C
and carry out operation O*

This implies for local operations:
T: Type of data on which the operation applies; in this case single elements.
C: The features of the element that are used for operations.
O: The type of local operations that is carried out.

Local operations are divided into:
- functions (1.-)
- operators (2.-)
and into:
- one dataset operations (-.1)
- more datasets operations (-.2)

Functions directly affect attribute- or cell values of one (1.1) or more datasets (1.2). Operators aim at the relations between the values of different thematic attributes within one dataset (2.1) or the values of thematic attributes of different datasets (2.2) with the same location.

Classes of functions are:
- trigonometric (e.g. sinus, cosines, etc.)
- exponential
- logarithmic
- statistic (max, min, average, etc.)
- miscellaneous, like:
  . determining absolute values
  . conditions

Classes of operators are:
- arithmetic (add, minus, divide, etc.)
- boolean (and, or)
- relational (lesser than, etc.)
- combinatorial
- miscellaneous, like:
  . binary operators
  . accumulative operators

All these local operations can be carried out with GIS-commands or with a query language of a Database Management System. It makes no difference which one is used, because there is no effect on the geometry. Mostly GIS's don't have all these possibilities (especially not in vector structure), so the step to a DBMS is easily taken. Examples of local operations are:
- which type of landuse occurs most, less or modal (type 1.1);
- what is the mean organic matter level in areas where the landuse is agriculture (type 1.2);

- calculate what the organic matter level is when 10 grams are added (type 2.1);
- calculate the organic matter value in areas where landuse is agriculture. When the existing organic matter level is less then 10 gram, add 5 gram. When it is higher than 10 gram, add 12 gram (type 2.2).

Local operations in a raster structure
In a raster structure local operations are quite easy to carry out. There is a difference between operations on one database and on more than one database.
Operations on one database calculate the new value of each raster cell based on a function or operator. For example sinus calculation of a cell value (type 1.1):
> resultgrid = SINUS (startgrid)

Or an addition (type 2.1):
> resultgrid = startgrid + 1

Operations on two or more databases search the raster cells on the same location. With those cells operations or functions are carried out. For example maximum cell value determination (type 1.2):
> resultgrid = MAX (startgrid1, startgrid2, startgrid3)

Or an addition (type 2.2):
> resultgrid = startgrid1 + startgrid2

Local operations in a vector structure
Vector structures can be processed as well by local operations. For this the geometric properties of the elements of both datasets should be fully georeferenced.

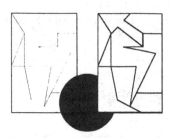

Figure 7: geometric similarity

This means that the examples of raster structure operations can be applied to a vector database.

4.2 Focus operations: filtering

With many operations on rasters people are interested in the surroundings of the element. This can be the searching of local changes of attribute values, like searching edges or isolated points. This can also be suppressing local variations in attribute values. This kind of operations is called filtering, because filters are used. The term filter is from the signal processing and has entered the field of GIS via image processing like it is used on satellite images.

Filters are applied on the attribute values of raster elements, which are in a certain window. This window is set up around a central element. The operation results in a new attribute value, given to the central element. Because the filters are introduced via image processing, most known filters are in ratio scale. Lower attribute scales are also possible. Besides filters the word window operations is used. Window operations can contain besides filters some other operations on attribute values.

A window V (figure 8) is a subset of the elements of raster R, write: $V \subset R$. This subset is identified by a central element with subscripts (i,j) and by two moving subscripts (r,s), while:

$$r \in D(r) = \{-n, ..., -1, 0, 1, ..., n\}$$

and

$$s \in D(s) = \{-m, ..., -1, 0, 1, ..., m\}$$

The window V(i,j) is in a raster with attribute A:

$$V(i,j) = \{..., A(i-r,j-s), ...\} \quad (r,s) \in Dr \otimes Ds$$

When for example $r \in \{-1,0,1\}$ and $s \in \{-1,0,1\}$ than will the domain of (r,s) be:

$Dr \otimes Ds = \{(-1,1), (0,1), (1,1), (-1,0), (0,0), (1,0), (-1,-1), (0,-1), (1,0)\}$

This domain and the window that is based on it are show in figure 8.

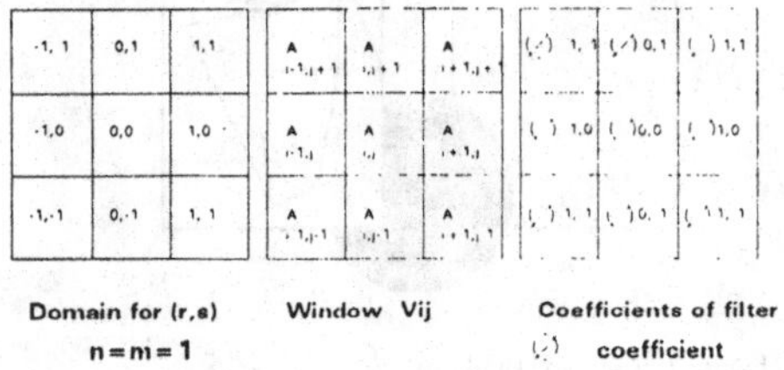

-1, 1	0,1	1,1	A $_{i-1,j+1}$	A $_{i,j+1}$	A $_{i+1,j+1}$			
-1,0	0,0	1,0	A $_{i-1,j}$	A $_{i,j}$	A $_{i+1,j}$			
-1,-1	0,-1	1,1	A $_{i-1,j-1}$	A $_{i,j-1}$	A $_{i+1,j-1}$			

Domain for (r,s)   **Window Vij**   **Coefficients of filter**
n = m = 1   (:)   coefficient

Figure 8: Definition of filter and window.

A filter is an operator that is applied on the elements of window V(i,j). The result will be one value F(i,j) assigned to position (i,j), like this:

$$F(i,j) = O(Vij) = O(..., A(i-r,j-s), ...)$$

Mostly a filter is carried out on all possible windows in a raster. Only the elements R will occur as central element of V(i,j) when the following is true:

$$i(o) + n < i < i(b) - n$$

and

$$j(o) + m < j < j(b) - m$$

This is because of the size of the window and the demand that the window should fall entire within the raster.
When a filter operation is carried out over a whole raster, a new raster is made with the elements F(i,j) under above mentioned constraints for i and j.
Operation $O(Vij)$ is not yet specified. In the following part of this paper some examples of filters will be given.

Linear filters

There is a special group of filters that have special properties; they are referred to as linear filters (figure 8) and have the following structure:

$$F(i,j) = \Sigma r \Sigma s \; \varphi r, s \times A(i-r,j-s)$$

or also:

$$F = \phi * R$$

When the sum of the coefficients of the filter is 1, we call this a normalised filter. At the later examples the normalisation takes place by multiplying the filter with a factor inversely proportional to the coefficients.
Linear filters are also called convolution filters, like convolution in signal and system theory.

The following rules apply to linear filters:
When the sum of R_1 and R_2 is defined, than will for a linear filter ϕ count:

$$\phi * R_1 + \phi * R_2 = \phi * (R_1 + R_2)$$

or: the sum of two filtered rasters is equal to the filtering of the sum of two original rasters.
When the sum of $\phi 1$ and $\phi 2$ is defined, than is:

$$\phi 1 * R + \phi 2 * R = (\phi 1 + \phi 2) * R$$

or: the sum of two filter operations of one rasters gives the same result as the compound filtering of one raster.

$$\phi * c.R = c.\phi * R$$

or: when we filter a raster whose attribute values are multiplied with a scalar c, the result will be the same as when the filtered raster will be multiplied with the scalar.

Some examples of linear filters are: moving average filter, gradient filters and Laplace filters The moving average filter calculates the mean value of the attribute values within a window. Gradient filters give an indication of how the central element in a window is situated in an area where attribute values increase in a certain direction.

A gradient filter determines the first order derivative of the attribute values in a certain direction. Slopes are detected in this way. If one should take the result of a gradient filter and apply this filter again, the result will be a second order derivative of the attribute values. A filter that carries out these two operations in once is called a Laplace filter. Figure 9 gives an example of two forms of a Laplace filter, which determine the second order derivative in different directions.

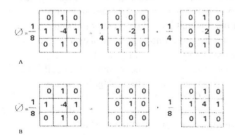

Figure 9: Laplace filters in two directions

Not linear filters
Ranking filters put the attribute values of window V(i,j) in sequence. When you've determined in advance which number in the sequence you want to select, the attribute value of that position is taken and assigned to element (i,j). The most important ranking filters are (see figure 10):
- median filter, which takes the median attribute value of the window;
- minimum filter, which takes the lowest attribute value of the window;

- maximum filter, which takes the highest attribute value of the window.

The median filter is insensible to exceptional values. It is therefore quite suitable to find edges between big compound areas without being disturbed by local irregularities in the attribute values. The maximum and minimum filter on the contrary look for the exceptional values.

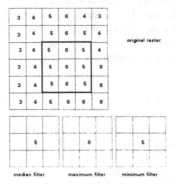

Figure 10: examples of ranking filter

Frequency filters count the number of times that certain attribute values occur in a window. The filter selects an attribute value based on the frequency and assigns this value to element (i,j). Frequency filters are:
- majority filters, which selects the value with the highest frequency (or the value which occurs most);
- minority filters, which select the value with the lowest frequency.

The majority filter is suitable to remove irregularities from edges of areas and to remove rare attribute values from rasters. The minority filters detect rare values.

Segment filters look for a continuous segment (region) with the same attribute value. These segments are ordered in size and based on this size one segment is selected. The attribute value of this segment is assigned to element (i,j). We distinguish:
- largest segment filter, which looks for the largest segment within the window;
- smallest segment filter, which looks for the smallest segment within the window.

The largest segment filter eliminates little isolated regions from the raster, while the smallest segment filter looks for them.

4.3 Focus operations: zoning

Objects in the real world often have a direct or indirect influence on their surroundings, or have to be protected against the influences of their surroundings. Development city plans or noise pollution are examples of the first. Buffering zones around cities to protect the area from urban activities is an example of the second. In GIS operations are developed to generate such influence zones. Buffer operations can be carried out for raster and vector structures. A buffer in a vector structure will always be an area; a buffer in a raster structure will always be a region.

The principle of the buffer operation is quite simple. First one or more elements are selected which form the source from where the buffer has to be calculated. This selection takes place by means of a query. After that the buffer is calculated from the selected object. This is done according to the form of the element, with consideration of inward or outward distance, and in relation to the border of the source. The result of that calculation is dependent of the determined distance and the form of the source element.

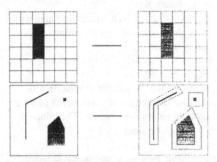

Figure 11: examples of zoning

4.4 Global operations

Global operations aims on the thematic and the geometric properties of the elements that are described in a database. The results of these operations are always new elements.
The starting point of the operation can be one database on the basis from which new elements are defined. The starting point can also be two databases, which are connected to each other. This connecting is only possible when georeferences of both databases are identical.

Global operations on one database can aim at determination of new elements based on geometric properties. Examples of this are:
- the conversion from a set of point elements to a set of line- or are element and vice versa;
- the conversion of a set of line elements to a set of area elements and vice versa;
- the conversion of a set of raster cells into regions.

Global operations on more than one database can aim at determination of new elements based on geometric properties of both databases. An example of this is determination of new elements by means of calculating intersections and topological relations.

When global operations are carried out on thematic properties of one or more databases, the new elements will always be new thematic attributes.

global operations in raster

Although a raster structure strictly speaking has no objects, there are still a great number of problems that have the necessity to divide a raster database into a large number of independent objects. An object in a raster database consists of one or more raster cell with the same value and with a neighbourliness relation (Molenaar, 1993). This last condition carries one of the problems of the difficulty of talking about objects. What kind of neighbourliness relation is meant?

Distances and weighted distances.

Because a raster cell represents a geometric and thematic unit, the real distance from one raster cell to another can be calculated. This is called the Euclidean distance. The calculation of the Euclidean distance can also take barriers or obstacles into account. It is also possible to calculate distance not based on the real distance, but based on social, economic, physic or technical nature. These cases are called weighted distance determinations.

Euclidean distance operations determine for every raster cell the distance to, the direction to and the allocation of the not-source cell to the source cells. Source cells are the raster cell that are interpreted as attraction or target points. When this shortest distance condition is valid for more than one source cell, the first found source cell is assigned to that raster cell. Distance calculations in raster structures are based on the Pythagorean theorem. The middle of the source cell and the not-source cell is used to construct the calculation.

Direction calculations in raster structures are based on degree-indication from 1 to 360. The north direction is indicated with 360 degrees, the south with 180 degrees, the west 270 degrees and the east with 90 degrees.

Allocation calculations are based on the results of the distance calculations combined with the allocating rule.

Weighted distance operations have great similarity with Euclidean distance determination. The only difference is that not real distances, but weighted distances (or accumulated travel cost) are calculated. Operations like this requires a raster database with source cells and a raster database with the values of the costs or barriers.

The calculation is based on a conceptual node-link representation of the raster. In this representation is the centre of the cell the node and the connection between node a link. Every node has 8 links; this means 8 possible displacing directions.

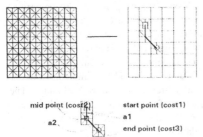

mid point (cost2)
a2
start point (cost1)
a1
end point (cost3)

Figure 12: node-link concept

In horizontal and vertical direction the costs or barriers are calculated with the following formula:

$$a1 = \frac{\text{costs}(x1, yj) + \text{costs}(x(i+d), y(j+c))}{2}$$

where a1 : the calculated costs or barrier is
i,j: give the position of x,y
e,d: give the displacement of +1 or -1
and e≠d and c=0 if d≠0 or d=0 if c≠0

In diagonal direction are the costs or barriers calculated with the same formula, but multiplied with √2 (square root of 2).

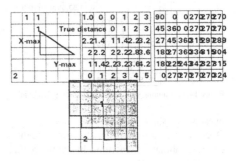

Figure 13: calculating distance in a raster

The calculation process is:
- the source cells are traced and made 0;
- the nearest neighbour cells are traced and the calculation is carried out for every neighbour cell. This means that both the horizontal/vertical and diagonal calculation is carried out. The lowest result value is assigned to the neighbour cell;
- the neighbour cell with the lowest value is traced and becomes the new source cell. From this cell are all the neighbour cells, which don't have a value yet, calculated. The result of the calculation is added to the value of the source cell and assigned to the neighbour cell.

The result of the calculation is a database with source cells and raster cells with a cost- or barrier value. This database does not give direction or allocation. For those two results some additional steps have to be taken.

In order to indicate how to move from a cost cell to another cost cell and eventually to a source cell back-link positions are used. Those back-link positions are based on the eight directions, that are used with the cost calculation, but the other way round (back-linking). The back-link positions are coded in a scale of 0 to 8. Zero refers to the source cell, 1 to the right neighbour cell, 2 to the right-under neighbour cell, etc. (see figure 14). According to these back-link positions the allocation and direction assignment takes place.

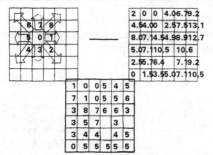

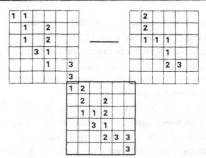

Figure 14: determining shortest cost path through back-linking

Figure 16: determining maximum value in zone

overlay in rasters

Overlay operations in raster structures are the so called combination functions. In set theory this is called a product of two sets; in raster terminology the product of two rasters. Every occurring combination gets a unique code. This means that this function gives a raster with zones, and an attribute-file where the original values of each code are described.

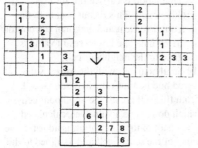

Figure 15: combination of two rasters

zonal functions

A zone in a raster is an area where all values have the same value. So called zonal functions carry out a certain function for a certain zone. Examples are:
zonalmean, calculation of mean value in a zone;
zonalsum, calculation of the sum;
zonalvariety, calculation of the number of different values in a zone.
With regard to the zones some geometric operations are possible, like determine the area, perimeter, thickness per zone. Thickness points at the highest distance between boundaries of a zone. This distance is calculated with a circle and the raster cell size.

Global operations in vector

In this paragraph some examples of global operations in vector structures are given. The determination of area objects (i.c. Thiessen polygons) based on a set of point objects is described. So called 'overlay' operations in a vector structure are described.

Thiessen polygons

Thiessen polygons are (boundaries of) area objects based on point elements. These polygons are constructed in such a way that the distance from any point within the polygon to the centre point of that polygon is always shorter than the distance to any other centre point.
Suppose there are two point objects (A and B). The perpendicular bisector of the connecting line between those two points marks the border between the areas that are closest to A or B. When three points are at stake, there are three connecting lines and three perpendicular bisectors. The three perpendicular bisectors intersect in one point. This point is also the middle of a circle described by A, B and C. The construction of the triangles between the three points A, B and C is called Delaunay-triangulation.
In order to construct a Thiessen polygon, the following two operations have to be carried out:

*   construction of a triangle network between point objects by means of a Delaunay-triangulation. This means that maximum three points lay on the border of the circle and no points within the circle. A triangle can be constructed from those three points. The angles of the triangles have to be as similar as possible.
*   The centres of these circles (or the middle points of the Delaunay triangles) form the angular points of the so called Thiessen polygon or Voronoi-diagram. To construct a

Thiessen polygon the nearest middle point has to be found. Between two points a connecting line is drawn, which may intersect only once with the Delaunay-triangle. This connecting line forms a border of a Thiessen polygon.

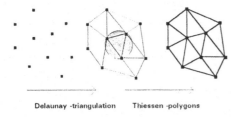

Delaunay -triangulation   Thiessen -polygons

Figure 17: construction Thiessen polygons

overlay
An overlay in vector structures means that from data from at least two databases derived elements can be constructed. This construction is based on the geometric data types and their specific topological relations. like coinciding lines and points, intersecting or crossing of lines. On the basis of these relations new objects are constructed.

5. Literature
This paper is written according to the lecture notes of:

v. Lammeren, R
    1995
    GIS-tools
    Wageningen.

Further literature:
ESRI
    1991
    ARC/INFO: data model, concepts &
    key terms.
    ESRI
ESRI
    1992
    Cell-based modelling with GRID
    ESRI
Maguire, D., Dangermond J.
    1991
    The functionality of GIS
    in: Geographical Information Systems.
    Volume 1: principles.
    by Maguire, D., Goodchild, M.,
    Rhind, D.
    Longman
Molenaar, M.
    1990
    Formal data structures and query
    spaces.
    Wageningen.
Molenaar, M.
    1993
    Lecture notes Basic principles of GIS II
    Wageningen.
Tomlin, C.D.
    1990
    Geographic information systems and
    cartographic modelling
    Prentice Hall.

SPATIO-TEMPORAL DATABASES:
FROM MOVING TO ACTIVE GEOGRAPHIC OBJECTS

R. Laurini

University C. Bernard Lyon I, Villeurbanne, France

ABSTRACT

Several geomatic applications require to use three dimensional and temporal information. Existing Geographic Information Systems do not allow to correctly deal with those characteristics.

In this paper, we emphasise that a GIS will not also store historical information, but also some other types of spatio-temporal data such as for real time geographic applications and for simulations. In other words, only the attributes are evolving during time, but also positions and shapes. So, geographic objects can have trajectories and deform themselves along time. In addition, some applications need active spatio-temporal objects, that is to say objects having a programmed behaviour in order to influence their own environment.

We finish by stressing that certain geographic phenomena governed either by finite-state automata or differential equations require special database structures.

----

0 - INTRODUCTION

Presently, all existing marketed Geographic Information Systems (GIS) products are either 2D or time-independent. Of course, by some tricks, time may be emulated on some GIS, for instance by storing the same coverage at different times. However, some applications need volumetric 3D data, not only staticly, but also dynamically. In these cases, fully spatio-temporal characteristics must be considered (3D+T).

Examples can range from demography, ecology, environmental planning, pollution control, urban and regional planning to utility network management, geology, geomorphology, transportation, traffic management, urban civil engineering (working site management), archaeology, tourism, virtual reality and so on.

But for those applications, time is a very varying dimension, ranging from milliseconds and lower to millennia and eras. As preliminary examples, let us examine some spatio-temporal queries such as : *What was the population of Venice in 1645 ? What will be the vehicles going to this direction ? Give the aerial photos of this zone in 1976 ? What is the level of noise in the Washington City Hall on October 23, 1991 ? What are the sheep flocks which have crossed a radio-active zone ? When this volcano will erupt ? What will be the damages in the next flooding ? Map the evolution of the Poland shape during the 5 last centuries*, etc.

We can distinguish several kinds of time for GIS:

- **historical GIS**, in this case, historical data are stored, for instance censuses with their dates,

- **real time GIS** for which information are stored at seconds level, for instance for traffic control or pollution monitoring and also for project management in civil engineering and geotechniques,

- **simulated time GIS** for which time does not really exist within a calendar, but it is a very important simulation parameter, such as in flooding simulation; in this case, a program is run until some calibration criterion is achieved; and when so, some forecasts could be made.

In this paper, only dynamic data will be considered, that is real and simulated time in connection with a GIS.

As a matter of facts, the database people say that "*The truth is within the database*", meaning that the contents of the database must mimic the real world as soon as possible. From that we can easily distinguish between the entity-time, i.e. the time when the entity has evolved, from the registration-time, i.e. the time when the modification is updated or entered in the database. Sometimes this time-lag can run from some microseconds to weeks or worst months or years[1], and this fact stresses the difference between the evolution of objets and the evolution of the knowledge we have about objects.

In this paper, general concepts of temporal databases will be firstly presented. Then, we will particularly examine issues regarding the evolution of attributes, the case of moving objects and then finish with the problems of active geographic objects.

---

[1]In the French cadastre, any modification is registered approximately four years later ...

I. TEMPORAL CONCEPTS

In the domain of databases, research regarding temporal databases has a long history. In this paragraph, we will only stress some of them, such as the temporal database concepts and the versions. And we will finish this paragraph by giving the specificities we have to add to temporal databases in order to define a spatio-temporal database system which can be use for geomatic applications.

1.1. Time and databases concepts

For key-elements, please refer either (Soo, 1991) for a bibliography on the problem or (Jensen et al. 1992) for a glossary of temporal database concepts.

One of the major feature is to distinguish the *valid time* (entity-time) and the *transaction time*. By *valid time*, we mean the date the object is evolving, whereas *transaction time* refers to the date the evolution was registered in the database. Of course, in temporal databases, some attributes are time-invariant whereas some others are time-varying. In Table 1, an example is given showing some time-varying (Population) and time-invariant (Capital City) attributes. However, to be very general, attributes such as Capital City can also vary along time. For instance, in Germany, Berlin, Bonn and now Berlin again.

Country	Capital City	Population	Population-Date
France	Paris	55	1990
France	Paris	43	1940
Italy	Rome	57	1995
Italy	Rome	53	1985
USA	Washington	250	1990
USA	Washington	230	1950
...		...	...

Table 1. *Example of time-invariant and time-varying attributes*

1.2. Versions of tuples, versions of object instances, versions of schemata

A first possibility is to use the version concept. By version, one means that there are different tuples with the same identifier (key), but for different dates, so giving versions of tuples. More generally, after versions of attributes, versions of tuples, there can be versions of objects and versions of schemata. By means of the version concept, it is easy to store historical information for any kind of topics.

Time-stamping is the main mechanism which can be used by giving dates of validity to attributes or to tuples. Tables 2 and 3 emphasise different solutions for versioning attributes and tuples

Country	Capital, date
Brazil	Rio de Janeiro, --> 1960
Brazil	Brasilia, 1960 --->
Germany	Berlin, --> 1945
Germany	Bonn, 1945-1990
Germany	Berlin, 1990 -->
......	

Table 2. *Attribute versioning*

Parcel Number	Landowner	Starting Time	Ending Time
33	Peter	12	24
33	Jack	25	35
45	Mary	28	37
45	Tom	38	42
....	....	....	....

Table 3. *Tuple versioning*

But in other cases, versions of objects must be defined. For instance, during a certain period, Germany was split into West-Germany and East-Germany: i.e., we can use different versions of object instances, each of them having different geometries. More generally, in the object version, the attribute values can evolve, new instances can be created. In some particular cases we will further examine, an object can change its classes without changing its identifier. Let us take the example of a man coming from the Bachelor Class after that he can enter to the Married Class and so on, without changing his identity. See Figure 1 to illustrate this example by using a finite-state automaton giving only the possible transitions.

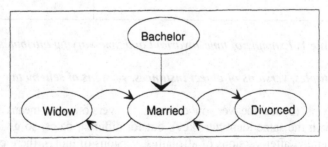

Figure 1. *Different possible states and transitions*

In addition, it could be interesting, during the life cycle of a database to modify its schema in order either to create some new object classes, or to modify the attribute list of an already existing object class, or to create some new relationships between objects. In this case, we deal with versions of schemata that we have to take into account when querying against the database about some contents at different times.

1.3. Specificities of dynamic spatio-temporal database systems

Starting from an existing temporal database system, several features must be integrated in order to specify a fully spatio-temporal database system.

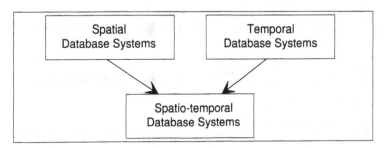

Figure 2. *Spatio-temporal databases must integrate concepts issued from both spatial and temporal databases systems*

a/ the majority of actual GIS are 2-dimensional; but sometimes, not only two but three dimensions are need in addition to time (3D+T);
b/ time is not only seen as historical parameter, perhaps leading to different versions of objects at different times, but rather a structuring parameter leading to the evolution of spatio-temporal patterns and phenomena;
c/ time must be used as a parameter shaping real or simulated evolution (numeric simulation or virtual reality);
d/ time must evolve by either increments in the case of finite-state automata or continuously when the phenomena are governed by differential equations.

II - ATTRIBUTE EVOLUTION OF SPATIO-TEMPORAL DATA

Firstly, we will examine attribute evolution of spatio-temporal data. In this case, information may be stored for different times especially by versionizing tuples as previously explained. But sometimes, information concerning intermediate dates must be retrieved; consequently one must use some extrapolation techniques. One important features concerned the integrity of evolution. For several applications, information concerning object's behaviour must be stored so that objects will follow exactly their planned evolution.

Back to table 1, one can imagine a query such as *"what was the population of France in 1950?"*. In this case, it is clear that to answer such a query, one need to interpolate between two dates, the corresponding tuples of which are stored in the database (namely 1940 and 1990). However, for a query against table 2, *"what was the capital of Germany in 1960 ?*, no interpolation is need. Figure 3 shows different kinds of possible interpolation (nearest value, linear, spline, stochastic and model-based interpolation).

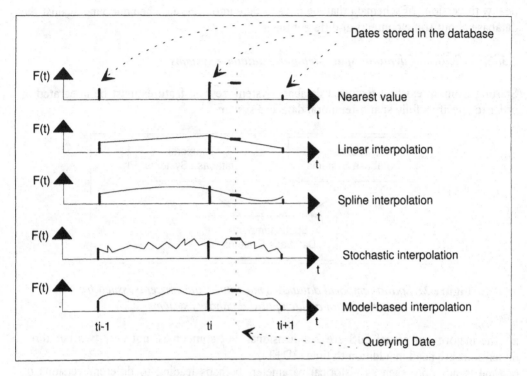

Figure 3. *Different possibilities for interpolation*

Among attributes, identifiers are very important because several database systems are constructed on identifiers. And identifiers can evolve along time: for instance Saint Petersburg, Petrograd, Leningrad, and now back to Saint Petersburg; and different objects can have the same identifier, e.g. Mississipi (river and state), Congo (river and country), etc..

To tackle this problem, two identifiers are currently maintained, an **internal identifier** which is always the same, and perhaps one **external identifier**. The internal identifier is only used inside the computer in order to distinguish objects and to follow them along time without interruption.

III - MOVING AND DEFORMABLE OBJECTS

Among attributes, geometric information has a very important role in spatial databases. This information corresponds to position and shape. When position, (i.e. co-ordinates change continuously along time), one speaks about moving objects with trajectories, and when the shape also changes continuously, one speaks about deformation. Let us first examine those two cases. After that, we present some data structures in order to store this kind of information along time.

3.1. Trajectories

Two main cases must be examined, trajectories in simulated time and trajectories for real time.

a/ Simulated trajectories

For moving objects (Li and al. 1992), one has to know or to forecast its trajectories. For some objects such as trains, the trajectories can be supposedly a priori known due to the railways tracks. For cars, boats or projectiles, missiles, it is also possible to forecast trajectories.

When the trajectory is predictable (for instance by using the gravity law), future positions can be forecast by using functions such as:

$$x_i(t) = f_x(O_i, t)$$
$$y_i(t) = f_y(O_i, t)$$
$$z_i(t) = f_z(O_i, t)$$

In some other cases, when the moving objects are colliding (e.g. snooker balls), interferences between all trajectories must be taken into account. So the trajectory of object i is the consequence of the trajectories of all other objects (namely j) :

$$x_i(t + \Delta t) = F_x(O_i, O_j, t) \forall j = 1, n, (j \neq i)$$
$$y_i(t + \Delta t) = F_y(O_i, O_j, t) \forall j = 1, n, (j \neq i)$$
$$z_i(t + \Delta t) = F_z(O_i, O_j, t) \forall j = 1, n, (j \neq i)$$

b/ Real time acquisition of trajectories

The previous equations are necessary especially during simulation. Another aspect is when the database must follow moving objects in real time, for instance fleets of vehicles, flocks of animals using positioning systems such as GPS, ARGOS as depicted in Figure 4. In this case, the dynamic information is in reality real time information and the database maintains similarities with a temporal database, except for updating issues.

A very important parameter is the frequency for data acquisition which is linked to accuracy; as a common rule, the higher is the frequency, the lower is the accuracy.

A subsequent problem is the object's map positioning. After having acquired the co-ordinates, objects can be stored in the database and be used for cartography. Here due to error and inaccuracies, a driven car may be positioned not inside a street, but inside a "building": due to that difficulty, some tricks must be used so to force-fit the car within the more neighbouring street: the problem is known as object's re-positioning.

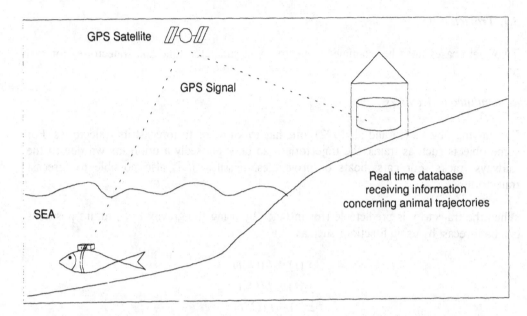

Figure 4. *Animal following by means of GPS (continuous trajectories)*

3.2. Deformations

After having examined the evolution of attributes and position, let us examine shape. Two kinds of deformations can be distinguished, discrete or continuous.

a/ Discrete deformations

In a cadastre, parcels evolve. Indeed, they can fusion, be split into two or several smaller parcels to give allotments. Sometimes, in some villages, parcels are totally reorganised under the process of consolidation. Figure 5 illustrates the story of some parcels at different times.

In this case, one must consider a set of discrete deformations and for each of them, an event can be registered within the database.

b/ Continuous deformations

Some objects like clouds have continuous deformations. For continuous deformations, two cases must be taken into account:
    - same shapes, but with a regular variation (for instance a square which extends itself bigger and bigger, and always being a square);
    - a starting shape, and ending shape and several intermediate shapes (for instance a regular transformation from a circle into a square);

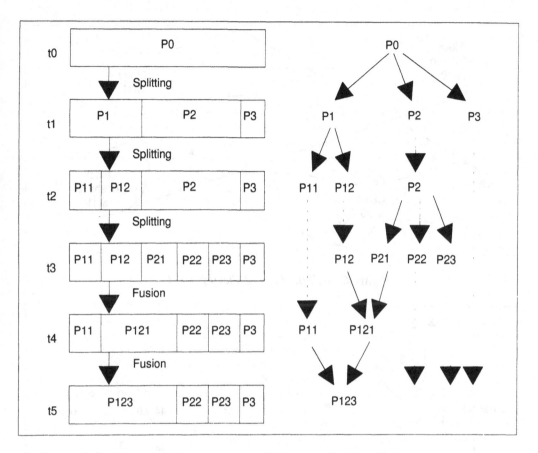

Figure 5. *Example of discrete evolution of parcels*

In other words, some objects can have continuous deformations, but always keeping the same topological characteristics. So we speak about homeomorphisms and warpings.

3.3. Deformations and mobility

But many geographic objects are both moving and deformable. Let us quote some of them: clouds, sandhill, lagoon, soil erosion, flooding, flock of sheep, continents, etc.

In this case, it is possible to answer a query such as : *Will the flock of sheep get wet* ? (Yeh-de Cambray 1995) by studying the deformations of the clouds and of the flock (Figure 6).

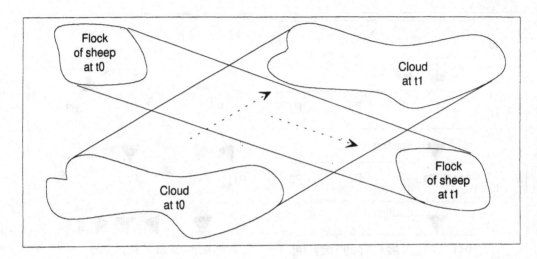

Figure 6. *Will the flock of sheep get wet ?*

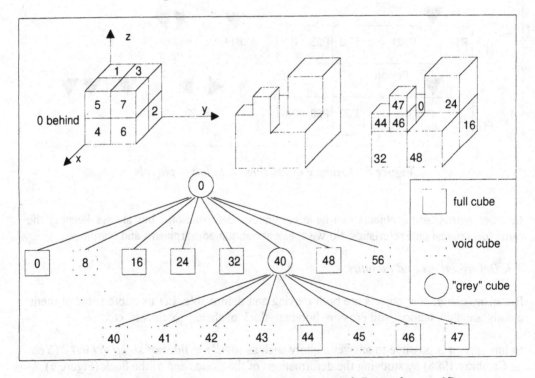

Figure 7. *Representing an object with octrees with Peano keys at 3D.
For 4D, same structure, but within sixteen pointers.*

3.4. Data structures

One of the major problem is to propose a data structure able to store this kind of information, namely 3D and time-dependent. In other words a 4D model is necessary. A possibility is to use an extension of quadtrees. Quadtrees are defined by the recursive subdivision of a square, and similarly, octrees correspond to the recursive subdivision of a cube (See Figure 7). In our case, we need to tackle 4D structure whereas quadtrees and octrees are only 2D and 3D respectively, and 2^4-tree structure can be defined with 16 pointers. A possibility is to use Peano keys in order to encode our 2^4-tree. A Peano key is obtained by interleaving x, y, z and t in binary. For more details, see (Laurini, 1987, Samet, 1990, Laurini-Thompson, 1992). Figure 8 illustrates this process in order to compute a 4D Peano key:

$$(x,y,z,t) \dashrightarrow p(x,y,z,t).$$

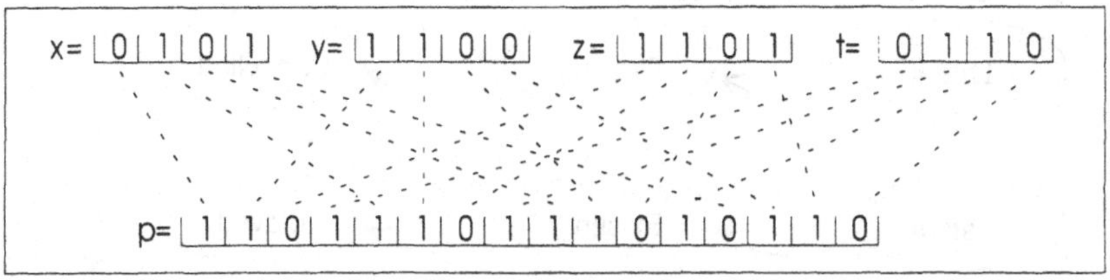

Figure 8. *Interleaving x, y, z and t in order to get a spatio-temporal Peano key (4D)*

So, any spatio-temporal moving and deformable object can be modelled by those 2^4-trees, either directly with 16 pointers, or by means of Peano keys. With the relational model, we can define a Peano relation (Laurini, 1987, Laurini-Thompson, 1992):

ST-OBJECT (#object, 4D_Peano_key, size, attributes)

Back to the flock query, the sheep flock and the clouds can be modelled with this technique giving:

CLOUD (4D_Peano_key, size)
FLOCK (4D_Peano_key, size).

In order to answer the query: *Will the flock of sheep get wet ?*, we have to transform everything in 2D+T (x, y and t) by eliminating the z bit in the Peano keys, giving

CLOUD (2D+T Peano_key, size)
FLOCK (2D+T Peano_key, size)

and then to perform a Peano tuple intersection of the two previous Peano relations. If the answer is null, it means that no sheep were wet.

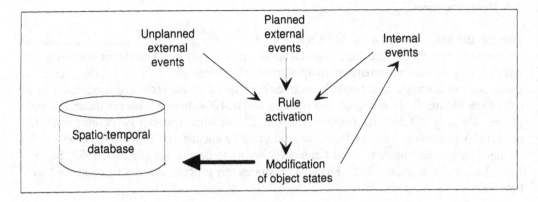

Figure 9. *Events and rule activation*

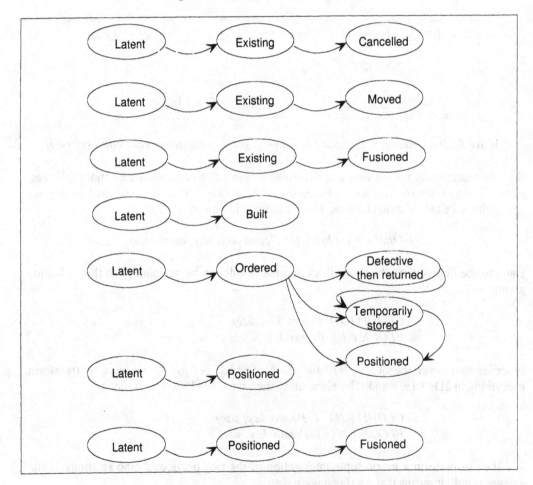

Figure 10. *Example of states and transitions for some active objects in urban civil engineering.*

IV- ACTIVE SPATIO-TEMPORAL OBJECTS

In the previous cases, objects were considered as passive, i.e. having no influence on their environment. However, in a lot of geomatic applications, it is worthwhile to deal with objects having actions over their environments, for instance in project management for civil engineering. In this section, we will first define active objects and active databases, and then examine the implications to geographic databases.

4.1. Active objects

According to Lemoigne (1977), within a system, an active object does something, it intervenes, it has a function in its environment.

A database management system (DBMS) is said active when it has the possibility to automatically execute some actions or programs in response to events which are either generated by the database itself or outside (Collet et al. 1994).

Presently, active databases are overall used in all domains in which some events can have an influence on the running of the database (Sistla-Wolfson, 1995).

By active rules, we mean rules that can monitor the database evolution during time. The general organisation is given in Figure 9 in which one can see the differences between internal and external events which will govern the behaviour of database active objects.

4.2. Active spatio-temporal objects

Let us take an example in civil engineering. During a project such as a new road construction, the contents of the database must evolve from the actual situation (nothing in the countryside, or some buildings in the city center) to the very end of the project. Another examples can be the transformation of a crossroad into a roundabout, the construction of an urban tunnel, etc. In these cases, one can define some active objects having different states as illustrated in figure 10. Based on those concepts, a prototype was developed in Lyon (Chabal, 1995) in which rules were modelled by triggers such as:

```
when e if C then activate process P.
```

in which e is a database event, C, a condition or a set of conditions to check and P a process to be run.

Another prototype was also developed in our laboratory in hydrology. In each region, a different system of differential equations (model) has to be selected as illustrated in Figure 11 (Kettal, 1994). Taking the environment parameters into account, rules are used in order to choose the best hydrological model for each zone at each time, for instance to simulate a flooding. Due to the evolution, the parameters are changing and so the model. At each iteration, rules are fired in order to select the best choice as depicted in Figure 12.

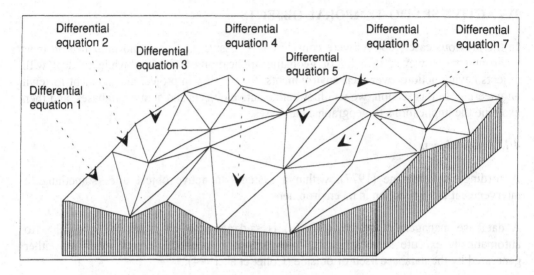

Figure 11. *In each zone, a different hydrological model is selected, based on differential equations*

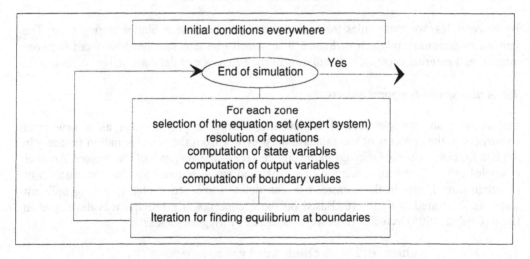

Figure 12. *Selection of rules and boundary value equilibrium*

One of the main problem is to ensure flow continuity in spite of the spatiall splitting into several zones with different behaviour (model). In other words, the continuity must be solved at boundaries since the values generated by different models are not the same (state variables and their derivatives). In order to match those values, another iteration is necessary, because two neighbouring sets of equations do not necessarily give the same values at one side and at the other side of the boundary as depicted in Figure 13. In the mentionned prototype, only first-order and second order derivatives were matched.

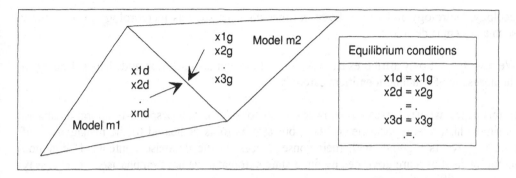

Figure 13. *Necessity of finding an equilibrium at boundaries so that boundaries conditions match (boundary continuity).*

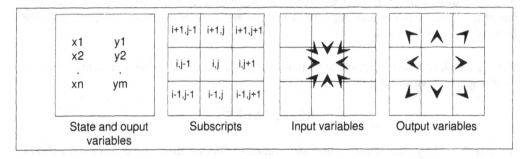

Figure 14. *Elements of the cellular automaton for solving the hydrological problem*

Another possibility is to use, a cellular automaton is used as illustrated in Figure 14, each cell having a different set of equations for its behaviour, whiich is obviously linked to the neighbours in order to simulate the water flows.

For this application, rules were as follows:

```
For zone z if p1, p2, p3 then select model m
```

in which z stands for a named zone, $p1$, $p2$, $p3$ some parameters of the zone and m a model, that is to say a special set of differential equations

V- CONCLUDING REMARKS

Presently, no marketed GIS are able to efficiently solve the spatio-temporal problems. Of course, by some tricks, those problems can be solved in certain cases. However a general system dealing with spatio-temporal databases is need, especially for storing information concerning some phenomena in geomatics such as geomorphology, civil engineering, geology, hydrology and so on. Even in other domains such as meteorology or astronomy such a system is desirable

We saw that not only attributes can vary along time, but also position and shapes, leading to the necessity of handling identifiers carefully

In this paper, we have stressed the necessity to consider not only spatio-temporal databases as simple historical repositories of data, but also as tools dedicated to the management of time and events together with their consequences on the database contents. If common temporal databases are governed by finite-state automata, we have emphasised the necessity to use also differential equations for some classes of phenomena.

The goal of this paper was to give some conceptual basis in order to build such a software product. But in addition to database concepts, some other features are desirable especially linked to visual interfaces and mapping. Among them, let us quote, animated cartography in order to visualise the results.

Having such a tool, new applications can be envisioned such as problem in spatio-temporal reasoning and in virtual reality.

REFERENCES

Amghar Y., Flory A.: *Modélisation de systèmes d'information au travers du concept d'objet actif*, Revue d'Ingénierie des Systèmes d'Information, Vol. 2, n°3, p 293-315, 1994.

Chabal E.: *Gestion de projets en génie civil urbain et bases de donnnées actives*. Mémoire de DEA. June 1995. 90 p

Collet C., Coupaye T, Svensen T.: *NAOS : Efficient and modula reactive capabilities in an object-oriented database system*, Proceedings of the 20th Very Large Data Bases (VLDB) Conference Santiago, Chile, 12 p, 1994.

Jensen CS, Clifford J., Gadia SK, Segev A., Snodgrass RT: *A Glossary of Temporal Database Concepts*. ACM SIGMOD Record, Vol 21,3, September 1992, pp 35-43.

Kettal E.: *Expert Spatial Knowledge: an Expert Geographical Information System in Diffusion of Water*. Proceedings of the 5th International Conference DEXA, Athens, Grèce, September 7-9, 1994 . Edited by Springer-Verlag by D. Karagiannis (LNCS 856), pp 453-464.

Laurini R.: *Manipulation of Spatial Objects by a Peano Tuple Algebra*. Computer Vision Laboratory Technical Report, University of Maryland CAR-TR 311, July 1987. 34p.

Laurini R., Thompson D.: *Fundamentals of Spatial Information Systems*. Academic Press. 1992.680 p.

Lemoigne JL: *La Théorie du Système Général, Théorie de la Modélisation*. Presses Universitaires de France. 1977.

Li KJ., Badji N., Laurini R.: *Objets mobiles : Vers les bases de données spatio-temporelles*. 8èmes Journées Bases de Données Avancées. Trégastel, 15-18 Septembre 1992, pp 102-120. Edited by INRIA, Rocquencourt, France.

Samet H.: *The Design and Analysis of spatial Data Structures*. Addison-Wesley, 1989.

Sistla AP, Wolfson O: *Temporal Triggers in Active Databases*. IEEE Transactions on Knowledge and Data Engineering. Vol 7, 3 pp 471-486. June 1995.

Soo MD: *Bibliography on Temporal Databases*. ACM SIGMOD Record, Vol 20,1 March 1991 pp 14-23.

Yeh TS, de Cambray B: *A Model for the Management of Highly Variable Spatio-temporal Data*. Joint European Conference on Geographical Information (JEC-GI), The Hague, March 26-31, 1995. Volume 1 pp 126-131.

COMPUTER GRAPHICS - PRINCIPLES AND PRACTICE

T. Ertl

University of Erlangen-Nuremberg, Erlangen, Germany

Abstract. These are the short notes for a two hour tutorial on principles and practice of computer graphics and scientific visualization. They are intended to summarize the contents of the tutorial transparencies and slides but they cannot completely replace them since restrictions in space and print quality do not permit the inclusion of figures and example images. For further reference the following standard text should be consulted: [3, 8, 5, 1, 6, 2, 9]

1 History

The notion of what is considered to be computer graphics has changed quite dramatically from simple 2D line drawings to interactive walk throughs of 3D architectural scenes rendered for example using the radiosity method. In order to put the development into perspective, a few important milestones are listed here, some of which are explained later-on :

The 50's: the first computers equipped with a cathode ray tubes (MIT Whirlwind) and interactive light pen input (SAGE). 1963: Ivan Sutherland's revolutionary thesis Sketchpad - the theory and implementation of an interactive CAD package. The 60's: the vector display epoch with the foundation of Evans&Sutherland, Tektronix's storage tube and the IBM 2250 CAD console. The 70's: start of the raster area with Gouraud's shading, Phong's lighting, Catmull's textures and z-buffer. The 80's: the proliferation of PC's and workstations, software standards like GKS, X11, PHIGS, global illumination algorithms such as ray tracing (Whitted) and radiosity (Cohen). The beginning of the 90's: Silicon Graphics defines high-end graphics (RealityEngine) and software standards (OpenGL, OpenInventor).

2   Raster graphics hardware

Most of what we consider as advanced computer graphics today is intrinsically based on the success of raster scan display technology, which replaced vector displays during the 80's due to enormous decrease in memory cost. Almost alll personal computers and workstations on the market today are equipped with a color monitor, a cathode ray tube (CRT) with three electron guns for the primary colors red, green and blue that excite the corresponding phosphor dots arranged in a triangular or in-line pattern on the screen. The electron beam scans the screen in a regular raster, line after line, while the intensities of the beams are regulated to generate the desired color combination at each of the picture elements (pixels) and darkness during the horizontal and vertical retrace. As opposed to standard television technology, where each of two interlaced half images are presented at a rate of 25 Hz, the refresh rate of a modern non-interlaced display is at least 60 Hz, with 72 Hz considered to be the ergonomical minimum. Screen resolutions range from 640x480 pixels for the PC VGA standard to 1280x1024 for almost all modern Unix workstations and even up to 1600x1200 for special applications. Since the corresponding pixel time, on the order of 10 nanoseconds, is obviously not long enough to derive the color of each pixel from even a simple scene description, intermediate storage for all the pixel values is required for any raster scan display.

This memory is called the frame buffer and is arranged as a stack of bit-planes storing the color representation for each pixel. There are essentially two classes of frame buffer depth: the true color systems with 24 bits per pixel (some PC's use 3x5 bits in 2 bytes) allowing for about 16 million different colors in a single image by combining 256 different shades of red, green and blue and the color table systems with a depth of 8 bits per pixel which are used as indices into a lookup table of 3-byte color definitions allowing for the display of images with 256 different colors from a palette of 16 million. The memory requirements for the frame buffer range from 256 Kbyte for 16 color VGA resolution to 4 Mbytes for a true color workstation. Newer low-cost systems employ hardware-based color dithering in order to make true color software to run on 8-bit frame buffers. Double buffering splits the frame buffer in two halves (back and front) decoupling the rendering rate (writing to the back buffer) from the screen refresh rate (reading from the front buffer), thus allowing for smooth animations of even complex scenes.

3 Applications

The enormous progress made in the area of graphics hardware would not be sufficient to make computer graphics a business with an estimated revenue of about 50 billion US Dollars world wide a year without an enormous variety of application areas, some of which are listed here: graphical user interfaces (GUIs) with window systems APIs, point&click desktop tools and page description languages (Postscript) for WYSIWYG publishing (DTP), business and presentation graphics, plotting and scientific visualization, image manipulation and digital painting, photo-realistic rendering, animation, simulation and virtual reality, computer-aided design (CAD), geographic information systems (GIS), computer aided instruction (CAI) and last, but not least, computer games.

4 Computer graphic fields

Computer graphics as it has evolved during the last years splits into a number of related fields each of which has developed its own methods, software packages and research specialists. At the heart of computer graphics is, undoubtly, the synthesis of photo-realistic images of three-dimensional scenes, a process called rendering, which ranges from hardware assisted polygonal z-buffer rendering to ray tracing and radiosity software. It is supplemented by the field of geometric modeling describing the scene by means of mathematical curves and surfaces, the field of animation and simulation specifying the temporal behavior of objects, the field of imaging dealing with the manipulation, storage and compression of digital images and video sequences and the field of scientific visualization assisting in the graphical presentation of measured and computed data. Graphical user interfaces, although still carrying the word graphics in it, have become such a vital part of any operating system that many consider it as a field separate from computer graphics. Multimedia, however, seems to have reversed this trend by extending user interfaces to 3D-graphics, imaging and animation. In the following, all of these fields will be discussed in greater detail putting special emphasis on rendering and visualization.

5 GUIs and 2D raster graphics

While the PC world is dominated by proprietary GUI systems like MS Windows, IBM OS/2 Presentation Manager or Apple Macintosh System 7, there is only one de-facto standard GUI in the Unix world: X-Windows (in its current release X11R6) is a device and operating system independent software architecture consisting of a low level client API for 2D raster graphics and window and event handling (Xlib), a server implementation controlling the display and the input devices (Xserver) and a protocol specification allowing network transparent client server communication. The Xtoolkit, built on top of Xlib, provides a policy-free basis for building abstractions of user interface elements (widgets) like buttons, menus and scrollbars. OSF/Motif is by now the most successful widget library available defining a consistent look&feel for applications and window management operations across many platforms.

Several protocol extensions were implemented in order to overcome the functional restrictions of the X-protocol while maintaining compatibility with the general concepts : the display postscript extension (DPS) allows a resolution independent page description layout to be transmitted to and rasterized by the X-Server, the X image extension (XIE) provides an efficient means for transfering compressed images. With respect to 3D graphics, the two competing protocol extensions are PEX (once an acronym for Phigs extension to X) and GLX, describing the transmission of OpenGL rendering commands to an X-Server exploiting local graphics hardware support. While graphics packages designed in the pre-X epoch had their own functions for handling input events and devices, using text fonts and generating windows and menus, the newer ones (OpenGL and PEXlib) are designed as pure 3D rendering APIs, leaving all other functionality to X/Motif.

6   Modeling

Before even the simplest scene (e.g. composed of cubes and spheres) can be rendered, all objects and their spatial relationship have to be modeled. This is mainly done by describing objects using curves and surfaces (surface graphics as opposed to volume graphics) by means of low degree polynomials. Linear approximation connects vertices by straight lines segments (polylines) or flat surface elements (polygons). In order to ensure flatness, only simple polygons like triangle meshes or quadrilaterals are used. Polygonal models, however, require a dense spacing of vertices to achieve acceptable smoothness of curved lines and surfaces.

The number of control points can be reduced by using parametric representations of 3rd degree polynomials called splines. In contrast to interpolating splines, where the curve passes exactly through all vertices, the interior control points of a Bezier spline are only approximated. The geometric meaning of the control parameters greatly increases the ease of interactive design especially with the local control property of the B-spline curves and patches. The extension to non-uniform rational B-splines (NURBS) even supports the exact representation of conic sections like circles. However, with polygonal rendering still being the base of hardware accelerated graphics, parametric curves and surfaces have to be tessellated (e.g. by recursive subdivision) before they can actually be drawn.

Natural objects like trees or natural phenomena like clouds are obviously not very well or efficiently represented by surface models. Fractals or volumetric descriptions, where a voxel-based description of objects is derived as in constructive solid geometry (CSG), become more acceptable as progress is made in volume rendering methods.

In all cases, complicated models are built from simple ones in a hierarchical manner. Each of them is described in its own local coordinate system and it is put into place by an appropriate coordinate transformation usually composed of translations, rotations and scaling operations. This hierarchy can be in a thought of as a directed acyclic graph (DAG) with the basic elements at the leaves and the modeling transformations at the interior nodes. The traversal of the graph can be done efficiently by employing a matrix stack for the assembly of the current modeling transformation. Graphics packages like PHIGS, where this scene graph is actually stored in memory and rendering is done by starting a graph traversal are called retained mode packages. Immediate mode packages, like OpenGL, draw the scene by calling a functional hierarchy and explicitly manipulating the transformation matrix stacks.

7 Animation and simulation

The geometric modeling phase only determines the static appearance of a scene possibly viewed from different camera positions. If a more lively scenario is desired the various objects have to change position, shape or visual appearance with time. In order to produce a flicker-free realtime animation sequence, more than 20 frames per second have to be generated. This process requires a lot of CPU time for the rendering of the single frames and a lot of disk space (more than 1 Gbyte for an uncompressed one-minute clip) if it has to be stored digitally. Computer animation began with the conventional animation techniques like keyframing and image interpolation and gradually moved into more advanced parametric and kinematic interpolation schemes specified in scripting languages. It was realized that in many cases realistic looking motion can only be generated by simulating the correct dynamic behavior and by integrating the laws of physics. Thus, physically-based animation spread from particle systems to autonomously moving, legged figures.

8 Color

Besides specifying the geometric and temporal behavior of objects one has to define the visual appearance, which is done mainly by assigning colors to the vertices or control points (also see the section on texture). Based on physiological results (tri-stimulus theory) color is assumed to be representable in a three-dimensional space. Well-known color systems are the hardware oriented color systems RGB (red, green, blue, additive) for CRTs and CMY (cyan, magenta, yellow, subtractive) for printing on white paper. Easier to manipulate when designing colors are the perception-oriented color spaces HSV and HLS, with the independent variables hue (the dominant frequency), saturation (color purity) and lightness or value (the luminance). Color systems based on the CIE standards offer the desired uniform perception variation, however, they are still seldomly used because of their complexity. When dealing with the RGB triples, usually a fourth component A is added describing the opacity of the color. The default value of one means that this color is completely visible, whereas a opacity of zero represents a fully transparent and, thus, completely invisible color. Intermediate value can be used in compositing, a blending operation producing transparency effects.

9 Coordinate systems, perspective projection and clipping

After the various parts of a scene are constructed in their respective modeling coordinate systems, they are assembled in a common coordinate system called world coordinates. In order to derive a 2D image from the 3D scene, those world coordinates have to be projected onto a virtual screen. This can be done by means of an orthographic projection. For the more realistic appearance of close-by objects, however, perspective projection is to be prefered. The best way to deal with the nonlinear character of this projection, which can be seen in the foreshortening effect, is to introduce homogeneous coordinates, where the additional component is used to represent the perspective division. Now, all transformations including perspective projection can be written as a 4x4 matrix. Following the standard graphics pipeline, the world coordinate vertices and normals (required for lighting calculations) are first converted into eye coordinates by means of the viewing transformation. Then, the viewing frustrum defined by the field-of-view, the front and the back clipping planes is transformed into a unit cube (perspective transformation). Since no objects outside of the viewing frustrum are visible, lines and polygons are efficiently and correctly clipped in these nomalized homogeneous coordinates. Thereafter, the perspective division results in coordinates which can be transformed into window coordinates by means of the viewport transformation.

10 Scan conversion and anti-aliasing

Once the vertices describing a line or a polygon are transformed into the final 2D coordinates, all pixels on the line or inside the polygon have to be computed and set to the defined color. This process is termed scan conversion since it usually proceeds from one scan line intersecting the object to the next. For each scan line, an incremental algorithm (i.e. Bresenham for lines and Active Edge Table for polygons) computes the right span of pixels from the slope of the lines and from previous results. Obviously, a staircase pattern is produced which is quite visible for slopes close to a multiple of 90 degrees. These artifacts, together with other similar effects, are called aliasing and result from the fact that the spatial sampling resolution is not good enough for the high frequency components introduced by the sharp edges. Anti-aliasing either increases the effective sampling rate by rendering with sub-pixel precision and appropriate averaging to generate the final pixel color, or by removing the high frequencies with a low-pass filter resulting in blurring.

11   Hidden surface removal

If the list of polygons to be rendered is not pre-sorted, the color of the polygon drawn last instead of the one being closest to the eye would determine the color of the pixels of the final image. Nowadays, the problem of hidden-surface removal is almost exclusively solved by the z-buffer approach. Here, an additional buffer with the same spatial resolution as the frame buffer, either in hardware or in main memory, is used to store for each pixel its corresponding distance from the eye in the form of a normalized z clip coordinate (in a precision of 2 to 4 bytes). If another polygon is rendered, the z-value of each generated fragment is compared to the respective z-buffer value. The pixel color is only set to the new value if the polygons z-value is smaller than the stored z-value, thus being in front of the old pixel. The depth buffer is updated accordingly.

12   Lighting and shading

The reflection of light from one or more light sources off the surfaces of the objects is a major clue for the correct perception of 3D scenes, i.e. an unlit sphere with uniform color cannot be distinguished from a flat circle. Therefore, any graphics package used for the realistic rendering of 3D scenes has to support some form of lighting calculation. The correct way to do this would again be the numerical simulation of the physics of light transport together with the light-matter-interaction defining the reflection properties of the surfaces. With some remaining restrictions, this has become computationally feasible only very recently. Traditional computer graphics is based on many crude approximations and heuristics in order to emulate the lighting calculations with reduced complexity. Reflection is classified to be either diffuse (Lambert's cosine law) or specular or a combination thereof. Local lighting models, that take into account the influence of only directly visible light sources and neglect shadow effects or light scattered from intermediate diffuse or specular reflecting surfaces, form the basis for hardware supported lighting calculations of packages, like OpenGL or PHIGS.

The Phong lighting model determines the color of a pixel as a combination of an emissive color, an ambient color (a substitute for multiple scattering effects), and a diffuse and a specular color contribution of a light source. The last two terms are summed for all light sources in the scene possibly taking distance attenuation into account. The reflection behavior of the surface is characterized by ambient, diffuse and specular reflection coefficients which scale each of the appropriate terms. The diffuse component depends on the angle between the surface normal and the light source vector and is therefore view independent, whereas the specular component is also dependent on the viewing direction.

For a smooth color variation across a surface, the lighting computation should be performed for each pixel. Thich is, however, prohibitively expensive. Therefore, various shading methods were developed to reduce this complexity by means of interpolation. Gouraud shading evaluates the lighting equation only at the polygon vertices and derives the color of interior pixels by linear interpolation. This process can be efficiently combined with scan conversion and is available in many graphics hardware accelerators. However, it fails to reproduce highlight effects inside of polygons. Phong shading (not to be confused with Phong lighting) does not interpolate the colors, but interpolates the normal vectors and evaluates the lighting equation for each pixel, thus, correctly finding all highlights. Since it has a much higher computational cost, it is usually not supported in hardware.

13   Texture mapping and other fragment operations

Phong lighting and shading produce very smooth surfaces with a plastic-like appearance. Real surfaces, however, exhibit characteristic color variations and a certain degree of roughness. Instead of modeling e.g. a brick wall by small brick-sized polygons and still missing the characteristic brick surface structure one could just scan the photograph of a real brick wall and glue this image onto a wall-sized polygon. This technique is called texture mapping. Besides specifying the texture, the user is required to assign texture coordinates to each vertex and to define a wrapping behavior for the case where a small texture is used to tile a large polygon. Since the assignment of texture coordinates to vertices is independent of the viewing conditions, situations might easily arise where one image pixel covers either only a small fraction of a texel or several hundreds of texels. Sophisticated filtering methods are to be employed in order to avoid strong aliasing effects. A common approach is the mipmap technique where the texture map is stored together with a pre-filtered hierarchy of coarser resolutions. High-end graphics workstations perform all texturing operations like trilinear mipmap interpolation and pixel-wise perspective division in real-time without any performance penalty.

The decline in cost of memory allowed for more bitplanes to be allocated for the frame buffer. The accumulation buffer is an additional color buffer, which can be used for multi-pass rendering algorithms, like anti-aliasing, depth-of-field, motion blur, or soft shadows. These algorithms compute a complete series of slightly different images, which are successively blended into the accumulation buffer. Stencil planes support the masking of arbitrary shaped frame buffer regions and can be used for the windshield of flight simulators, for hidden line removal, or for the rendering of overlapping translucent polygons.

14   Ray tracing

Local illumination algorithms such as Phong lighting will fail to generate realistic images, e.g. of a bathroom with mirrors in it, since the reflection of light from a surfaces is obviously a non-local illumination effect. Ray tracing is a physically motivated rendering technique that takes into account the effects of multiple scattering of light from perfectly specular surfaces, automatically producing correct shadows (direct illumination). For each pixel in the image a ray is traced from the eye into the scene, making it a view-dependent algorithm. If the ray hits a non-reflective object, shadow rays are sent out from the intersection point to all light sources. If one of the shadow rays hits another object, the corresponding light source can be neglected, since it is obscured. All remaining light sources are used to derive a diffuse color for this pixel. For each hit with a reflective object, new rays are recursively launched in the direction of reflection and in the direction of refraction (if the object is partially transparent). The colors accumulated by those rays are scaled by additional coefficients and are combined to form the final pixel color. Despite the fact that the recursion depth is usually restricted, ray tracing remains a very compute-intensive procedure with most of the time spent in ray intersection tests. Many hierarchical approaches have been implemented in order to speed up these calculations. The recursive nature of the algorithm, however, has limited the development of efficient hardware support, and ray tracing of complex scenes still remains a non-interactive rendering approach.

15   Radiosity

In contrast to ray tracing, radiosity attempts to model multiple diffuse reflections between objects (global illumination), completely ignoring specular reflection. The method is based on physical principles modeling radiation transfer in a closed environment of diffuse reflecting surfaces by means of a finite element approach. The scene is split into a number of surface patches and the interaction of each patch with all other patches is considered. This leads to a system of linear equations for the energy leaving each patch (radiosity) and it is solved using an iterative approach. The matrix elements are essentially determined by the form factor, which describes the geometric relationship (size, orientation, visibility) between two respective patches. Radiosity allows the synthesis of images of a new level of realism with extended light sources, soft shadows, and color bleeding effects. Recent advances have introduced hierarchical schemes, wavelet decompositions, and clustering strategies. However, radiosity is still a very compute-intensive technique where rendering times are measured in minutes and hours. One important aspect of radiosity is that it computes view-independent entities, which can later-on be loaded into a hardware accelerated polygonal renderer, allowing for interactive walk-throughs of the computed scene.

16   Software packages

As a result of its availability on most Unix platforms and even on PCs, OpenGL seems to be becoming the de-facto standard for network transparent, immediate mode, 3D rendering APIs. It does not define any subsets, thus, the programmer can rely on the availability of Phong-lit, Gouraud-shaded, mipmap-textured, and z-buffered polygon rendering. The second widely used API is PHIGS PLUS, a retained mode package defined as an ISO standard. Since it relies on its own input events and mechanisms, it interfered with the surrounding window system. This spawned the development of the PEX protocol and the PEXlib interface. In its current version this package also allows for immediate mode rendering, texture support is planned for a future release. An interesting high-level API is OpenInventor, an object-oriented class library built on top of OpenGL also defining a scene description language. Ray tracing packages are available in the public domain (rayshade, POV-ray) and in almost all commercial animation packages (Alias, Wavefront, Softimage).

17   Scientific visualization

The advances in high performance computing and in the area of numerical methods over the last couple of years made possible the simulation of time-dependent, three-dimensional problems in science and engineering. These computations produce an immense amount of data, which, in many cases, can be interpreted only visually by means of computer graphics generating images and animation sequences. The necessary methods, tools and devices are the research field of scientific visualization, which evolved from the combined activities of scientists, mathematicians and engineers into a discipline of it own.

The current focus of research activities [4, 7] is on volume visualization, fluid flow visualization, perception and user interfaces, data modeling and multivariate datasets. The areas of direct volume rendering of 3D scalar fields and the visualization of vector and tensor fields, especially, have introduced many new representation techniques, which will step-by-step become available in modular visualization environments like AVS, NAG Iris Explorer, IBM Data Explorer, or IDL.

References

[1] Cohen, M., Wallace, J. Radiosity and Realistic Image Synthesis. Academic Press (1993).

[2] Farin, G. Curves and Surfaces for Computer Aided Geometric Design. Academic Press (1990).

[3] Foley, J., van Dam, A., Feiner. S., Hughes, J. Computer Graphics: Principles and Practice. Addison-Wesley (1990).

[4] Gallagher, R. Computer Visualization. CRC Press (1995).

[5] Glassner, A. Ray Tracing. Academic Press (1989).

[6] Neider, J., Davis, T., Woo, M. OpenGL Programming Guide. Addison-Wesley (1993).

[7] Rosenblum, L. Scientific Visualization. Academic Press (1994).

[8] Watt, A., Watt, M. Advanced Rendering Techniques. Addison-Wesley (1992).

[9] Young, D. The X-Window System - Programming and Applications with Xt (OSF Motif Edition). Prentice Hall (1994).

Printed in the United States
By Bookmasters